Index of Applications

CALCULUS FOR THE MANAGEMENT
AND SOCIAL SCIENCES

Calculus for the Management and Social Sciences

John Hegarty *Bentley College*

ALLYN AND BACON, INC.
Boston · London · Sydney · Toronto

To Marie

Library of Congress Cataloging in Publication Data

Hegarty, John.
  Calculus for the management and social sciences.

  Includes index.
  1. Calculus. 2. Management—Mathematics.
3. Social sciences—Mathematics. I. Title.
QA303.H46   515   79-20561
ISBN 0-205-06886-3

Printed in the United States of America.

Preparation buyer: Patricia Hart
Series editor: Carl Lindholm

Contents

Preface

This book has been written for a one-semester or two-quarter introductory course in calculus for students of business or the social sciences who have had one to two years of algebra. Applications in the fields of accounting, economics, finance, and management have been emphasized. Because most students take calculus in the freshman or sophomore year, basic applications have been chosen.

Traditionally, textbook applications of calculus have been restricted to the natural sciences and engineering. However, the contents of journals such as *Accounting Review, American Economic Review, Decision Science,* and *Management Science* provide convincing evidence of the widespread use of calculus in business areas.

A number of features have been incorporated into the text to provide an intuitive and supportive approach for students with diverse backgrounds:

1. Where possible, calculators and graphs are employed to convince and motivate the student. For example, limits are introduced and discussed in Section 2.1 by the use of tables and graphs alone.

2. Each example is worked out completely. Including an additional step or two has been found to pay enormous dividends in reducing student anxiety and frustration.

3. Whenever a problem requires a graph, an appropriate coordinate system is provided. Students using a prepublication edition found this feature very helpful because the availability of the coordinate systems enabled them to carry out assignments more efficiently and quickly. The availability of the completed graphs in the book was a great asset for studying and reviewing material for examinations.

4. In Exercises 3.2–3.5, 6.1, and 6.2, the student is not only required to find the equation of the tangent line at a point, but is also asked to plot the line on an accompanying coordinate system that contains the graph of the function. Problems of this kind are very effective in convincing a student that the first derivative represents the slope of the tangent line.

5. Two chapters (5 and 6) are devoted to exponential and logarithmic functions and their derivatives. I became convinced that it was desirable to allot more space and therefore time to these topics after continually finding that some students were being frustrated in their attempts to find critical points because they were unable to solve equations such as $\ln(x - 2) + 1 = 0$, or even worse, writing $de^x/dx = xe^{x-1}$.

6. Appendix A contains a review of sets and set algebra. Although set notation appears only in Chapters 1 and 10, a review is included for those students who may have forgotten some of the basic concepts. Appendix B contains a review section on algebra that emphasizes those topics a student will repeatedly encounter.

For quarter- or semester-length courses where abbreviated coverage is desired, choices can be made from the many applications. Additional flexibility is provided by the optional nature of Chapter 1 and Sections 6.3, 7.2, 8.5, 9.3, 9.4, and 9.5.

I would like to take this opportunity to thank publicly the many people who assisted in the publication of this book.

Colleagues at Bentley College were not only generous with their time and advice but also very gracious in suggesting changes. Karen Schroeder, Richard Swanson, Larry Dolinsky, Ralph Johanson, Nikolaos Kondylis, and Philip Laurens read all or portions of the original manuscript and contributed many valuable ideas. Bruce Aborn, Michael Epelman, Nelson Hartunian, Harold Perkins, Harold Rice, and Michael Saxe used prepublication editions in their classes and offered many recommendations for improvement.

The following reviewers are gratefully acknowledged: Professor William Perrizo of North Dakota State University, Professor Charles Lewis of Monmouth College, Professor L. M. Larsen of Kearney State College, Professor John Roberts of the University of Louisville, Professor Dorothy Crepin of Lower Columbia College, Professor Arthur Clemmons of Southern Oregon State College, Professor Alan Olinsky of Bryant College, Professor R. Smith of Bryant College, Professor Joyce Longman of Villanova University, Professor Richard D. Ringiesen of Indiana University, Professor Gerald White of Western Illinois University, Professor John A. Pfaltzgraff of the University of North Carolina, Professor Boyd Benson of Rio Hondo College, Professor Paul Banks of Boston College, Professor David Cullen of Loyola Marymount University, and Professor William Fuller of Lower Columbia College.

Special thanks are extended to Grace Chaffee, Nora Krafian, Kathi Krajewski, and Mary Trimble, who spent many hours typing the original manuscript and many revisions of the same.

Last, but by no means least, my wife Marie and children John, Anne, Susan, and Karen were enthusiastic and supportive during each stage of development.

1

Functions and Their Graphs

CHAPTER 1 ঌ INTRODUCTION

One of the primary goals of calculus is to determine the direction and size of the change in one quantity or variable produced by a known change in a second quantity or variable. For example, the management of a fast food chain contemplating a 10 percent increase in the price of its best-selling cheeseburger would be very interested in determining what change the increase will produce in sales and thus in revenue. The change can be predicted if the mathematical relationship between the two variables (price and sales) is known. The relationship usually takes the form of an equation in two variables known as a *function*. The emphasis in this chapter is placed on describing the concept of a function and its visual representation, the graph of a function.

ঌ 1.1 Equations in Two Variables and Their Graphs

Solutions to equations in one variable can be represented graphically by showing them as points on a real number line. For example the equation

$$5x + 1 = 3x + 9$$

has as its only solution

$$x = 4$$

This solution can be represented as a point on the following number line.

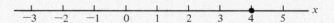

Equations in two variables are far more useful and important than those in one variable. For example, a company selling frozen pizzas at \$2 apiece finds that its revenue depends on the number of pizzas it sells. In words, the relationship can be stated as

Revenue = (\$2 per pizza) × (number of pizzas sold)

If we denote the quantity of pizzas sold by the symbol q and the revenue by R, the relationship can be expressed as the equation

$$R = 2q \tag{1.1.1}$$

This equation allows us to calculate the revenue quickly and efficiently for a given value of q.

For each equation in two variables, a solution requires specifying two pieces of information, that is, a value for each variable. For example, a solution to the equation

$$y = 3x - 2 \tag{1.1.2}$$

requires assigning values to both x and y. Generally, a solution is written as an *ordered* pair (x, y) where the first element indicates the value assigned to x, while the second element represents a corresponding value of y found from the equation. For example, the ordered pairs $(0, -2)$, $(1, 1)$, $(2, 4)$, and $(\frac{4}{3}, 2)$ are solutions to Equation 1.1.2 because the equations

$$-2 = 3(0) - 2 \qquad 1 = 3(1) - 2 \qquad 4 = 3(2) - 2 \qquad 2 = 3(\tfrac{4}{3}) - 2$$

are true statements. One the other hand, ordered pairs such as $(1, 5)$, $(2, -3)$, and $(4, -7)$ are not solutions because the equations

$$5 = 3(1) - 2 \qquad -3 = 3(2) - 2 \qquad -7 = 3(4) - 2$$

are false statements.

Any ordered pair of real numbers (x, y) can be represented geometrically as a point in a two-dimensional coordinate system called the *cartesian coordinate* system. The coordinate system is formed by aligning two perpendicular number lines so that the point of intersection O, called the origin, represents the number 0 on both lines as shown in Figure 1.1. The horizontal number line is called the x axis, the vertical line the y axis. Points on the x axis to the right of the origin represent the set of positive numbers, points to the left the set of negative numbers; likewise points on the y axis above the origin represent the set of positive numbers, points below the set of negative numbers.

The ordered pair of numbers (x, y) associated with any point P in the plane is found by drawing two lines through P, one horizontal, the second vertical. The value of the first member in the ordered pair, x, is found by noting the point where the vertical line intersects the x axis; in the same manner, the value of the second member, y, is found by noting the point where the horizontal line in-

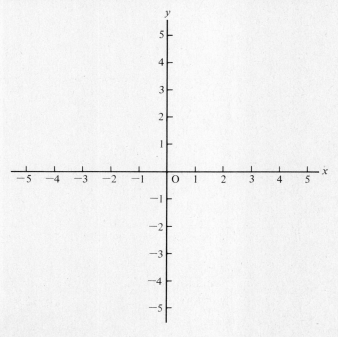

Figure 1.1

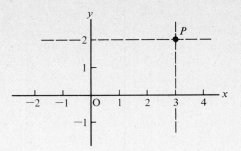

Figure 1.2

tersects the y axis. Thus, in Figure 1.2, the point P represents the ordered pair of numbers $(3, 2)$. The origin O represents the ordered pair $(0, 0)$.

Problem 1 Find the ordered pair of numbers associated with each of the points Q, R, and S on the following coordinate system.

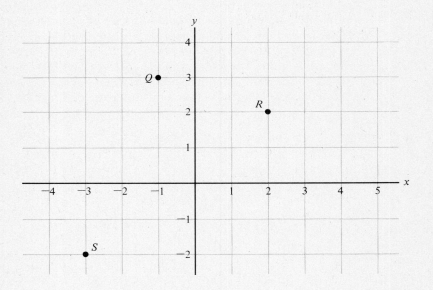

Answer $Q(-1, 3)$, $R(2, 2)$, $S(-3, -2)$

In the same way, each ordered pair of numbers (x, y) can be represented geometrically as a point in the cartesian coordinate system. The point is located by drawing a vertical line through the point x on the x axis and a horizontal line

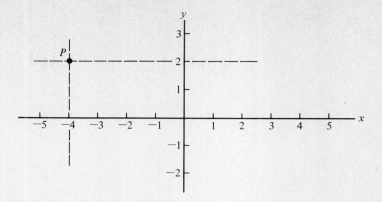

Figure 1.3

through the point y on the y axis; the point of intersection of the two lines defines $P(x, y)$. For example, the ordered pair $(-4, 2)$ is represented by the point P shown in Figure 1.3.

Problem 2 On the following coordinate system, find the points corresponding to the ordered pairs $(1, 2)$, $(-3, -3)$, $(4, 0)$, and $(2, -1)$.

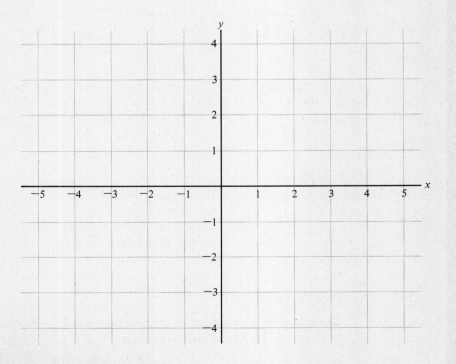

Answer

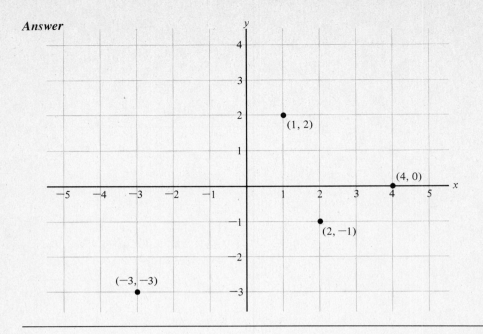

Figure 1.4

QUADRANT	x	y
1	+	+
2	−	+
3	−	−
4	+	−

The first number x is an ordered pair (x, y) is called the *x coordinate* of the point P, while y is called the *y coordinate*. The x and y axes divide the plane into four quadrants, as shown in Figure 1.4, where the signs of the x and y coordinates are also shown for each quadrant.

GRAPHS

As mentioned earlier in this section, a solution to an equation in the two variables x and y is represented by an ordered pair (x, y) of numbers that satisfies the equation. Generally, equations in two variables have an infinite number of solutions so that listing all the solutions is impossible. However, a picture of the equation can be obtained by the technique of graphing.

DEFINITION

The *graph* of an equation in two variables is the set or collection of all points (x, y) that satisfy the equation.

For simple equations, the graph is sketched by finding a representative group of points by selecting values of one variable and determining the corresponding values of the second variable. These points are then connected by means of a smooth curve.

Example 1 | Sketch the graph of the equation

$$y - 2x = 1$$

Solution | By setting up a table such as the following one and plotting the resulting points, we can sketch the graph by connecting them.

x	y
-3	-5
-2	-3
-1	-1
0	1
1	3
2	5

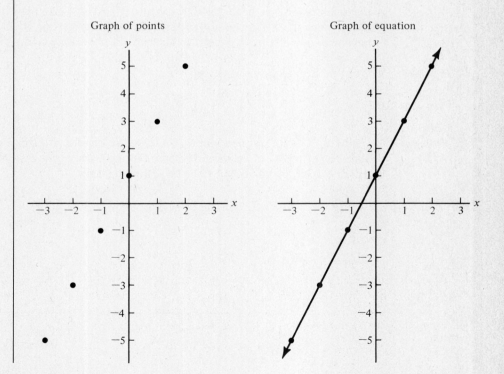

Graph of points   Graph of equation

Example 2 | Sketch the graph of the equation

$$y = x^2 - 3$$

Solution Proceeding as we did in Example 1, we set up a table containing enough ordered pairs to enable us to plot a graph of the equation.

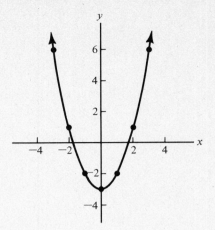

x	y
-3	6
-2	1
-1	-2
0	-3
1	-2
2	1
3	6

When plotting the graph of an equation, it is often necessary to use *nonintegral* values of one or both variables in order to get a complete picture of the graph, especially when the equation is *not* defined for one or more values of either variable. The next example illustrates a situation of this kind.

Example 3 Sketch the graph of the equation

$$y = \frac{2}{x}$$

Solution The equation is not defined when $x = 0$, so we cannot use this value when we construct our table. If we limit ourselves to only a few integral values of x, we find that our graph turns out like the following one.

x	y
-4	$-\frac{1}{2}$
-3	$-\frac{2}{3}$
-2	-1
-1	-2
1	2
2	1
3	$\frac{2}{3}$
4	$\frac{1}{2}$

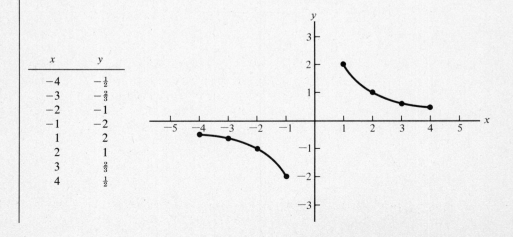

The graph is not complete because it does not indicate what the curve looks like in the region between $x = -1$ and $x = 1$. This deficiency can be remedied by selecting a few nonintegral values of x in the intervals $-1 < x < 0$ and $0 < x < 1$. In addition, selecting one or two values of x beyond the scale on the x axis is often helpful in sketching the "tails" of the graph (see the following graph).

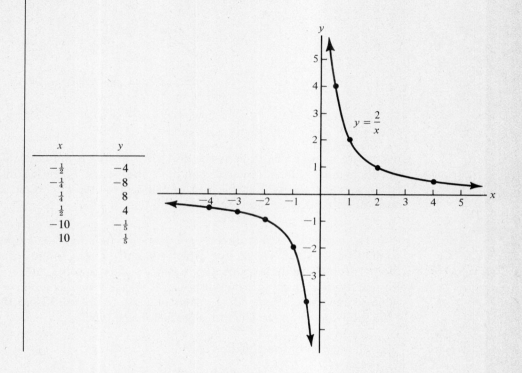

x	y
$-\frac{1}{2}$	-4
$-\frac{1}{4}$	-8
$\frac{1}{4}$	8
$\frac{1}{2}$	4
-10	$-\frac{1}{5}$
10	$\frac{1}{5}$

In applications, the symbols x and y are replaced by symbols more descriptive of the quantities under study as illustrated by the following example.

Example 4   For a house call, customers of the Reliable Service Company are billed according to the following schedule: service call—$20; labor—$15 per hour (prorated). The equation which describes the charge C (in dollars) in terms of the length of time t (in hours) to repair a major appliance has the form

$$C = 20 + 15t$$

Sketch the graph of this equation.

Solution   Noting that the equation does not hold for negative values of t, a table such as the following enables us to sketch a graph of the equation.

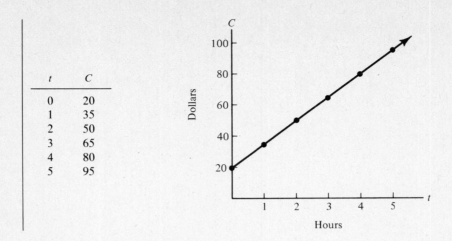

t	C
0	20
1	35
2	50
3	65
4	80
5	95

NOTE: In many cases such as the preceding one, it is not convenient to have identical units of length or scales on the x and y axes. Selecting an appropriate scale on each axis will enable you to highlight all the important features of a curve in a small area.

The algebraic structure of an equation often indicates which variable should be assigned values and which should then be found from the equation. This means that it is sometimes easier to carry out the calculations needed to find a representative group of points if values are assigned first to the variable y, and the corresponding values of x are then found from the equation. The following example illustrates a case of this type.

Example 5 | Sketch a graph of the equation
$$y^2 = x + 4$$

Solution | If values are assigned first to the variable x, finding the corresponding values of y requires computing the square root of $(x + 1)$. On the other hand, finding values of x from the equation is quite simple once values are assigned to y.

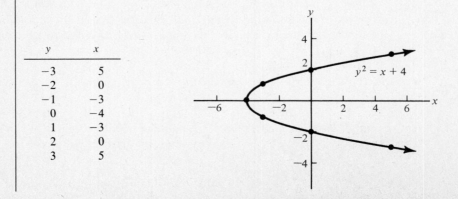

y	x
−3	5
−2	0
−1	−3
0	−4
1	−3
2	0
3	5

EXERCISE 1.1

1. From among the following ordered pairs (x, y), find those that satisfy the equation $2x^2 + y = 7$.
$(1, 5)$, $(-1, 3)$, $(0, 5)$, $(2, -1)$, $(-4, -20)$, $(-2, -1)$, $(7, 85)$, $(\frac{3}{2}, \frac{5}{2})$

2. From among the following ordered pairs (x, y), find those that satisfy the equation $x^2 y - y = 2$.
$(0, 2)$, $(0, -2)$, $(\frac{1}{2}, -4)$, $(-\frac{1}{2}, -\frac{8}{3})$, $(2, -\frac{2}{3})$, $(-1, 5)$

3. Find the coordinates associated with each of the points $P, Q, R, S, T,$ and V shown on the coordinate system.

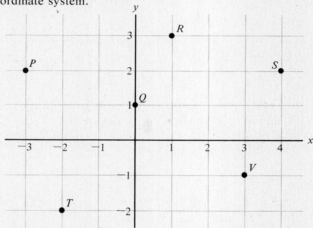

4. On the accompanying coordinate system, find the points corresponding to the ordered pairs $(2, 1)$, $(-3, 4)$, $(1, -4)$, $(0, -1)$, $(-1, 0)$, $(-2, -1)$.

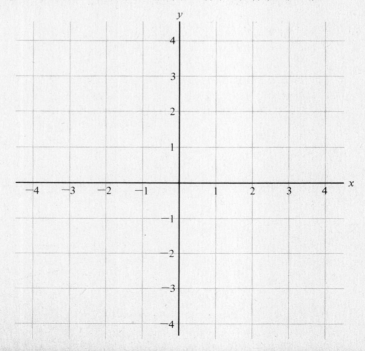

Sketch the graph of each equation in Exercises 5–16 on the accompanying coordinate systems.

5. $y = x - 2$

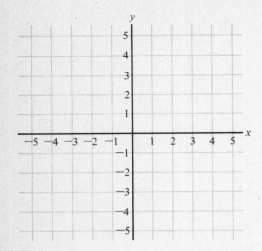

6. $2x + 3y = 6$

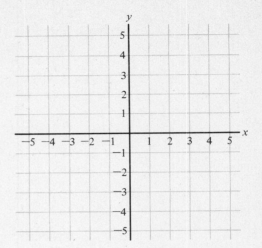

7. $2x + y = -4$

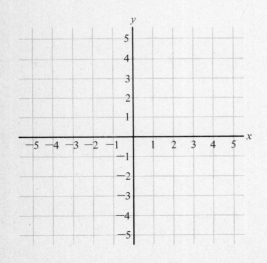

8. $2y - x = 4$

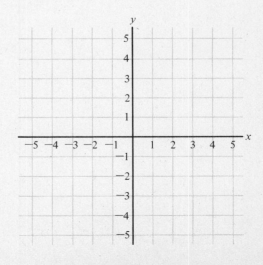

9. $y = 4 - x^2$

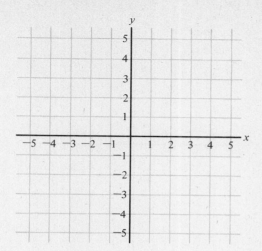

10. $y = x^2 - 4x$

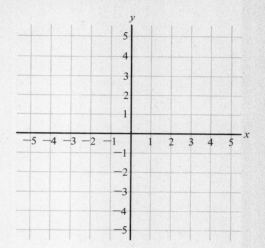

11. $y = x^2 + 4x$

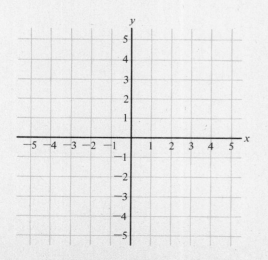

12. $y^2 + x = 3$

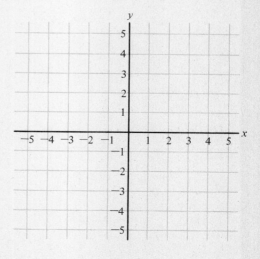

13. $y^2 - x = 4$

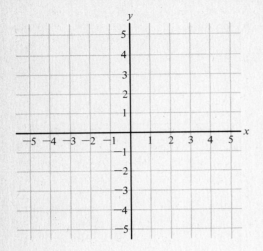

14. $y = \dfrac{1}{x^2}$

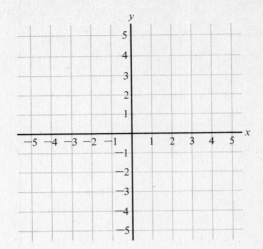

15. $y = \sqrt{x + 4}$

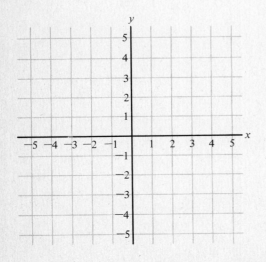

16. $y = 3 + 2x - x^2$

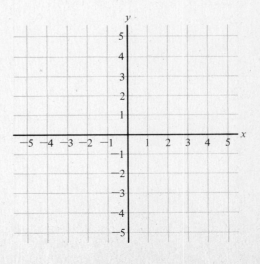

17. A part-time cashier in a supermarket earns \$4.00 per hour. Write the equation that describes her weekly salary s in terms of h, the number of hours worked each week.

18. The monthly salary of a sales clerk in a large department store consists of two parts: (1) a base salary of \$400 per month, plus (2) a 3 percent commission on the dollar value of all items he or she sells during the month. Write an equation that describes the monthly salary S in terms of d, the dollar value of all items sold.

1.2 Functions and Their Graphs

Finding and describing the relationship between two variables is an important activity in many fields. For example, the weekly salary S of most workers depends on or is a function of the number of hours h worked each week; the daily output O of a machine is a function of its age t in years; the amount of sunlight S reaching the earth's surface is a function of the pollution level P in the earth's atmosphere.

In mathematics, when the word *function* is used, it has a very special meaning. In its simplest form, a function is a rule that enables us to associate with each given value of one variable, call it x, a *unique* value of a second variable, call it y. The resulting association or matching is then represented as a set of ordered pairs (x, y) of real numbers. For example, suppose that the variable x represents a real number belonging to the set D defined as

$$D = \{x \mid x = 1, 2, 3, 4, 5\}$$

and that each value of x is paired with a *unique* value of a second variable y by means of the equation

$$y = x^2$$

For each value of x, the equation determines one and only one value of y

$$x = 1, \ y = (1)^2 = 1$$

$$x = 2, \ y = (2)^2 = 4$$

$$x = 3, \ y = (3)^2 = 9$$

$$x = 4, \ y = (4)^2 = 16$$

$$x = 5, \ y = (5)^2 = 25$$

This pairing of each value of x with a unique value of y is an example of a function; it can also be represented as the following set of ordered pairs (x, y)

$$\{(1, \ 1)(2, \ 4)(3, \ 9)(4, \ 16)(5, \ 25)\}$$

The set of all first elements x is called the *domain* of the function; in the preceding example, the domain equals

$$\{x \mid x = 1, 2, 3, 4, 5\}$$

The set of all second elements y is called the *range* of the function; in the preceding example, the range equals

$$\{y \mid y = 1, 4, 9, 16, 25\}$$

In summary, a function is a rule that pairs or matches each number x in one set called the domain with exactly one number y belonging to a second set called the range.

In most cases, the rule defining the pairing or matching of the two variables takes the form of an equation in two variables. In addition, the domain and the range are not always given. When this happens, it is assumed that they are the sets of all real numbers for which the equation is defined. To see what this means, let us consider the function defined by the equation

$$y = 2x + 1 \qquad\qquad (1.2.1)$$

Neither the domain nor the range has been given. To find them, we have to determine those values of x and y for which the equation is defined. This equation, like most we shall meet, is written with the variable y standing alone on one side and all the terms containing x on the other. To find the domain, we examine the right-hand side of Equation 1.2.1 and ask the question: Are there any values of x for which the right-hand side is *not* defined? Since the right-hand side contains only the operations of multiplication and addition, the expression $(2x + 1)$ produces a real number for all values of x. Therefore, we conclude that the domain is the set of all real numbers, or

Domain $= \{x | x$ is any real number$\}$

To find the range, we attempt to repeat this process by solving Equation 1.2.1 for x in terms of y, getting

$$x = \frac{y - 1}{2} \qquad\qquad (1.2.2)$$

In this case, we ask if there are any values of y for which the right-hand is *not* defined. If any value of y is given, the right-hand side always yields a real number, so we get for the range

$R = \{y | y$ is any real number$\}$

These conclusions are supported by the following graph of the function.

x	$y = 2x + 1$
-3.0	-5.0
-2.0	-3.0
-1.0	-1.0
0.0	1.0
1.0	3.0
2.0	5.0

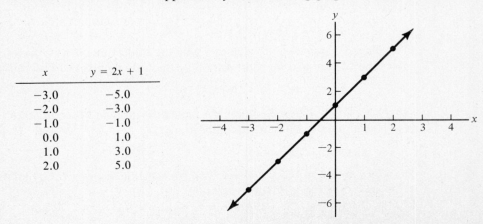

In most cases, the results do not work out so easily. The task of finding the range can be extremely difficult because inverting an equation and solving for x alone in terms of y can be not only challenging but often impossible. However, we will confine our study of functions in this section to those for which the domain and range are not difficult to find.

Example 1 | Find the domain, range, and graph of the function

$$y = x^2 - 2$$

Solution | Because the only operations appearing in the expression $(x^2 - 2)$ are multiplication and subtraction, the expression is defined for all values of x, so we can conclude that

$$D = \{x \,|\, x \text{ is any real number}\}$$

The range R can be found by noting that $x^2 \geq 0$ for all x and that $(x^2 - 2) \geq -2$, so we define the range as

$$R = \{y \,|\, y \geq -2\}$$

These conclusions are reinforced by the following graph.

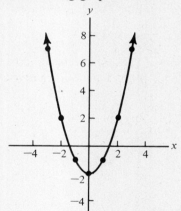

x	$y = x^2 - 2$
-3.0	7.0
-2.0	2.0
-1.0	-1.0
0.0	-2.0
1.0	-1.0
2.0	2.0
3.0	7.0

For the remainder of this section, our attention will be focused on defining the domain and sketching the graph of some simple functions. For these functions, the two operations that may yield values of x for which the given function is not defined are: (1) **division by zero,** and (2) **finding even roots when the radicand is negative.** These cases are illustrated in the next two examples.

Example 2 | Find the domain and sketch the graph of the function

$$y = \frac{3}{2 - x}$$

Solution | The expression $[3/(2 - x)]$ is defined for all values of x, except $x = 2$, so that the domain D is

$$D = \{x \,|\, x \neq 2\}$$

The graph can be sketched by selecting values of x, taking care to include non-integral values close to 2, and calculating the corresponding values of y. This procedure is illustrated in the accompanying table and graph.

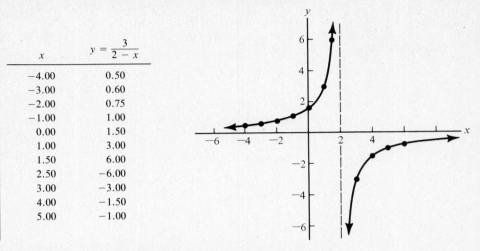

x	$y = \dfrac{3}{2 - x}$
-4.00	0.50
-3.00	0.60
-2.00	0.75
-1.00	1.00
0.00	1.50
1.00	3.00
1.50	6.00
2.50	-6.00
3.00	-3.00
4.00	-1.50
5.00	-1.00

Example 3   Find the domain and sketch the graph of the function

$$y = \sqrt{x + 1}$$

Solution   The equation indicates that real values of y are obtained when the radicand $(x + 1)$ is nonnegative, that is

$$x + 1 \geq 0$$

or

$$x \geq -1$$

Therefore, the domain is the set of all real numbers greater than or equal to -1 and can be written as

$$D = \{x | x \geq -1\}$$

Following is the graph of the function.

x	$y = \sqrt{x + 1}$
-1.0	0.0
0.0	1.0
3.0	2.0
8.0	3.0

Not all equations in two variables represent functions. If the equation has the property that, for some x, there are two or more corresponding values of y, then the equation does not represent a function. A simple illustration of this situation is the equation

$$y^2 = x - 1$$

NOTE: Two values of y are paired with each value of x, except $x = 1$, for which a solution can be found. For example, when $x = 5$, two solutions for y are generated from the resulting equation, $y^2 = 4$,

$$y_1 = +2 \qquad y_2 = -2$$

We can sketch the graph of the equation by using the following table and coordinate system.

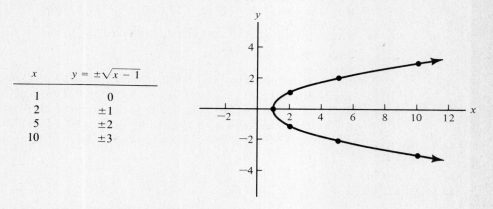

x	$y = \pm\sqrt{x - 1}$
1	0
2	± 1
5	± 2
10	± 3

NOTE: The equation cannot be solved when $x < 1$.

The graph of an equation can often indicate whether or not the equation represents a function. If a vertical line drawn through the coordinate system cuts the curve at two or more points, the curve cannot represent a function; the intersection of a vertical line with two or more points on the curve shows that there is a value of x for which there are two or more values of y. An illustration of this situation is shown in the accompanying figure.

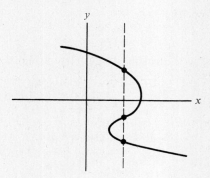

$f(x)$ NOTATION

The second element y in an ordered pair (x, y) is often written as $f(x)$ to be read "f of x". This symbolism may appear rather strange when first encountered, but can perhaps be clarified by means of a simple example. Suppose that the rule matching elements of the domain with corresponding elements of the range is given by

$$y = f(x) = 8x^6 + 5x^5 - 9x^4 + 3x^3 - 11x^2 + 17x - 9$$

Then the element of the range, corresponding to the element $x = 2$ of the domain, is written $f(2)$ and is easily found by substituting 2 for x on both sides of the equation, giving

$$y = f(2) = 8(2)^6 + 5(2)^5 - 9(2)^4 + 3(2)^3 - 11(2)^2 + 17(2) - 9$$

or, after carrying out the arithmetic, we have

$$y = f(2) = 533$$

Note that on the last line, both elements of the ordered pair, that is, 2 and 533, are displayed explicitly; the notation enables one to keep track of specific elements in the ordered pairs easily. The technique just illustrated is also used when x in $f(x)$ is replaced by any quantity or expression; the same quantity or expression is substituted for x on the right-hand side of the equation.

Example 4 If $y = f(x) = 3x^2 + 2$, find

a. $f(4)$ b. $f(\frac{1}{2})$ c. $f(4 + h)$
d. $f(4 + h) - f(4)$ e. $f(a + b)$

Solution a. $f(4) = 3(4)^2 + 2$
                 $= 50$
b. $f(\frac{1}{2}) = 3(\frac{1}{2})^2 + 2 = \frac{3}{4} + 2$
   $= \frac{11}{4}$
c. $f(4 + h) = 3(4 + h)^2 + 2$
             $= 3(16 + 8h + h^2) + 2$
             $= 50 + 24h + 3h^2$
d. Using the results from parts a and c, we have

$$f(4 + h) - f(4) = (50 + 24h + 3h^2) - 50$$
$$= 24h + 3h^2$$

e. $f(a + b) = 3(a + b)^2 + 2$
             $= 3(a^2 + 2ab + b^2) + 2$
             $= 3a^2 + 6ab + 3b^2 + 2$

When writing a functional relationship as $y = f(x)$, the variable x is called the *independent* variable because arbitrary values are assigned to it first; the variable y is called the *dependent* variable because its value is determined once a value has been assigned to x.

In applications, symbols or letters more descriptive of the quantities being examined are used in place of x and y, as illustrated in Example 5.

Example 5 | A company installs a new stamping machine whose output O (units per hour) is a function of its age t (years)

$$O = f(t)$$

The relationship between the variables O and t is described by the equation

$$O = f(t) = \frac{60,000}{300 + t^2} \qquad 0 \leq t \leq 10$$

where the inequality indicates the domain of the function. Sketch a graph of the function.

Solution | The table together with the graph indicate that output diminishes slowly at first but more rapidly as the machine ages.

t	O
0	200
2	197
4	190
6	179
8	165
10	150

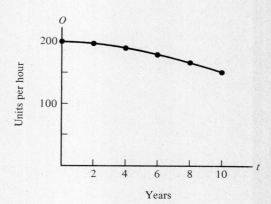

PIECEWISE FUNCTIONS

Quite often, a function must be defined by two or more equations, where the domain is "broken up" into pieces, one for each equation as shown in the following examples.

Example 6 | Sketch the graph of the function

$$y = f(x) = \begin{cases} x + 2 & x < 1 \\ 3 & x \geq 1 \end{cases}$$

Solution | The graph is found by first plotting the equation $y = x + 2$, restricting the variable x to values less than 1. The graph is shown in Figure 1.5 (a) where the open circle $\bigcirc$ indicates that (1, 3) is *not* a point on this part of the graph. Next the graph of the equation $y = 3$ is plotted, using only values of x greater than or equal to 1; this piece is shown in Figure 1.5(b). The graph of the entire function can be plotted by superimposing the two pieces as the third graph shows.

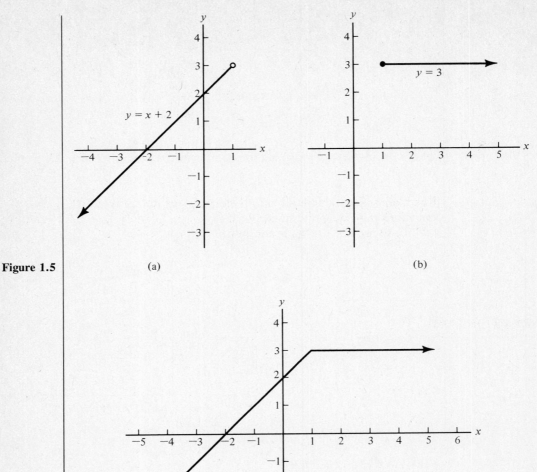

Figure 1.5

(a)   (b)

Many piecewise functions display gaps or breaks at those values of x where the rule or equation changes form. The next example illustrates an example of this type.

Example 7 Graph the function

$$y = f(x) = \begin{cases} x + 2 & x < 0 \\ 1 & x = 0 \\ x & x > 0 \end{cases}$$

Solution Proceeding as we did in the previous example, we obtain the graph shown in the following figure.

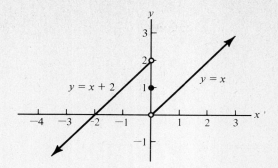

Remember that open circles $\bigcirc$ indicate points that are *not* on the graph; closed circles $\bullet$ indicate points that are on the graph. In the preceding graph, (0, 2) and (0, 0) are *not* on the graph but (0, 1) is.

EXERCISE 1.2

1. For each of the following functions, find the range and then write each function as a set of ordered pairs of numbers.

   a. $y = f(x) = 8 - 4x$; domain = $\{-5, -3, -1, 1, 3\}$
   b. $y = f(x) = 3x^2 - 6x + 2$; domain = $\{2, 3, 4\}$
   c. $y = f(x) = 5 - x$; domain = $\{0.5, -2, -10\}$
   d. $y = f(x) = \dfrac{7x + 1}{2x - 3}$; domain = $\{0, 1, 3\}$

In Exercises 2–6, the equations that are the rules for matching elements of the domain x to corresponding elements in the range y are shown for various functions. In each case, find the domain and sketch a graph of the function on the adjacent coordinate system.

2. $y = f(x) = 3x - 1$

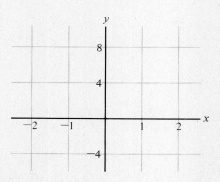

3. $y = f(x) = 8 - 2x^2$

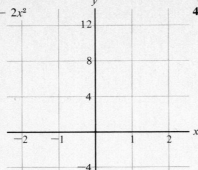

4. $y = f(x) = \dfrac{1}{x^2}$

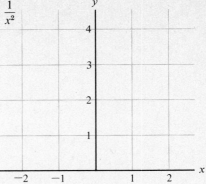

5. $y = f(x) = \dfrac{x^3}{2} + 1$

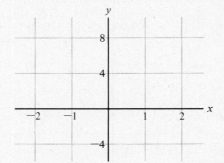

6. $y = f(x) = \sqrt{x + 4}$

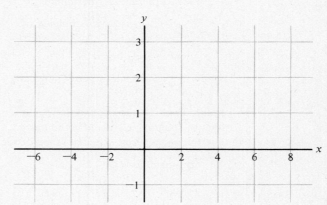

7. If $y = f(x) = 2x + 3$, find

   a. $f(7)$ **b.** $f(\frac{3}{4})$ **c.** $f(-3)$
   d. $f(a)$ **e.** $f(a + 2)$ **f.** $f(a + 2) - f(a)$

8. If $y = f(x) = \dfrac{12}{x - 2}$, find

   a. $f(4)$ **b.** $f(3)$ **c.** $f(4 - 3)$

   d. $f(4) - f(3)$ **e.** $f(\frac{4}{3})$ **f.** $\dfrac{f(4)}{f(3)}$

   g. $f(a)$ **h.** $f(a + 1)$ **i.** $f(a + 1) - f(a)$

9. If $y = f(x) = x^2 - x + 4$, find

   a. $f(0)$ **b.** $f(-1)$ **c.** $f(6)$

   d. $f(1)$ **e.** $f(1 + h)$ **f.** $f(1 + h) - f(1)$

10. If $y = f(x) = 2x^2 + 1$, find

   a. $f(1)$ **b.** $f(-1)$ **c.** $f(1 + h)$

   d. $f(1 + h) - f(1)$ **e.** $\dfrac{f(1 + h) - f(1)}{h}$

11. If $y = f(x) = \dfrac{2}{x - 2}$

   a. $f(0)$ **b.** $f(1)$ **c.** $f(-2)$

   d. $f(\tfrac{1}{2})$ **e.** $f(1 + h)$ **f.** $f(1 + h) - f(1)$

12. If $y = f(x) = \sqrt{x - 1}$, find

   a. $f(2)$ **b.** $f(5)$ **c.** $f(\tfrac{5}{4})$ **d.** $f(\tfrac{25}{16})$

13. The gasoline mileage of an automobile is a function of the speed at which the car is driven. The equation describing this function for a particular automobile is

$$M = \frac{100\,V - V^2}{125} \qquad 0 \le V \le 75$$

where V is the speed of the car (mph) and M is the gasoline mileage (miles per gallon of gasoline).

a. What is the gasoline mileage when the car is moving at 30 mph?

b. On the accompanying coordinate system, sketch a graph of the function.

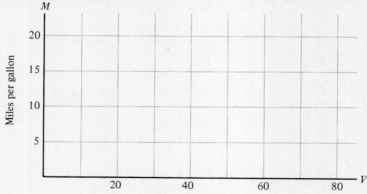

Miles per hour

Plot the functions in Exercises 14 and 15 on the accompanying coordinate systems.

14. $y = f(x) = \begin{cases} 1 - x & x < 2 \\ x - 3 & x \ge 2 \end{cases}$

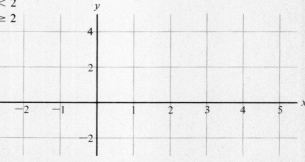

15. $y = f(x) = \begin{cases} x + 1 & x < 0 \\ -1 & x = 0 \\ x^2 & x > 0 \end{cases}$

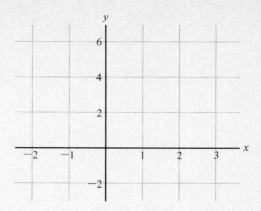

16. The operator of a small self-service gasoline station pays 90¢ per gallon for gasoline; he sells it for $1.10 per gallon at the pump. If his monthly overhead (rent, utilities, insurance) is $500, find the equation that describes his monthly profit P as a function of x, the number of gallons of gasoline sold each month.

17. Functions defined in a piecewise fashion are not always mere mathematical curiosities, as perhaps the following problem will demonstrate.

The Pilgrim Parking Company owns a centrally located parking lot in downtown Boston. Its daily rates for parking are:

$1.00 for 1st hour or fraction thereof
Additional 50¢ for 2nd hour or fraction thereof
Additional 50¢ for 3rd hour or fraction thereof
Additional 50¢ for 4th hour or fraction thereof
$2.50 for all-day parking

a. Write a function that represents mathematically the parking fee as a function of the length of time (hours) a car is left in the lot.

b. Plot the function on the following coordinate system.

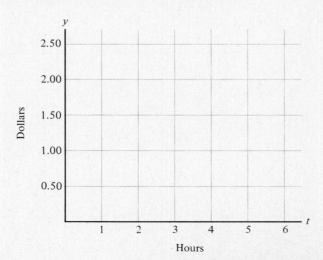

ৰ্জ্জ 1.3 Linear Functions and Straight Lines

In the previous section, an attempt was made to examine the properties of a function and to see what the graphs of some simple functions look like. The functions were restricted to those with a fairly simple mathematical form for two reasons:

1. In order to save time and energy in plotting a sufficient number of points to graph the function, and
2. Because there are more efficient methods for accomplishing the same task when more complex functions are encountered; these methods will be demonstrated in detail later in the text.

Even at this stage, however, the graphs of certain simple functions can be plotted once their algebraic structures have been recognized and selected points found. Two of the easiest to handle, linear and quadratic functions, will be treated in detail in this chapter. Before getting deeply involved with linear functions, it is necessary to introduce the concept of the slope of the line segment connecting any two points, (x_1, y_1) and (x_2, y_2), in the plane.

DEFINITION

The *slope m* of a nonvertical line segment connecting two distinct points (x_1, y_1) and (x_2, y_2) is defined as

$$m = \frac{y_2 - y_1}{x_2 - x_1} \qquad\qquad \textbf{(1.3.1)}$$

Example 1 Find the slope of the line segment connecting $(-2, -1)$ and $(4, 3)$.

Solution Using Equation 1.3.1 with $(x_1, y_1) = (-2, -1)$ and $(x_2, y_2) = (4, 3)$, we get

$$m = \frac{3 - (-1)}{4 - (-2)} = \frac{4}{6} = \frac{2}{3}$$

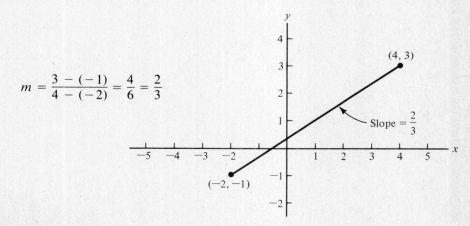

We get the same result if we let $(x_1, y_1) = (4, 3)$ and $(x_2, y_2) = (-2, -1)$, because the algebraic signs of both the numerator and denominator are reversed in the process

$$m = \frac{-1 - 3}{-2 - 4} = \frac{-4}{-6} = \frac{2}{3}$$

In Equation 1.3.1 the quantity $(y_2 - y_1)$ is referred to as the *change in y* and $(x_2 - x_1)$ as the *change in x*, so the slope m can also be written as

$$m = \frac{\text{Change in } y}{\text{Change in } x}$$

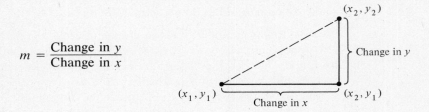

This form of the definition indicates that for each unit change in x, the slope represents the corresponding change in y.

For the sake of consistency, let us assume that whenever we move along a line or any curve, we always move from left to right so that the change in x, that is, $(x_2 - x_1)$, will always be considered positive. If the corresponding change in y, $(y_2 - y_1)$, is

1. *Positive,* the slope m is also positive and the line is said to be *rising*.
2. *Negative,* the slope m is negative and the line is said to be *falling*.
3. *Zero,* the slope m is 0 and the line is *horizontal*.

These three situations are depicted graphically in Figure 1.6 [(a)–(c)]. NOTE: The slope of a *vertical* line segment [Figure 1.6(d)] for which $x_1 = x_2$ is *not* defined.

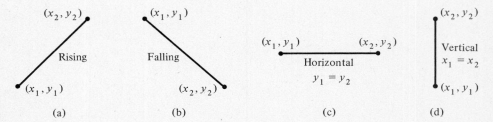

Figure 1.6   (a)   (b)   (c)   (d)

GRAPHING A LINEAR FUNCTION

Throughout this book, the *linear function*

$$y = f(x) = mx + b$$

will be encountered often, so it is important to understand and remember all its properties and characteristics. The graph of a linear function is a straight line for which

1. m represents the *slope.*
2. b represents the *y intercept,* that point on the line where it crosses the y axis.

Graphing a linear equation is quite easy because only two points are needed to construct a straight line. For the sake of accuracy, it is recommended that the two points *not* be situated close together; otherwise a slight error made in connecting the two points can cause a large deviation between the true line and your graph.

Example 2   Graph the linear function

$$y = f(x) = 2x - 3$$

Solution   If we arbitrarily select $x_1 = -1$ and $x_2 = 3$ as the x coordinates of the two points to be plotted, the corresponding y coordinates are found easily from the equation; the results are $y_1 = -5$ and $y_2 = 3$. Using these points as guides, the graph of the straight line can be drawn easily (see Figure 1.7). NOTE: The slope of the line segment connecting $(-1, -5)$ and $(3, 3)$ equals 2, the coefficient of x, and the line passes through $(0, -3)$ where -3 is the y intercept.

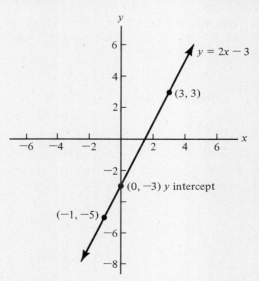

Figure 1.7

It is not difficult to show that the slope of the line segment connecting any two points (x_1, y_1) and (x_2, y_2), which satisfy the equation $y = mx + b$, equals m. Because (x_1, y_1) and (x_2, y_2) satisfy the equation $y = mx + b$, the following hold

$$y_1 = mx_1 + b \qquad\qquad\qquad\qquad\qquad \textbf{(1.3.2)}$$

$$y_2 = mx_2 + b \qquad\qquad\qquad \textbf{(1.3.3)}$$

Subtracting the left-hand side of Equation 1.3.2 from the left-hand side of Equation 1.3.3, repeating the same for the right-hand side of two equations, and then equating the two results yields

$$y_2 - y_1 = m(x_2 - x_1)$$

or

$$m = \frac{y_2 - y_1}{x_2 - x_1}$$

As noted earlier, the quantity b in the equation $y = mx + b$ represents the y intercept of the line. This can be verified quite easily by noting that the ordered pair $(0, b)$ satisfies the equation $y = mx + b$.

The x *intercept*, if it exists, is denoted as a and is defined as that point where the line intersects the x axis; it is represented by the ordered pair $(a, 0)$. Substituting this ordered pair into the equation $y = mx + b$ and solving for a enables one to write a in terms of b and m

$$0 = ma + b$$

yielding

$$a = -\frac{b}{m}, \ m \neq 0$$

The characteristics of the graph of the equation $y = f(x) = mx + b$ are shown in Figure 1.8. It should come as no surprise if we tell you that the equation

$$\boxed{y = mx + b} \qquad\qquad\qquad \textbf{(1.3.4)}$$

is also called the *slope-intercept* form of the equation of a straight line. NOTE: If the linear equation defining the function is not written in the form $y = mx + b$, it can be put into that form very easily.

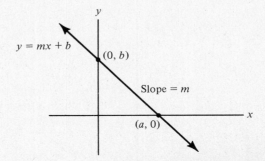

Figure 1.8

Example 3 | Write the linear equation $4x - 3y = 7$ in the slope-intercept form; determine the slope and the x and y intercepts.

Solution | The equation can be put into the slope-intercept form by subtracting $4x$ from both sides of the equation and then dividing every term by the coefficient of y, that is, -3. The result is

$$y = \frac{4x}{3} - \frac{7}{3}$$

from which the following values can be derived

$$m = \tfrac{4}{3} \qquad b = -\tfrac{7}{3} \qquad a = \tfrac{7}{4}$$

The graph of this function is shown in Figure 1.9.

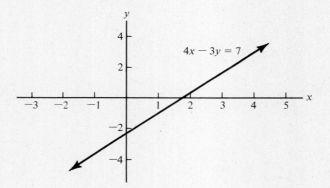

Figure 1.9

FINDING THE EQUATION OF A STRAIGHT LINE

So far, it has been assumed that we have been given the linear equation from which certain characteristics of the line such as the slope and the intercepts, together with the graph, could be deduced. Now we want to look at the problem in reverse: Given certain characteristics of a line or its graph, find the equation of the line.

Just as the graph of a straight line can be drawn if you are given either the coordinates of two points on the line or the slope together with the coordinates of one point, so too can the equation of the line be found if either of the following is given or can be found:

1. The slope m of the line and the coordinates of one point (x_1, y_1).
2. The coordinates of two points (x_1, y_1) and (x_2, y_2).

To see how this information is used, consider the following.

Example 4 | Find the equation of the line whose slope equals 2 and which passes through the point $(1, -3)$.

Solution | First, it is an easy task to construct the line by locating the point $(1, -3)$ and drawing through it a line whose slope equals 2; such a graph is shown in Figure 1.10. To obtain the equation satisfied by all points (x, y) on the line itself, we select an arbitrary point on the line and label its coordinates (x, y) as shown in Figure 1.10. We know that the slope of the line segment connecting (x, y) and $(1, -3)$ equals 2, a condition that can be expressed in the following way, using Equation 1.3.1:

$$\frac{y + 3}{x - 1} = 2$$

or

$$y + 3 = 2(x - 1)$$

yielding

$$y = 2x - 5$$

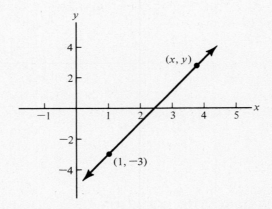

Figure 1.10

This technique can be generalized to the case in which the line passes through the fixed point (x_1, y_1) and has slope m as shown in Figure 1.11. The same procedure can be followed as in the previous example. An arbitrary point (x, y) is selected, and again noting that the slope of the line segment joining (x, y) to (x_1, y_1) is m, we have

$$m = \frac{y - y_1}{x - x_1}$$

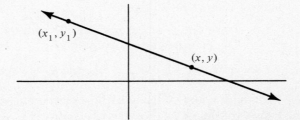

Figure 1.11

Multiplying both sides of the equation by $(x - x_1)$, we have the formula

$$\boxed{y - y_1 = m(x - x_1)}$$

(1.3.5)

Equation 1.3.5 is called the *point-slope* formula because it indicates the basic information needed to employ the formula.

The point-slope formula can also be used when the coordinates of two points on the straight line are given.

Example 5 | Find the equation of the straight line that passes through $(6, -1)$ and $(1, 4)$.

Solution | First, it is possible to find the slope of the line

$$m = \frac{4 - (-1)}{1 - 6} = -1$$

and then to use the point-slope formula. The question then arises: Which of the two points is substituted for (x_1, y_1) in the point-slope formula? The answer: Use either because they both lie on the straight line whose equation we want to develop

$$x_1 = 6 \quad y_1 = -1 \qquad\qquad x_1 = 1 \quad y_1 = 4$$
$$y - (-1) = -1(x - 6) \qquad y - 4 = -1(x - 1)$$

yielding
$$y = -x + 5$$

Thus, the same equation results when we substitute either ordered pair into the point-slope formula.

The point-slope formula, Equation 1.3.5, can be cast into a second well-known form in the following way: substitute into the equation

$$y - y_1 = m(x - x_1)$$

the ordered pair $(0, b)$ for (x_1, y_1), getting

$$y - b = mx$$

from which we obtain

$$y = mx + b$$

the slope-intercept equation.

Because the point-slope and slope-intercept equations are equivalent, it makes no difference which one is used to obtain the equation of a straight line.

Example 6   Find the equation of the straight line that passes through the points $(1, -2)$ and $(3, -8)$.

Solution   a. *Point-slope formula 1.3.5*

The slope of the line can be found easily

$$m = \frac{-8 - (-2)}{3 - 1} = -3$$

Substituting one of the points, say $(1, -2)$, into the point-slope formula gives

$$y - (-2) = -3(x - 1)$$

from which we get

$$y = -3x + 1$$

b. *Slope-intercept formula 1.3.4*

The slope is known from part a, so we have

$$y = -3x + b$$

Because either ordered pair must satisfy this equation, we arbitrarily select $(1, -2)$ and substitute this pair, giving

$$-2 = -3(1) + b$$

or

$$b = 1$$

so the equation again reads

$$y = -3x + 1$$

The equations of horizontal and vertical lines such as those shown in Figure 1.12 have a very simple form. Because a horizontal line represents a set of ordered pairs (x, y) for which the y coordinate is constant and equal to the y intercept b, its equation assumes the simple form

$$\boxed{y = b}$$

(1.3.6)

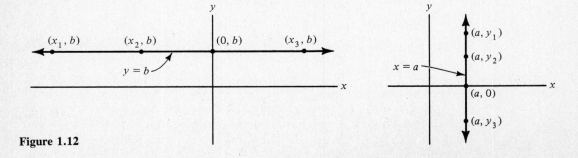

Figure 1.12

In the same way, because a vertical line represents a set of ordered pairs (x, y) for which the x coordinate is constant and equal to the x intercept a, its equation assumes the simple form

$$x = a$$

(1.3.7)

The equations of the horizontal and vertical lines shown on the following graphs are $y = 3$ and $x = -2$, respectively.

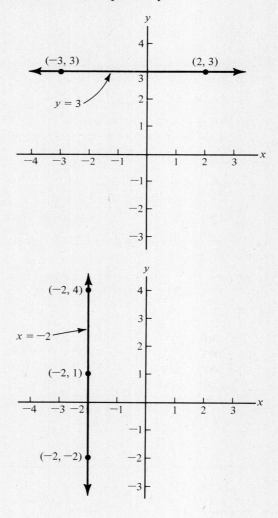

EXERCISE 1.3

1. On the following coordinate system, draw the graph of each of the following lines.
   a. The line has slope equal to $+2$ and passes through $(1, 3)$.
   b. The line passes through $(-2, 4)$ and $(1, -1)$.

c. The line passes through $(3, -1)$ and has slope equal to $-\frac{3}{4}$.

d. The line passes through $(-2, 4)$ and $(-2, -1)$.

e. The line passes through $(1, -3)$ and $(-1, -3)$.

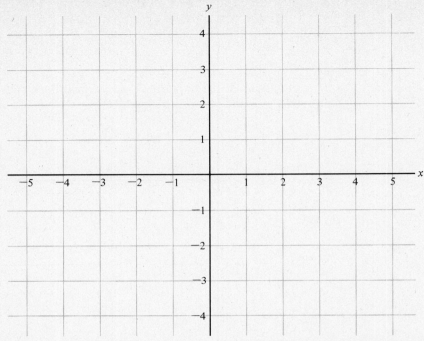

For each linear equation in Exercises 2–5, sketch the graph and find the slope and intercepts of the line representing each.

2. $y = 2x + 5$

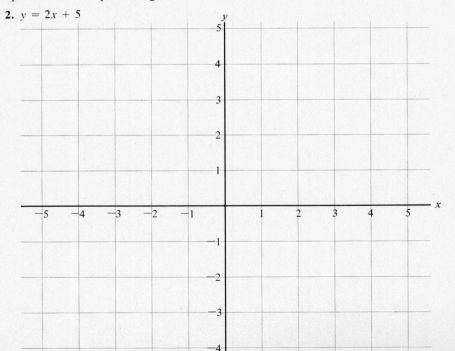

3. $y = -3x - 2$

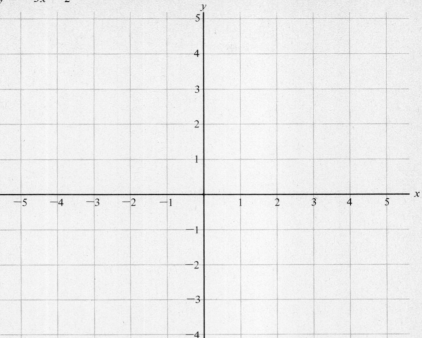

4. $3x + 5y = 10$

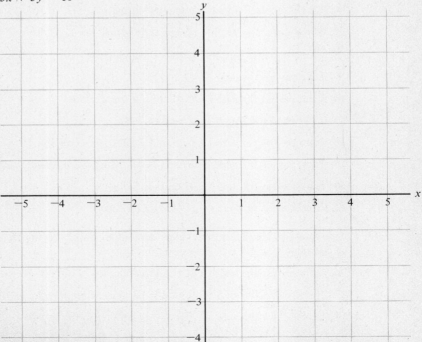

5. $8y - 2x = 5$

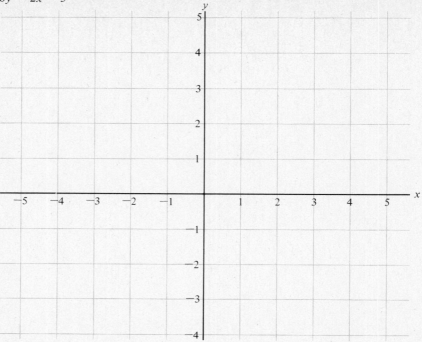

Find the equation of each straight line described in Exercises 6–15.

6. Slope equals -4 and line passes through $(1, 5)$.

7. Slope equals $\frac{2}{3}$ and line passes through $(6, -4)$.

8. Slope equals 3 and line passes through $(\frac{1}{2}, \frac{3}{2})$.

9. Line passes through $(1, 1)$ and $(2, 0)$.

10. Line passes through $(3, -2)$ and $(2, 3)$.

11. Line passes through $(3, \frac{1}{2})$ and $(4, -\frac{1}{2})$.

12. Line passes through $(7, -3)$ and y intercept equals 2.

13. Line passes through (c, d) and $(-d, -c)$, where c and d are constants.

14. Line passes through $(3, 1)$ and $(3, 5)$.

15. Line passes through $(-2, 4)$ and $(3, 4)$.

Two lines are parallel if their slopes are equal. Using this information, find the equation of each line described in Exercises 16–18.

16. Line passes through $(-2, 1)$ and is parallel to the line $6x + 3y = 7$.

17. Line passes through $(3, 3)$ and is parallel to the line $2x - 3y = 5$.

18. Line passes through $(\frac{3}{2}, -\frac{1}{2})$ and is parallel to the line $4x - 2y = 3$.

19. a. The Speedy Rent-a-Car Agency charges $12 per day plus 20¢ per mile for renting their compact autos. Find the equation that gives the dollar charge D as a function of the mileage M driven each day. Graph the function on the following coordinate system.

   b. The Drive-away Rent-a-Car Agency charges $5 per day plus 25¢ per mile for renting their compact autos. Again, find the equation that gives the dollar charge D as a function of the mileage M driven each day, and graph the function on the adjacent coordinate system.

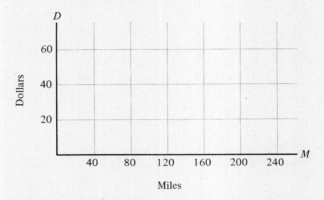

   c. How many miles per day would a person have to drive in order to make the Speedy offer more attractive than that of the Drive-away Agency.

20. a. What is the slope of a horizontal line segment?

   b. What is the equation of the following horizontal line?

   c. What is the equation of the following vertical line?

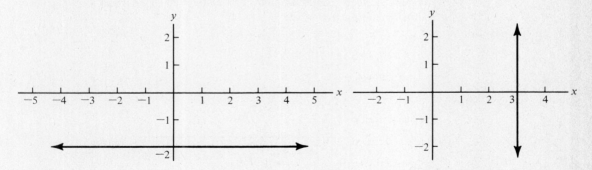

21. When borrowing or lending money for a short period of time, usually less than one year, the interest I paid on the principal P (the amount borrowed or loaned) is often calculated according to the simple interest equation

$$I = Prt$$

in which r is the annual interest rate and t is the time expressed in years. The amount or value of the loan A grows with time according to the equation

$$A = P + I = P(1 + rt)$$

a. If a person takes out a six-month note for $1000 at 18 percent per year from a local finance company, what is the equation that gives A as a function of t where t is expressed in years?

b. If the variable t is expressed in units other than years, say for example months, then the interest rate r must be expressed as a monthly rate. For the problem in part a, express A as a function of t where t is expressed in months.

c. On the following coordinate system, plot a graph of the equation found in part b.

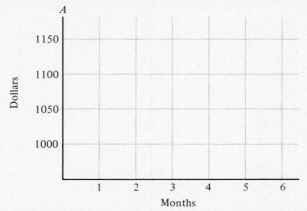

d. If the individual decides to pay off the note in four months, how much money does she have to pay the finance company?

22. The United States is in the process of adopting the metric system. One of the changes this entails is measuring temperature in degrees Celsius (°C) instead of degrees Fahrenheit (°F). If the relationship between the two systems is linear, find the equation that expresses the Celsius temperature T_C as a function of the Fahrenheit temperature T_F if 32°F equals 0°C and 212°F equals 100°C and graph the equation on the following coordinate system.

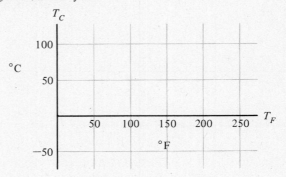

❧ 1.4 Applications of Linear Functions

Linear functions can be applied to the analysis of many elementary situations commonly encountered in the fields of business and economics; some of these applications are illustrated in this section. They are by no means meant to be complete and exhaustive but are intended only to demonstrate that linear functions can be applied in meaningful ways.

STRAIGHT-LINE DEPRECIATION

When a firm purchases a piece of equipment with a useful life of many years, the total cost of the item cannot be charged as an expense at the time of purchase; instead, this expense is spread out over the useful life of the asset. As this expense is applied periodically, it is recorded on the income statement as a cost known as *depreciation*. In addition to "spreading out" the expense, depreciation also indicates how rapidly the value of the asset, recorded in the balance sheet, is decreasing.

One of the more common depreciation techniques is the *straight-line* method in which the decline in value is uniform from one period to the next. To illustrate this technique, suppose we use the following notation:

C—original cost of the item (in dollars)
T—useful life of the item (in years)
S—salvage or resale value of the item (in dollars)

The annual depreciation D, expressed in dollars per year, is given by the equation

$$D = \frac{C - S}{T}$$

(1.4.1)

All the quantities introduced thus far are constants for any item. To develop an equation describing the worth or value of any item in terms of the length of time from the date of purchase, let us introduce two variables:

1. The independent variable t, which expresses the length of time— usually in years—from the date of purchase ($t = 0$) to the date at which the value of the item is determined.
2. The dependent variable V, which gives the value of the item at any time t.

Assuming that the item decreases in value at a uniform rate, the relationship between V and t is given by the equation

$$V = -Dt + C$$

(1.4.2)

Example 1 | Amalgamated Products purchased an executive jet for $1,500,000. The company plans to use the plane for 10 years at which time they expect to sell it for $500,000. What is the form of the equation expressing the relationship between book value V and time t from the date of purchase? Draw a graph of the function.

Solution | The annual depreciation D is given by

$$D = \frac{1,500,000 - 500,000}{10} = 100,000 \text{ dollars/year}$$

so

$$V = -100,000t + 1,500,000 \qquad 0 \le t \le 10$$

when

$$t = 2, \ V = -100,000(2) + 1,500,000 = 1,300,000 \text{ dollars}$$

$$t = 7, \ V = -100,000(7) + 1,500,000 = 800,000 \text{ dollars}$$

Following is the graph of the function.

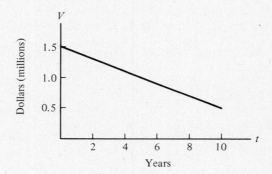

LINEAR DEMAND EQUATIONS

Although you may not yet have been formally exposed to the concept of a demand equation, you have been exposed to the basic ideas underlying it by virtue of living in a competitive business environment. You know that the price of an item plays an important part in your decision to purchase or not to purchase the item; the increased attractiveness of an item as its purchase price decreases forms the personal aspect of the demand phenomenon. As an example, the recent large price drops in competing calculator models have resulted in a tremendous surge in the numbers purchased by the general public; the low price of some very exotic calculators has all but made the slide rule a relic of the not-too-distant past.

The demand equation represents an attempt to present the unit price p of an item as a function of the number of items q (for quantity) that can be sold.* If a demand equation is linear, we would expect the slope of the line to be negative because high prices are generally associated with low sales and low prices with high sales (see following graph).

---

* Economists generally draw demand curves with the horizontal axis reserved for q and the vertical axis for p. For this reason, we treat q as the independent variable and p as the dependent variable even though it is more logical to reverse their roles.

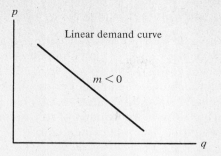

Linear demand curve

$m < 0$

Example 2   Suppose the accompanying table contains the results of a market research study conducted by the MacDougall Hamburger Co. to determine the relationship between unit price and weekly sales for its new Gluttonburger (one pound of meat in a six-inch roll); the results are shown graphically on the adjacent coordinate system. Find the linear equation that describes the relationship between p and q.

q(thousands)	p(dollars)
50	1.00
40	1.25
30	1.50
20	1.75

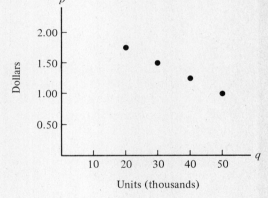

Units (thousands)

Solution   The linear equation that gives p in terms of q can be developed using the methods described in Section 1.3. Using the extreme values given in the preceding table, we get for the slope of the line

$$m = \frac{p_2 - p_1}{q_2 - q_1} = \frac{1.00 - 1.75}{50 - 20} = -0.025$$

Next, using the point-slope formula, Equation 1.3.5, in the form

$$p - p_1 = m(q - q_1)$$

we get using $(50, 1.00)$ for (q_1, p_1)

$$p - 1 = -0.025(q - 50)$$

$$p = -0.025q + 2.25 \qquad 20 \le q \le 50$$

Caution would have to be exercised if you were to attempt to extend the validity of the equation to values of q beyond the domain. For example, the equation implies that the company could charge nothing for the burger ($p = 0$) and yet only 90 thousand would be "sold" in this unusual situation.

LINEAR SUPPLY EQUATIONS

Just as a demand equation describes mathematically how consumers react to price changes, a *supply* equation describes mathematically how sellers or producers of goods react to price changes. The supply equation gives the unit price p as a function of the number of items q produced or made available for sale. If the supply equation is linear, we would expect its slope to be positive because producers are more inclined to step up output as unit prices rise in order to increase profits (see following graph).

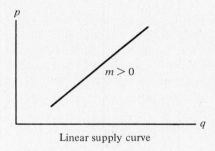

Linear supply curve

Example 3 | When the price per bushel of wheat was $2.50, 75 million bushels per month were offered for sale. When the price increased to $3.00 per bushel, the number of bushels of wheat rose to 100 million per month. Find the supply equation, assuming that it is linear.

Solution | The supply curve is shown on the following figure. Using the methods developed in Section 1.3, we get the equation by first finding the slope

$$m = \frac{3.00 - 2.50}{100 - 75} = 0.02$$

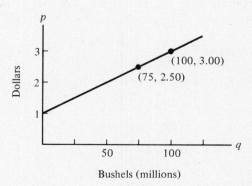

Bushels (millions)

Again using the point-slope formula, Equation 1.3.5, in the form

$$p - p_1 = m(q - q_1)$$

we get using (100, 3) for (q_1, p_1)

$$p - 3 = 0.02(q - 100)$$

$$\boxed{p = 0.02q + 1}$$

BREAK-EVEN ANALYSIS

In the operation of any enterprise, one is always concerned about income or revenue generated through sales and with the costs associated with operating the enterprise. The operation generates a profit when revenue exceeds the costs, whereas it is operating at a loss when the reverse is true. The profit P can be written in terms of the revenue R and total cost C as

Profit = Revenue − Total cost

or

$$P = R - C \tag{1.4.3}$$

In most applications, the goal is to express R and C and thus P as functions of x, the number of items sold. The revenue R can be written as

$$R = px$$

where p is the unit price. In many situations the total cost C can be written as the sum of two terms, that is

$$C = F + V$$

where F represents what are known as *fixed costs,* costs independent of the number of units sold, such as rent, office equipment, and insurance. V represents the *variable costs,* such costs as labor, raw material, transportation and machinery, which are functions of the number of units produced. Like the revenue function, V can often be written as

$$V = vx$$

where v is the unit variable cost. The profit function P can then be written in terms of x as

$$P = px - (F + vx)$$

$$\boxed{P = (p - v)x - F} \tag{1.4.4}$$

Example 4   A company manufacturing specialty T-shirts sells the shirts to a wholesaler for $2 apiece. The company's annual fixed costs are $70,000, and the variable unit cost is $0.60. Find the profit function P as a function of x, the number of T-shirts produced and sold.

Solution | Noting that $F = 70{,}000$, $p = 2.00$, and $v = 0.60$, Equation 1.4.4 becomes

$$P = (2.00 - 0.60)x - 70{,}000 = 1.40x - 70{,}000$$

Following is the graph of the profit function.

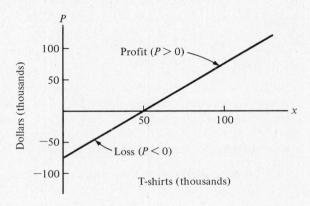

The x intercept on a profit curve is called the *break-even* point, that is, the point where $P = 0$; the corresponding value of x represents the number of items that must be produced and sold to cover costs. The break-even point is found by setting P equal to zero in Equation 1.4.4 and solving the resulting equation for x. Denoting the solution as x_{BE}, we get the following:

$$0 = (p - v)x_{BE} - F$$
$$F = (p - v)x_{BE}$$

or

$$\boxed{\;x_{BE} = \frac{F}{p - v}\;}$$

Example 5 | Find the break-even point for the company producing specialty T-shirts (Example 4).

Solution | For the break-even value of x, we get

$$x_{BE} = \frac{70{,}000}{2.00 - 0.60} = 50{,}000$$

Equation 1.4.3 indicates that the break-even point ($P = 0$) occurs when revenue R equals total costs C. This is shown for the T-shirt manufacturer on the following graph where both the revenue curve $R = 2.00x$ and total cost curve $C = 0.60x + 70{,}000$ are plotted. Break even occurs when x equals 50,000 units and $R = C = 100$ thousand dollars.

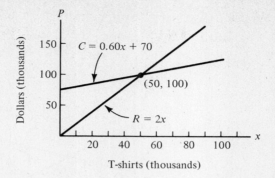

EQUILIBRIUM EQUATIONS

In a supply-demand situation, market equilibrium is achieved when the quantity offered for sale equals the quantity consumers plan to buy. Graphically, the point where the supply and demand curves intersect gives the equilibrium quantity q_E and price p_E, illustrated in the following figure.

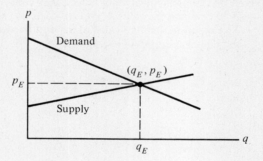

Example 6 | Find the equilibrium quantity and price for a commodity whose supply and demand equations are

$$p = 0.5q + 3 \text{ (Supply)}$$

$$p = -q + 6 \text{ (Demand)}$$

Solution | Because the ordered pair (q_E, p_E) satisfies both the supply and the demand equation, we have at equilibrium

$$p_E = 0.5q_E + 3$$

$$p_E = -q_E + 6$$

or equivalently

$$0.5q_E + 3 = -q_E + 6$$

Solving this equation for q_E gives

$$q_E = 2$$
$$p_E = 4$$

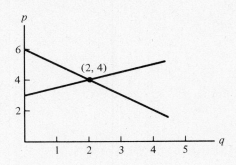

EXERCISE 1.4

1. A taxi company purchases a fleet of 10 cabs for \$35,000. If the useful life of the taxis is 3 years and the salvage value of each taxi is estimated to be \$500 at the end of the 3-year period, find the value of the fleet as a function of time and plot the function on the adjacent coordinate system.

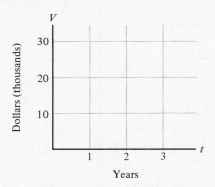

2. The owner of a photo studio purchased an enlarger for \$1500. If the lifetime of the enlarger is 12 years with a salvage value of \$300, find

   a. Its annual depreciation.
   b. Its book value when it is 5 years old.
   c. The equation showing the book value V as a function of time t in years.

3. A small engineering company recently purchased a portable computer terminal for \$4400. The lifetime was expected to be five years with a salvage value of \$400. Because of rapid technological developments taking place in the computer industry, the management of the company has revised downward the lifetime to four years and the salvage value to \$200. Find the equations that give the book value of the terminal as a function of time under the five- and four-year lifetime assumptions.

4. A company that manufactures lawn furniture finds it can sell 200,000 units if it charges $10 for its best-selling chaise longue. The company expects to sell 250,000 units when the price is decreased to $9 per unit. Find the demand equation and plot it on the following coordinate system.

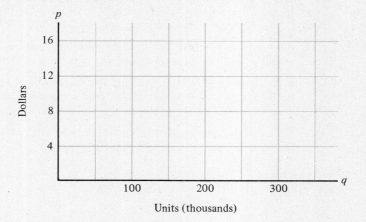

5. A group sponsoring a rock concert is trying to determine the ticket price it should charge. On the basis of past experience, it estimates that an audience of 10,000 people can be drawn if the ticket price is $15. For each dollar increase in the price of a ticket, 500 fewer people will attend the concert.

a. Find the demand equation, assuming that it is linear.
b. If the concert hall holds 8750 people, what is the maximum ticket price that will ensure a full house?

6. Because of a rapid increase in the price of silver, a film manufacturing company raised the price on its 20-exposure roll from $1.75 to $2.00. Immediately, monthly sales of the film decreased from 2 million to 1.8 million rolls. Find the linear equation that describes the price p in dollars in terms of monthly sales q expressed in millions of rolls.

7. The manufacturer of a new tennis racket produces 500 units per week, all of which are sold to a wholesaler for $20 each. When the racket proves to be very popular, the wholesaler increases the weekly order to 900. The manufacturer is unwilling to allocate additional resources to producing tennis rackets but agrees to do so when the wholesaler indicates a willingness to pay $25 for each racket. Find the linear supply equation for this situation.

8. In planning production for the coming year, the operations research group in a large office-equipment company recommends manufacturing 30,000 electric typewriters if the selling price remains constant at $150 per unit. They also recommend manufacturing 4000 additional typewriters for each $10 increase in the selling price. Find the supply equation for this product.

9. A retired carpenter rents a small shop where he makes and sells doll houses. The houses sell for $75 each while material for each costs $20. If rent and utilities cost $220 per month, find the equation that describes his monthly profit P in terms of x, the number of doll houses sold each month. How many houses must be sold to break even?

10. A small company manufactures an indoor-outdoor thermometer it sells to whole-salers for $6 each. Labor and material costs for each thermometer are $4 while fixed annual costs are $25,000.

a. How many thermometers must be sold for the company to break even?
b. Find the equation that describes the company's annual profit as a function of x, the number of thermometers sold each month.

11. A sales representative for the Data Base, Inc. company, an internationally known manufacturer of computer equipment, has been offered two salary plans

a. Base salary of $500 per month plus a commission of 3 percent on all sales, or
b. Base salary of $600 per month plus a commission of 5 percent on all sales in excess of $15,000 per month.

Write the functions representing the gross monthly salary under each plan and plot the functions on the adjacent coordinate system. For what range of sales is Plan a preferable to Plan b?

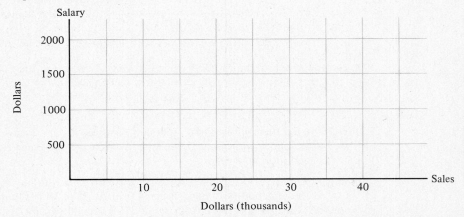

12. Sally Senior recently sold her home for $50,000 and moved into a one-bedroom apartment in an elderly citizens' housing complex. She has been advised to invest the $50,000 and use the income generated to supplement her social security pay-ments. After some study, the number of investment possibilities have been nar-rowed to the following: (1) AAA corporate bonds yielding 8 percent interest per year, and (2) bank certificates of deposit yielding 6.5 percent interest per year.

a. Assuming the entire $50,000 is to be invested in one or both of these, find her annual income I as a function of x, the amount invested in corporate bonds.
b. Plot the function on the following graph.

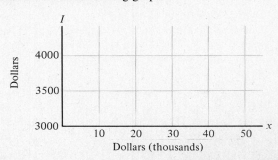

13. A nationwide manufacturer of calculators finds that its profits are $4 million when it sells 300,000 units. If the company's fixed costs are $800,000 and unit variable costs are $20 per calculator, find

    a. The unit price of each calculator.
    b. The profit P as a function of x, the number of calculators sold.
    c. The break-even point.

For Exercises 14–17, find the values of q and p at which market equilibrium is achieved.

14. $p = 0.5q + 3$ (supply)
$\quad\;\;p = -0.3q + 7$ (demand

15. $p = 0.02q + 5$ (supply)
$\qquad\;p = -0.04q + 8$ (demand)

16. $p = 0.35q + 1.3$ (supply)
$\quad\;\;p = -0.15q + 5.5$ (demand)

17. $p = 0.6q + 2.3$ (supply)
$\qquad\;p = -0.8q + 6.7$ (demand)

❧ 1.5 Quadratic Functions and Applications

The two preceding sections were devoted to studying linear functions and some of their applications. Although functions of this type are important, their usefulness is very limited. Nonlinear functions are generally more useful because they represent more accurately those situations in science, business, and economics where mathematics can be applied. To see how nonlinearity arises, consider the following example.

Example 1 A retail store finds that the relationship between the unit price p it charges for its best-selling CB radio and the number of units q it can sell per month is described by the equation

$$p = 300 - 10q \qquad 5 \le q \le 20$$

Find the monthly revenue function $R(q)$.

Solution The monthly revenue R that the company realizes from sales of the radio is related to the number sold via the equation

$$R(q) = pq = (300 - 10q)q = 300q - 10q^2 \qquad 5 \le q \le 20 \qquad \textbf{(1.5.1)}$$

The presence of the quadratic term, that is, $-10q^2$, indicates that the revenue R is not a linear function of q. The nonlinearity of the function can be seen vividly by plotting R as a function q; the results are shown in Figure 1.13. Inspection of

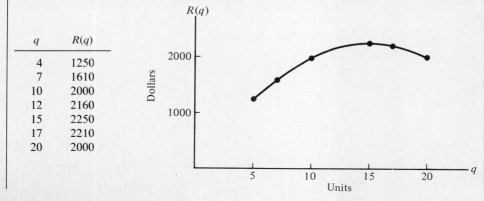

q	$R(q)$
4	1250
7	1610
10	2000
12	2160
15	2250
17	2210
20	2000

Figure 1.13

the graph indicates that revenue increases as sales increase when $q < 15$, but revenue declines as q increases when $q > 15$. From this result, it is easy to see that the monthly revenue is maximized when 15 radios are sold; to attain that goal, the unit price of each radio should be set at

$p = 300 - 10(15) = 150$ dollars

A *quadratic* function is defined as one having the form

$$y = f(x) = ax^2 + bx + c$$

(1.5.2)

where a, b, and c are constants and $a \neq 0$. The graph of a quadratic function is a curve called a *parabola,* which assumes one of the two shapes shown in Figure 1.14. The curve will have

1. The bowllike shape shown in Figure 1.14 (a) when a is positive.
2. The inverted bowllike shape shown in Figure 1.14 (b) when a is negative.

The graph of a parabola can be drawn quickly once the following selected points are found:

y **Intercept**   This is found by setting $x = 0$ in Equation 1.5.2; the result is $y = c$.

x **Intercepts** (if they exist)   They are found by setting $y = 0$ in Equation 1.5.2 and solving the resulting quadratic equation

$ax^2 + bx + c = 0$

(1.5.3)

for x. The solution(s) of Equation 1.5.3 can be found by factoring (where possible) or by using the quadratic formula

$$x = \frac{-b \pm \sqrt{b^2 - 4ac}}{2a}$$

(1.5.4)

Figure 1.14   (a)   (b)

The discriminant ($b^2 - 4ac$) determines the number of x intercepts as follows:

1. If $b^2 - 4ac > 0$, there are two real and distinct solutions to Equation 1.5.3 and the parabola intersects the x axis at two points whose x coordinates are designated as x_1 and x_2 (Figure 1.15).

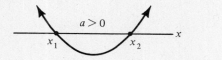
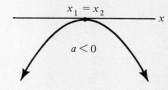

Figure 1.15

2. If $b^2 - 4ac = 0$, there is one real solution to Equation 1.5.3 and the parabola intersects the x axis at one point, that is, $x_1 = x_2$ (Figure 1.16).

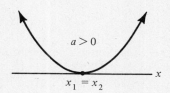

Figure 1.16

3. If $b^2 - 4ac < 0$, there are no real solutions and the parabola lies totally above or below the x axis as shown in Figure 1.17.

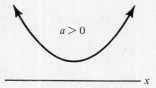
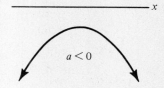

Figure 1.17

The Vertex   The vertex is the point on the parabola where the y coordinate is either a *minimum* ($a > 0$) or a *maximum* ($a < 0$); thus we look for either the *lowest* or *highest* point on the curve. If we label the x and y coordinates of the vertex as x_V and y_V, they can be found by the following formulas:

$$x_V = -\frac{b}{2a} \qquad y_V = \frac{4ac - b^2}{4a} \tag{1.5.5}$$

These points are shown in Figure 1.18 on page 54.

The formulas for x_V and y_V can be found by applying the method of comple-

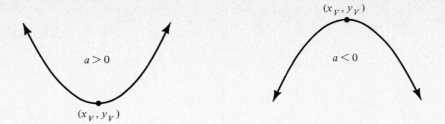

Figure 1.18

ting the square to the right-hand side of Equation 1.5.2. The first step is to rewrite the equation as

$$y = a \left(x^2 + \frac{b}{a} x\right) + c$$

Adding and subtracting $b^2/4a$ on the right-hand side gives

$$y = a \left(x^2 + \frac{b}{a} x + \frac{b^2}{4a^2}\right) + c - \frac{b^2}{4a}$$

The trinomial inside the parentheses becomes a perfect square and the equation can be written as

$$y = a \left(x + \frac{b}{2a}\right)^2 + c - \frac{b^2}{4a} \qquad\qquad \textbf{(1.5.6)}$$

Examination of Equation 1.5.6 indicates that

1. When $a > 0$, $a(x + b/2a)^2$ is either positive or zero so that the *minimum* value of y will occur when $x + b/2a = 0$, or $x = -b/2a$.
2. When $a < 0$, $a(x + b/2a)^2$ is either negative or zero so that the *maximum* value of y will occur also when $x + b/2a = 0$, or $x = -b/2a$.

Substituting $x = -b/2a$ into Equation 1.5.6 gives for the corresponding value of y

$$y = c - \frac{b^2}{4a} = \frac{4ac - b^2}{4a}$$

Example 2 Find the y intercept, the x intercepts (if any), and the vertex of the parabola whose equation is

$$y = f(x) = 2x^2 - 3x - 5$$

Solution a. The y intercept equals -5.
b. The x intercepts are found by solving the equation $2x^2 - 3x - 5 = 0$. By either factoring the left-hand side, that is, $(2x - 5)(x + 1)$, and solving the resulting equation $(2x - 5)(x + 1) = 0$ for x or by using the quadratic formula, and x intercepts are found to be

$$x_1 = \tfrac{5}{2} \qquad x_2 = -1$$

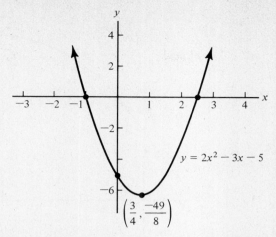

Figure 1.19

c. The coordinates of the vertex are found by using Equation 1.5.5, and the results are

$$x_V = \tfrac{3}{4} \qquad y_V = -\tfrac{49}{8}$$

Since the coefficient of x^2, that is, 2, is positive, the vertex $(\tfrac{3}{4}, -\tfrac{49}{8})$ is a minimum. These points together with a rough sketch of the parabola are shown in Figure 1.19.

Problem 1 Find the y intercept, the x intercepts (if they exist), and the vertex of the parabola whose equation is

$$y = f(x) = 6 + x - x^2$$

Answer y intercept: (0, 6)
$\qquad\quad$ x intercepts: (3, 0) and $(-2, 0)$
$\qquad\quad$ vertex: $(\tfrac{1}{2}, \tfrac{25}{4})$ maximum

---

APPLICATION: QUANTITY OR GROUP DISCOUNTS

Piecewise defined functions in which one equation is quadratic can arise in situations involving quantity discounts.

Example 3 A charter airline charges a flat rate of $200 per passenger for groups of 50 or fewer people. For each additional passenger over 50, the airline reduces by $1 the ticket price for all passengers in the group. Find the equation that gives the airline's revenue R as a function of x, the number of passengers in a group.

Solution First, the ticket price p can be written as

$$p = \begin{cases} 200 & x \le 50 \\ 200 - 1(x - 50) = 250 - x, & x > 50 \end{cases}$$

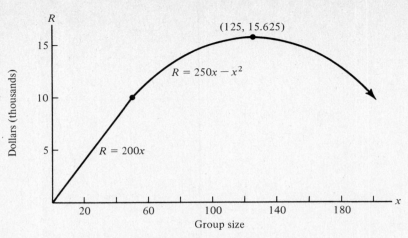

Figure 1.20

so that the revenue function $R(x)$ which has the form

$$R(x) = px$$

can be written

$$R(x) = \begin{cases} 200x & x \le 50 \\ 250x - x^2 & x > 50 \end{cases}$$

The graph of this function is shown in Figure 1.20. The maximum revenue is attained when the group size equals 25.

EXERCISE 1.5

1. Plot the graph of each of the following functions on the following coordinate system.

    a. $y = x^2$ **b.** $y = 2x^2$ **c.** $y = x^2/4$ **d.** $y = -x^2$

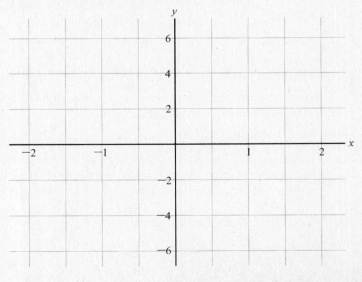

2. Plot each of the following functions on the following coordinate system.

   a. $y = x^2 - x$ **b.** $y = x^2 - 2x$ **c.** $y = x^2 + x$ **d.** $y = -x^2 + x$

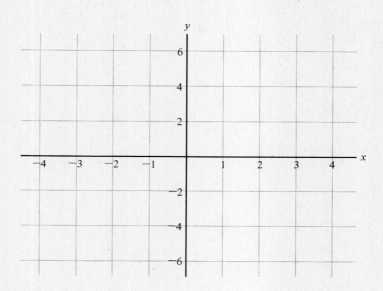

For each quadratic function in Exercises 3–7, find: (a) the y intercept, (b) the x intercepts (if any), and (c) the vertex. In addition, plot the graph of the function using the information obtained in parts (a), (b), and (c).

3. $y = 2x^2 - 3x + 1$

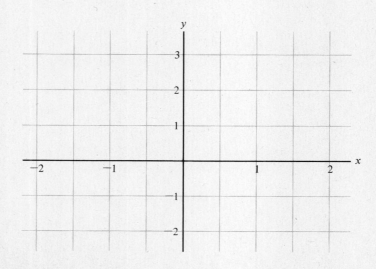

4. $y = -2x^2 - 4x + 6$

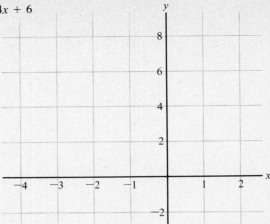

5. $y = x^2 - 8x + 16$

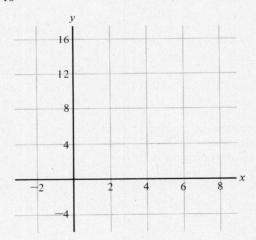

6. $y = 2x^2 - 5x + 5$

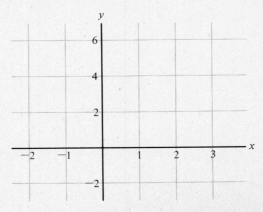

7. $y = -2x^2 - 4x + 1$

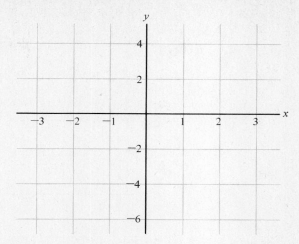

8. A small theater offers group discounts on tickets according to the following plan. For groups of 20 or fewer, ticket prices are $6 per person. For groups of more than 20, ticket prices for all members of the group are reduced by 10¢ per ticket for each additional person over 20. Find the revenue function $R(x)$ in terms of x, the number of people in a group. Find the size of the group for which $R(x)$ is maximized and plot the graph of $R(x)$ on the accompanying coordinate system.

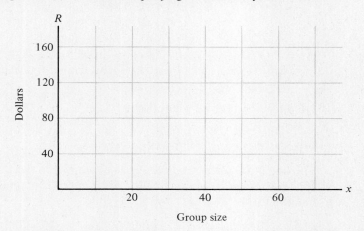

9. A resort hotel that has 100 rooms can rent all of its rooms if it charges $25 per day per room. For each dollar increase in the daily rate, the number of vacant rooms increases by 2. If the cost of cleaning and maintenance is $4 per day per room occupied, what room rate will maximize the profit function for the hotel?

10. Anyone who lives outside the Sun Belt knows and has experienced the additional discomfort produced by high winds on a cold winter day. The addition of the wind on top of the low temperatures produces what is known as a wind-chill factor, that is, the temperature feels lower than that recorded on a thermometer. If we let T_A represent the air temperature (°F), that is, the thermometer reading, V the wind speed in miles per hour, and T the wind-chill temperature (°F), the relationship between T and V can be expressed approximately by the equation

$$T = 0.03V^2 - 2.4V + 10 \qquad 0 \leq V \leq 40$$

when $T_A = 10°F$.

a. If the wind speed V equals 30 mph, what is the wind-chill temperature?
b. Plot a graph of the function on the accompanying coordinate system.
c. Wind speeds greater than 40 mph have little additional effect on the wind-chill temperature; how would you complete the graph to show this feature?

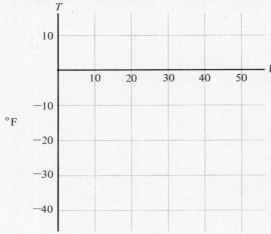

Miles per hour

11. For the group discount problem described in Example 3, find the revenue function R when the independent variable x represents the number of passengers over 50.

1.6 Other Types of Functions

POLYNOMIAL FUNCTIONS

The linear and quadratic functions examined in the three previous sections are special cases of functions called *polynomial* functions. A polynomial function has the form

$$y = f(x) = a_0 + a_1x + a_2x^2 + \cdots + a_nx^n \qquad (1.6.1)$$

where n is a nonnegative integer and $a_0, a_1, a_2, \ldots, a_n$ are real numbers. The following are examples of polynomial functions:

1. $y = f(x) = 2x^3 - 7x^2 + 3x + 6$
   $a_3 = 2 \qquad a_2 = -7 \qquad a_1 = 3 \qquad a_0 = 6$

2. $y = f(x) = 5 - 8x^4$
   $a_4 = -8 \qquad a_3 = a_2 = a_1 = 0 \qquad a_0 = 5$

3. $y = f(x) = \dfrac{x^3}{4} + \dfrac{7x^2}{3} - \dfrac{2x}{9} + 1$

$$a_3 = \tfrac{1}{4} \qquad a_2 = \tfrac{7}{3} \qquad a_1 = -\tfrac{2}{9} \qquad a_0 = 1$$

At this stage, no effort will be made to go into graphing polynomials because the point-by-point technique used so far is not the best method; more efficient techniques will be explored later in the text. However, simple power functions of the form

$$y = f(x) = ax^n \qquad\qquad\qquad\qquad\qquad \textbf{(1.6.2)}$$

can be graphed without much difficulty.

Example 1 | Plot a graph of each of the following functions:

$$y = f(x) = x^3 \qquad \text{and} \qquad y = g(x) = x^4$$

Solution | Using the accompanying table, we can plot each of the curves shown in Figure 1.21.

x	$y = x^3$	$y = x^4$
-3	-27	81
-2	-8	16
-1	-1	1
0	0	0
1	1	1
2	8	16
3	27	81

Figure 1.21

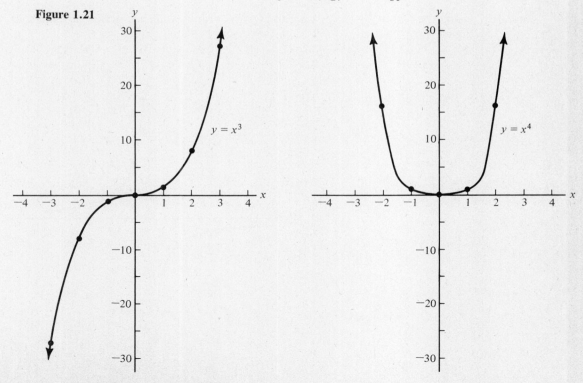

RATIONAL FUNCTIONS

The polynomial functions studied in the first part of this section are special cases of still another type of function, a *rational* function. A rational function is a function that can be written or expressed as the ratio of two polynomials, for example

1. $y = f(x) = \dfrac{x}{x + 3}$

2. $y = g(x) = \dfrac{3x^4 - 2x + 7}{8 - x^2}$

3. $y = h(x) = \dfrac{2 + x}{x^2 - 3x + 2}$

Again, we will not go into plotting rational functions at this time. However, it is possible to plot simple rational functions of the type

$$y = f(x) = \frac{a}{x^n} \tag{1.6.3}$$

Example 2 | Plot a graph of each of the following functions:

$$y = f(x) = \frac{1}{x} \quad \text{and} \quad y = g(x) = \frac{-1}{x^2}$$

Solution | Using the following table, we can plot the graph of each function as shown in Figure 1.22.

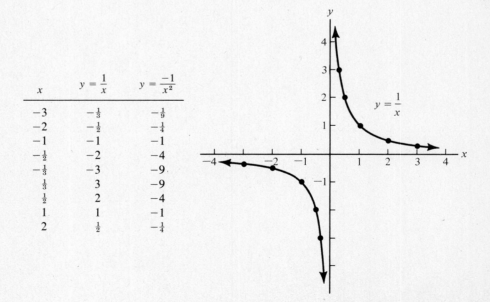

x	$y = \dfrac{1}{x}$	$y = \dfrac{-1}{x^2}$
-3	$-\frac{1}{3}$	$-\frac{1}{9}$
-2	$-\frac{1}{2}$	$-\frac{1}{4}$
-1	-1	-1
$-\frac{1}{2}$	-2	-4
$-\frac{1}{3}$	-3	-9
$\frac{1}{3}$	3	-9
$\frac{1}{2}$	2	-4
1	1	-1
2	$\frac{1}{2}$	$-\frac{1}{4}$

Figure 1.22

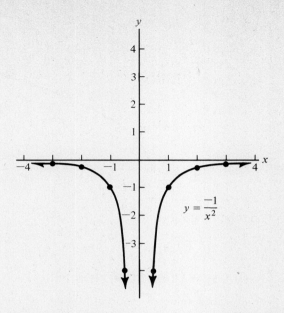

Figure 1.22
(continued)

One other feature should be pointed out regarding rational functions: the domain of a rational function does not include those values of x for which the denominator equals 0. For example, the function

$$y = f(x) = \frac{2x + 1}{x - 5}$$

has as its domain all real values of x except $x = 5$, whereas the function

$$y = f(x) = \frac{3}{x^2 - 4} = \frac{3}{(x + 2)(x - 2)}$$

has as its domain all real values of x except $x = 2$ and $x = -2$.

SQUARE ROOT FUNCTION

The next function we shall consider in this section is the square root function

$$y = f(x) = \sqrt{x}$$

whose graph is shown in Figure 1.23. The domain of this function is the set

$$D = \{x | x \geq 0\}$$

x	$y = \sqrt{x}$
0	0
$\frac{1}{4}$	$\frac{1}{2}$
1	1
4	2
9	3

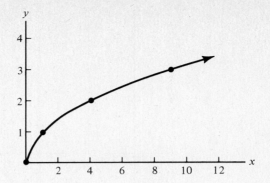

Figure 1.23

while the range R is

$$R = \{y|y \geq 0\}$$

ABSOLUTE-VALUE FUNCTION

The absolute value of any real number x, designated as $|x|$, is defined as

$$|x| = \begin{cases} x & \text{when } x \geq 0 \\ -x & \text{when } x < 0 \end{cases}$$

For example

$$|5| = 5 \qquad |-8| = -(-8) = 8$$

$$|4 - 11| = |-7| = (-7) = 7$$

The absolute-value function, defined as $y = f(x) = |x|$, can be written

$$y = f(x) = |x| = \begin{cases} x & \text{when } x \geq 0 \\ -x & \text{when } x < 0 \end{cases}$$

Using the following table, we can sketch a graph of the function $y = |x|$. The domain of the absolute-value function is the set of all real numbers; the range is the set of all nonnegative real numbers.

x	$y = |x|$
-3	3
-2	2
-1	1
0	0
1	1
2	2
3	3

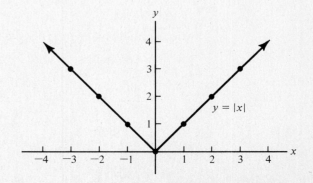

Figure 1.24

EXERCISE 1.6

1. Which of the following are polynomial functions?

   a. $y = f(x) = 8x^3 + 5x^2 - 2x + 1$ \qquad\qquad **b.** $y = f(x) = 6 - x$

   c. $y = f(x) = \dfrac{3}{x^2}$ \qquad\qquad **d.** $y = f(x) = \dfrac{x^4 - 7x + 2}{x^3 + x - 5}$

   e. $y = f(x) = \dfrac{x^2}{3} - \dfrac{x}{2} + \dfrac{2}{3}$ \qquad\qquad **f.** $y = f(x) = \sqrt{x + 2} + 7x$

   g. $y = f(x) = 2x^4 - \dfrac{5}{x^3}$

2. Find the domain for each of the following functions.

   a. $y = f(x) = \dfrac{2x}{x + 1}$ \qquad\qquad **b.** $y = f(x) = \dfrac{x^2 + 1}{x^2 - 3x + 2}$

   c. $y = f(x) = \dfrac{x - 5}{2x^2 - x - 3}$

Plot a graph of the functions in Exercises 3–9 on the accompanying coordinate systems.

3. $y = f(x) = \dfrac{3x^4}{4}$

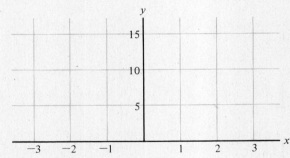

4. $y = f(x) = \dfrac{2}{x - 1}$

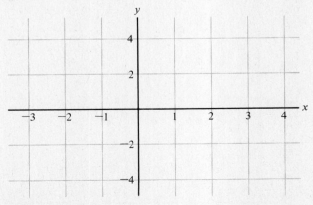

5. $y = f(x) = -3\sqrt{x}$

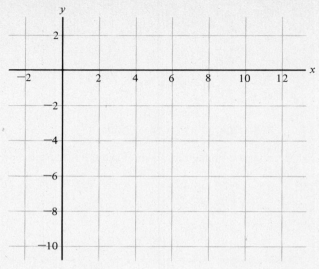

6. $y = f(x) = 2\sqrt{x - 1}$

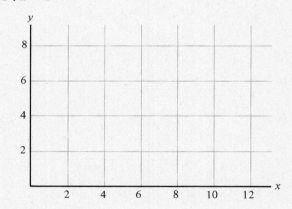

7. $y = f(x) = \sqrt{16 - x^2}$

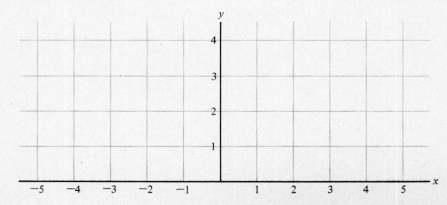

8. $y = f(x) = |x - 2|$

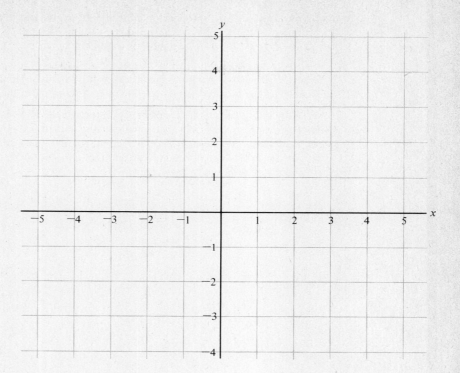

9. $y = f(x) = -|x + 1|$

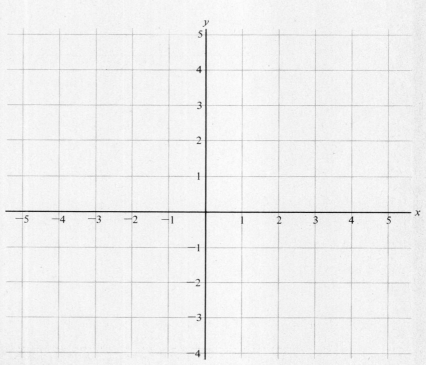

10. Suppose the income distribution for a nation is given by the equation

$$y = f(x) = \frac{2(10^{12})}{(\sqrt{x})^3} \qquad 1000 \leq x \leq 100{,}000$$

where x represents personal income measured in dollars and y gives the number of people whose incomes are greater than x. According to the equation

a. How many people have incomes greater than 1600 dollars?
b. How many people have incomes greater than 10,000 dollars?
c. How many people have incomes greater than 40,000 dollars?
d. On the following coordinate system, plot a graph of the equation over the set of values of x shown in the table.

x	y
1600	
6400	
10,000	
40,000	
90,000	

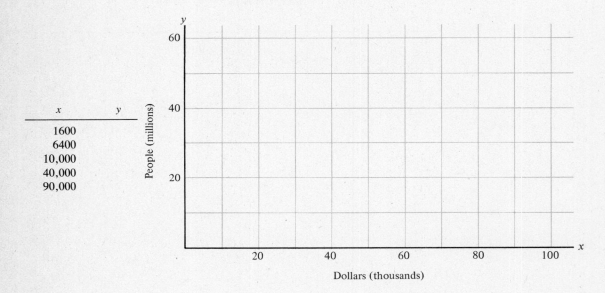

2

Limits and Continuity

The foundations of both differential and integral calculus are based on the concept of the *limit* of a function. However, a complete and thorough understanding of this concept is not easy to achieve; for this reason, an intuitive, descriptive approach to the subject is presented in Section 2.1. Standard techniques for finding limits are described next in Section 2.2. The concept of *continuity*, which ties together the concept of a function with that of a limit, is treated in Section 2.3. In what follows, enough of the essentials will be presented to enable you to operate comfortably with derivatives and integrals. No attempt at mathematical rigor will be made, but at the same time a purely mechanical approach will be avoided.

⮞ 2.1 Limits—A Descriptive Approach

Before taking up the idea of a limit, it is necessary to discuss the concept of the independent variable x ''approaching'' a given value. For example, when the variable x is approaching the number 1, it is understood that x is assigned values closer and closer to 1 but is *not* set equal to 1. An example of the process is shown in Table 2.1 and the accompanying figure. It is important to note that x approaches 1 by moving toward that value from both the left and right. Mathematically, the process whereby x approaches the number 1 is symbolized as

$x \to 1$ x approaches or gets closer and closer to 1

Table 2.1

FROM THE LEFT x	FROM THE RIGHT x
0.00000	2.00000
.50000	1.50000
.90000	1.10000
.99000	1.01000
.99900	1.00100
0.99990	1.00010
.	.
.	.
.	.

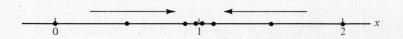

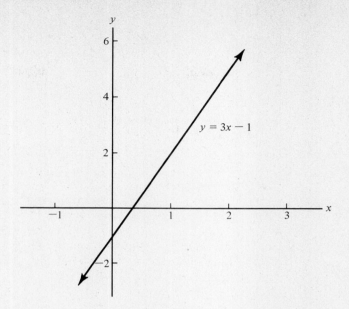

Figure 2.1

For any number a, the same notation is used

$\qquad x \to a \qquad$ x approaches or gets closer and closer to the number a

When studying a function, we are often interested in determining whether the dependent variable y approaches a unique value as x approaches a given value. For example, suppose we are given the function

$$y = f(x) = 3x - 1 \qquad\qquad\qquad (2.1.1)$$

whose graph is shown in Figure 2.1. In addition, suppose that x is approaching the number 1. We want to determine the value (if any) that the variable y is approaching. Table 2.2 illustrates the effect of "x approaching 1" on the variable y and Figure 2.2 shows the process graphically on the line $y = 3x - 1$.

Table 2.2

FROM THE LEFT		FROM THE RIGHT	
x	$y = 3x - 1$	x	$y = 3x - 1$
0.00000	−1.00000	2.00000	5.00000
.50000	0.50000	1.50000	3.50000
.90000	1.70000	1.10000	2.30000
.99000	1.97000	1.01000	2.03000
0.99900	1.99700	1.00100	2.00300
.	.	.	.
.	.	.	.
.	.	.	.

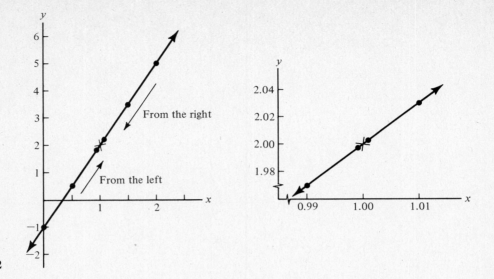

Figure 2.2

Columns 3 and 4 in Table 2.2 and Figure 2.2 indicate that, as x approaches 1, the variable y is also approaching or getting closer to some number; this number is called the *limit* of $f(x)$ as x approaches 1 and is described mathematically as

$$\lim_{x \to 1} f(x)$$

When $f(x) = 3x - 1$, we have

$$\lim_{x \to 1} f(x) = \lim_{x \to 1} (3x - 1) = 2$$

This result can also be written as

$$y \to 2 \qquad \text{as} \qquad x \to 1$$

In words, the variable y approaches 2 as the variable x approaches 1.

Now that a brief description of the limiting process has been presented, it probably seems to be an exercise in pomposity; if you had examined the equation

$$y = f(x) = 3x - 1$$

or its graph, you could have discovered the limit by substituting 1 for x in Equation 2.1.1. However, in an attempt to describe the process in as simple a manner as possible, you may have been deceived into believing that, operationally, finding the limit is a rather straightforward and trivial exercise. Let us assure you that, for many situations, it is not. Our only purpose and intention in the preceding example was to demonstrate the limiting process and the notation used in describing it.

The process can be applied to other types of simple functions such as

$$y = f(x) = 3x^2 \tag{2.1.2}$$

Table 2.3

FROM THE LEFT		FROM THE RIGHT	
x	$y = 3x^2$	x	$y = 3x^2$
1.000000	3.000000	3.000000	27.000000
1.500000	6.750000	2.500000	18.750000
1.900000	10.830000	2.100000	13.230000
1.990000	11.880300	2.010000	12.120300
1.999000	11.988003	2.001000	12.012003
1.999900	11.998800	2.000100	12.001200
1.999990	11.999880	2.000010	12.000120
.	.	.	.
.	.	.	.
.	.	.	.

And we can ask: What is the limit of the function $y = 3x^2$ as x approaches 2? Again, as you can see in Table 2.3, a limiting value of y appears to exist and to equal 12 (y approaches 12 as x approaches 2)

$$\lim_{x \to 2} f(x) = \lim_{x \to 2} 3x^2 = 12$$

At this point, your reaction to this whole process might understandably be one of frustration at being required to read about a process whose result can be computed apparently by substituting the limiting value of x into the equation that defines $f(x)$. To dispel any illusions you might have as quickly as possible, let us consider the function

$$y = f(x) = \frac{3x^2 - 3}{2x - 2} \qquad\qquad\qquad \textbf{(2.1.3)}$$

and ask the question: What is the limit of this function as x approaches 1? If you attempt to substitute $x = 1$ into Equation 2.1.3, you discover very quickly that you obtain $\frac{0}{0}$, which is not defined; this means of course that the function itself is not defined when $x = 1$. Surprisingly, however, it does *not* necessarily follow that the limit does *not* exist. To show you more explicitly, let us examine Table 2.4 on the next page.

Examination of the second and fourth columns seems to indicate that y is approaching a limit as x approaches 1. To investigate this more formally, let us look at the algebraic expression that defines $f(x)$

$$\frac{3x^2 - 3}{2x - 2}$$

This expression, as written, is not in its simplest form. Fractional expressions such as this can be reduced to simpler forms by factoring the numerator and denominator and canceling common factors, that is

$$\frac{3x^2 - 3}{2x - 2} = \frac{3(x + 1)(x - 1)}{2(x - 1)} = \frac{3(x + 1)}{2} \qquad x \neq 1$$

Table 2.4

FROM THE LEFT		FROM THE RIGHT	
x	$y = \dfrac{3x^2 - 3}{2x - 2}$	x	$y = \dfrac{3x^2 - 3}{2x - 2}$
0.00000	1.50000	2.00000	4.50000
.50000	2.25000	1.50000	3.75000
.90000	2.85000	1.10000	3.15000
.99000	2.98500	1.01000	3.01500
.99900	2.99850	1.00100	3.00150
0.99990	2.99985	1.00010	3.00015
⋮	⋮	⋮	⋮

so we can express $f(x)$ as

$$y = f(x) = \frac{3(x + 1)}{2} \qquad x \neq 1 \tag{2.1.4}$$

In this new form, it is easier to see why $\lim\limits_{x \to 1} f(x)$ exists even though $f(1)$ itself is not defined. In the limiting process, x is allowed only to approach 1 but *not to equal* 1; therefore, we can use Equation 2.1.4 for $f(x)$ to evaluate the limit and obtain

$$\lim_{x \to 1} f(x) = \lim_{x \to 1} \frac{3(x + 1)}{2} = 3$$

A visual representation of this result is shown in Figure 2.3. The graph of the function is a straight line with a pinhole or gap at $(1, 3)$ to indicate that $f(1)$ is not defined. At the same time it shows that y gets closer to 3 as x gets closer to 1.

At the risk of muddying up the waters too much, let us caution you that in

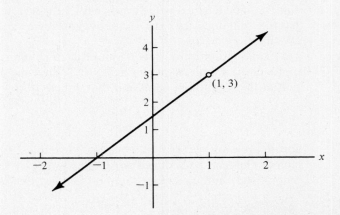

Figure 2.3

addition to difficulties sometimes encountered when attempting to find $\lim\limits_{x \to a} f(x)$, there are situations where

$$\lim_{x \to a} f(x) \qquad does\ not\ exist$$

To give a simple example, look at the function

$$y = f(x) = \frac{1}{x - 1}$$

and let us see what happens to this function as x approaches 1. As you can see from Table 2.5, the values of y in the second and fourth columns are not approaching a common value. In this case we have the result that

$$\lim_{x \to 1} \frac{1}{x - 1} \qquad does\ not\ exist$$

In addition, the function is not defined at $x = 1$. All of these features are illustrated in Figure 2.4 which contains the graph of the function. In dealing with the limit of a function, two items must be addressed

1. *Does the limit exist?*
2. *If the limit exists, find it.*

Fortunately, for most of the functions we will encounter, there are ways to answer these questions quickly and directly. These will be taken up in the next section.

When dealing with piecewise defined functions, it is instructive to look at the limit of the function as x approaches a value that indicates a change from one expression to a second expression defining the function.

Table 2.5

	FROM THE LEFT		FROM THE RIGHT	
x	$y = \dfrac{1}{x - 1}$	x	$y = \dfrac{1}{x - 1}$	
0.00000	-1	2.00000	1	
.50000	-2	1.50000	2	
.90000	-10	1.10000	10	
.99000	-100	1.01000	100	
.99900	-1000	1.00100	1000	
.99990	-10000	1.00010	10000	
0.99999	-100000	1.00001	100000	
.	.	.	.	
.	.	.	.	
.	.	.	.	

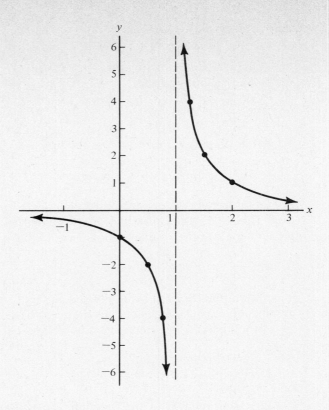

Figure 2.4

Example 1 Given the function

$$y = f(x) = \begin{cases} x - 2 & x \leq 3 \\ 7 - 2x & x > 3 \end{cases}$$

find $\lim_{x \to 3} f(x)$ if it exists.

Solution After completing Table 2.6 we see that the values of y in the second and fourth columns appear to be approaching a common value, namely 1. The graph in

Table 2.6

FROM THE LEFT		FROM THE RIGHT	
x	$y = x - 2$	x	$y = 7 - 2x$
2.0000	0.0000	4.0000	−1.0000
2.5000	.5000	3.5000	0.0000
2.9000	.9000	3.1000	0.8000
2.9900	.9900	3.0100	0.9800
2.9990	.9990	3.0010	0.9980
2.9999	0.9999	3.0001	0.9998
.	.	.	.
.	.	.	.
.	.	.	.

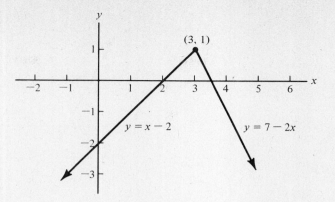

Figure 2.5

Figure 2.5 indicates that y approaches 1 as x approaches 3 from either the right or the left. So we can conclude that

$$\lim_{x \to 3} f(x) = 1$$

Example 2 | Consider the function

$$y = f(x) = \begin{cases} 3 - 4x & x \le 1 \\ x + 1 & x > 1 \end{cases}$$

find $\lim_{x \to 1} f(x)$ if it exists.

Solution | After completing Table 2.7, we see that the values of y in the second column appear to be approaching (-1) but those in the fourth column appear to be approaching 2 so we conclude that $\lim_{x \to 1} f(x)$ does *not* exist. This conclusion is reinforced by the graph in Figure 2.6 which shows a jump occurring at $x = 1$.

Table 2.7

FROM THE LEFT		FROM THE RIGHT	
x	$y = 3 - 4x$	x	$y = x + 1$
0.0000	3.0000	2.0000	3.0000
.5000	1.0000	1.5000	2.5000
.9000	−0.6000	1.1000	2.1000
.9900	−0.9600	1.0100	2.0100
.9990	−0.9960	1.0010	2.0010
0.9999	−0.9996	1.0001	2.0001
.	.	.	.
.	.	.	.

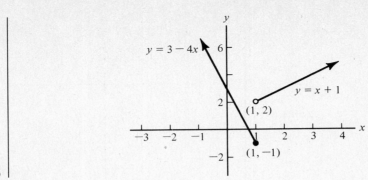

Figure 2.6

EXERCISE 2.1

The exercises given in this section will be easier to carry out if you have a calculator available.

For Exercises 1–13, complete the tables, expressing y to four decimal places, and determine: (1) whether the indicated limit exists, and (2) the value of the limit when it exists. In addition, plot a graph of the function in the vicinity of the limiting value of x.

1. $\lim\limits_{x \to 2} (5 - 4x)$

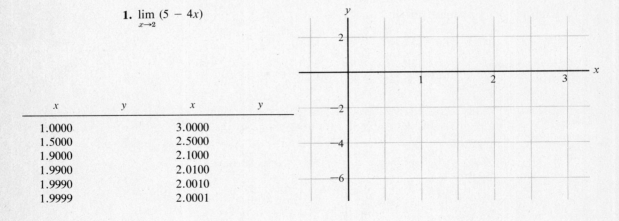

x	y	x	y
1.0000		3.0000	
1.5000		2.5000	
1.9000		2.1000	
1.9900		2.0100	
1.9990		2.0010	
1.9999		2.0001	

2. $\lim\limits_{x \to 1} \dfrac{x + 3}{x - 2}$

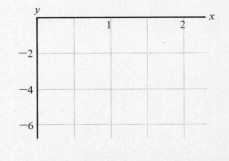

x	y	x	y
0.0000		1.9000	
.5000		1.5000	
.7500		1.2500	
.9500		1.0500	
.9950		1.0050	
0.9995		1.0005	

3. $\lim\limits_{x \to 3} \dfrac{x^2 - 3x}{x^2 - 9}$

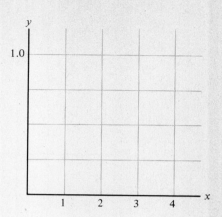

x	y	x	y
2.0000		4.0000	
2.5000		3.5000	
2.9000		3.1000	
2.9900		3.0100	
2.9990		3.0010	
2.9999		3.0001	

4. $\lim\limits_{x \to 2} \dfrac{3}{x - 2}$

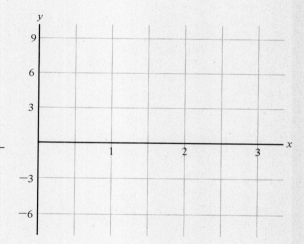

x	y	x	y
1.0000		3.0000	
1.5000		2.5000	
1.9000		2.1000	
1.9900		2.0100	
1.9990		2.0010	
1.9999		2.0001	

Use same table for Exercises 5, 6, and 7.

5. $\lim\limits_{x \to 0} \dfrac{x^2}{x}$

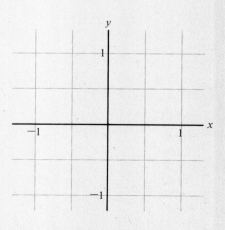

x	y	x	y
−1.0000		1.0000	
−0.5000		0.5000	
−0.1000		0.1000	
−0.0100		0.0100	
−0.0010		0.0010	
−0.0001		0.0001	

6. $\lim\limits_{x \to 0} \dfrac{x}{x}$

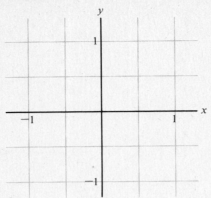

7. $\lim\limits_{x \to 0} \dfrac{x}{x^2}$

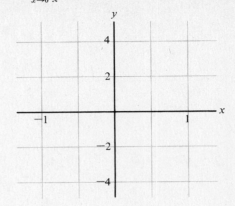

8. $\lim\limits_{x \to 1} \dfrac{2}{x^2 + 1}$

x	y	x	y
0.0000		2.0000	
.5000		1.5000	
.9000		1.1000	
.9900		1.0100	
.9990		1.0010	
0.9999		1.0001	

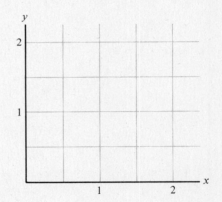

9. $\lim\limits_{x \to 1} \dfrac{x^3 - 1}{x - 1}$ (Simplify before carrying out calculations.)

x	y	x	y
0.0000		2.0000	
.5000		1.5000	
.9000		1.1000	
.9900		1.0100	
.9990		1.0010	
0.9999		1.0001	

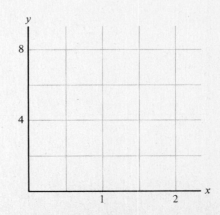

10. $\lim\limits_{x \to 1} \sqrt{x}$

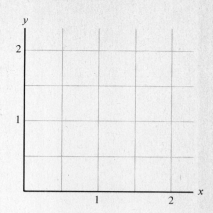

x	y	x	y
0.0000		2.0000	
.5000		1.5000	
.9000		1.1000	
.9900		1.0100	
.9990		1.0010	
0.9999		1.0001	

11. $\lim\limits_{x \to 0} \sqrt{x}$

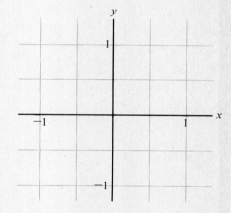

x	y	x	y
−1.0000		1.0000	
−0.2500		0.2500	
−0.0100		0.0100	
−0.0001		0.0001	

12. $\lim\limits_{x \to 2} f(x)$ where $y = f(x) = \begin{cases} 1 - x & x < 2 \\ x^2 - 5 & x \geq 2 \end{cases}$

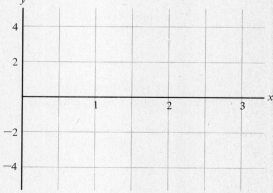

x	y	x	y
1.0000		3.0000	
1.5000		2.5000	
1.9000		2.1000	
1.9900		2.0100	
1.9990		2.0010	
1.9999		2.0001	

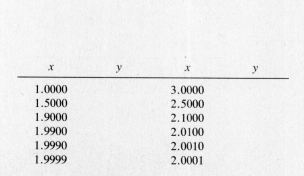

13. Find $\lim\limits_{x \to -1} f(x)$ where $y = f(x) = \begin{cases} x^2 & x \le -1 \\ x + 1 & x > -1 \end{cases}$

x	y	x	y
-2.0000		0.0000	
-1.5000		-0.5000	
-1.1000		-0.9000	
-1.0100		-0.9900	
-1.0010		-0.9990	
-1.0001		-0.9999	

For Exercises 14–18, construct tables that contain values of y corresponding to values of x close to the limiting value of x. From these tables, determine: (1) whether the indicated limit exists, and (2) the value of the limit if it exists. In addition, plot a graph of the function in the vicinity of the limiting value of x.

14. $\lim\limits_{x \to 0} \dfrac{x + 4}{x - 2}$

15. $\lim\limits_{x \to -1} \dfrac{x^2 - 1}{x + 1}$

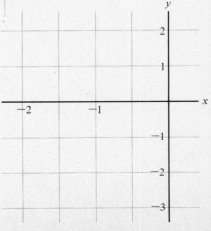

16. $\lim\limits_{x \to 2} \dfrac{5 - 3x}{x - 2}$

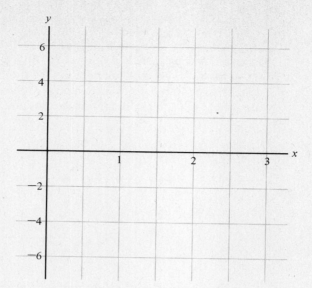

17. $\lim\limits_{x \to 4} \dfrac{x - 4}{\sqrt{x} - 2}$

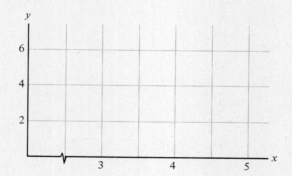

18. $\lim\limits_{x \to 2} \dfrac{2x^2 - 3x - 2}{3x^2 - 2x - 8}$

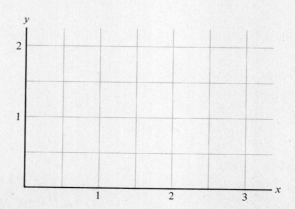

2.2 Techniques for Finding Limits

The procedures used in the preceding section for finding the limit were rather awkward and inefficient. Fortunately, for most functions we shall handle, there are more direct methods for finding the limit when it exists. Only the methods themselves and illustrations will be presented; no proofs will be given because the proofs rely on the definition of the limit, which has not been presented.

I. $\lim\limits_{x \to a} C = C$, where C is a constant.

Example 1 If $y = f(x) = 3$, find $\lim\limits_{x \to 5} f(x)$.

Solution $\lim\limits_{x \to 5} f(x) = \lim\limits_{x \to 5} 3 = 3$

II. $\lim\limits_{x \to a} x^n = a^n$, where n is a positive integer.

Example 2 If $y = f(x) = x^4$, find $\lim\limits_{x \to 2} f(x)$.

Solution $\lim\limits_{x \to 2} f(x) = \lim\limits_{x \to 2} x^4 = 2^4 = 16$

III. If $\lim\limits_{x \to a} g(x) = M$, then

$$\lim\limits_{x \to a} \sqrt[n]{g(x)} = \sqrt[n]{M}$$

provided $M > 0$ when n is even.

Example 3 Find $\lim\limits_{x \to 4} \sqrt{x^3}$.

Solution Because $\lim\limits_{x \to 4} x^3 = 4^3 = 64$

we get
$$\lim\limits_{x \to 4} \sqrt{x^3} = \sqrt{64} = 8$$

Example 4 Find $\lim\limits_{x \to -4} \sqrt{x^3}$.

Solution | Because $\lim\limits_{x \to -4} x^3 = -64$, and $\sqrt{-64}$ is not a real number

$\lim\limits_{x \to -4} \sqrt{x^3}$ does not exist

IV. The limit of a constant times a function equals the constant times the limit of the function. If $\lim\limits_{x \to a} f(x) = L$, then

$$\lim\limits_{x \to a} [Cf(x)] = C \left[\lim\limits_{x \to a} f(x)\right] = CL$$

where C is a constant.

Example 5 | Find $\lim\limits_{x \to 2} 4x^3$

Solution | $\lim\limits_{x \to 2} 4x^3 = 4(\lim\limits_{x \to 2} x^3) = 4(2^3) = 4(8) = 32$

V. The limit of a sum equals the sum of the limits. If $y = f(x) = g(x) + h(x)$ and $\lim\limits_{x \to a} g(x) = L$ and $\lim\limits_{x \to a} h(x) = M$, then

$$\lim\limits_{x \to a} [g(x) + h(x)] = \lim\limits_{x \to a} g(x) + \lim\limits_{x \to a} h(x) = L + M$$

Example 6 | If $y = f(x) = 5x + 3x^2$, find $\lim\limits_{x \to 2} f(x)$.

Solution | Because $\lim\limits_{x \to 2} 5x = 10$ and $\lim\limits_{x \to 2} 3x^2 = 12$, we can write

$\lim\limits_{x \to 2} (5x + 3x^2) = \lim\limits_{x \to 2} 5x + \lim\limits_{x \to 2} 3x^2$

$= 10 + 12 = 22$

VI. The limit of a product equals the product of the limits. If $y = f(x) = g(x) \cdot h(x)$ and $\lim\limits_{x \to a} g(x) = L$ and $\lim\limits_{x \to a} h(x) = M$, then

$$\lim\limits_{x \to a} [g(x) \cdot h(x)] = \left[\lim\limits_{x \to a} g(x)\right]\left[\lim\limits_{x \to a} h(x)\right] = LM$$

Example 7 | If $y = f(x) = (2x + 1)(3x^2 + 4)$, find $\lim\limits_{x \to 3} f(x)$.

Solution | Because $\lim\limits_{x \to 3} (2x + 1) = 7$ and $\lim\limits_{x \to 3} (3x^2 + 4) = 31$

$$\lim\limits_{x \to 3} (2x + 1)(3x^2 + 4) = 7(31) = 217$$

Example 8 | If $y = f(x) = \sqrt{x}\,(7 - x)$, find $\lim\limits_{x \to 4} f(x)$.

Solution | Because $\lim\limits_{x \to 4} \sqrt{x} = 2$ and $\lim\limits_{x \to 4} (7 - x) = 3$

$$\lim\limits_{x \to 4} [\sqrt{x}\,(7 - x)] = \left(\lim\limits_{x \to 4} \sqrt{x}\right)\left[\lim\limits_{x \to 4} (7 - x)\right] = 2(3) = 6$$

VII. The limit of a quotient equals the quotient of the limits. If $y = f(x) = g(x)/h(x)$ and $\lim\limits_{x \to a} g(x) = L$ and $\lim\limits_{x \to a} h(x) = M$, then

$$\lim\limits_{x \to a} \frac{g(x)}{h(x)} = \frac{L}{M} \qquad \text{provided } M \neq 0$$

Example 9 | If $y = f(x) = (3x + 2)/(1 - x^2)$, find $\lim\limits_{x \to 2} f(x)$.

Solution | Because $\lim\limits_{x \to 2} (3x + 2) = 8$ and $\lim\limits_{x \to 2} (1 - x^2) = -3$, then

$$\lim\limits_{x \to 2} \frac{3x + 2}{1 - x^2} = \frac{\lim\limits_{x \to 2} (3x + 2)}{\lim\limits_{x \to 2} (1 - x^2)} = -\frac{8}{3}$$

Example 10 | If $y = f(x) = (\sqrt{x} + 2)/(x + 3)$, find $\lim\limits_{x \to 1} f(x)$

Solution | Because $\lim\limits_{x \to 1} (\sqrt{x} + 2) = 3$ and $\lim\limits_{x \to 1} (x + 3) = 4$

$$\lim\limits_{x \to 1} \frac{\sqrt{x} + 2}{x + 3} = \frac{3}{4}$$

Technique VII cannot be used when the limit of the denominator, that is, M, equals 0. The existence or nonexistence of $\lim\limits_{x \to a} [g(x)/h(x)]$ in this case depends on the limit of the numerator, that is, L, as described in techniques VIII and IX.

VIII. If $y = f(x) = \dfrac{g(x)}{h(x)}$, $\lim\limits_{x \to a} g(x) = L \neq 0$, and $\lim\limits_{x \to a} h(x) = 0$, then

$$\lim\limits_{x \to a} \frac{g(x)}{h(x)} \qquad \text{does not exist}$$

Example 11 | If $y = f(x) = (3x + 2)/(x^2 - 1)$, find $\lim_{x \to 1} f(x)$.

Solution | Because $\lim_{x \to 1} (3x + 2) = 5$ and $\lim_{x \to 1} (x^2 - 1) = 0$

$$\lim_{x \to 1} \frac{3x + 2}{x^2 - 1} \qquad \text{does not exist}$$

The following partial graph of the function indicates why the limit does not exist. As x approaches 1 from the left, not only are the corresponding values of y decreasing, but they decrease more and more rapidly as we get closer to the vertical line $x = 1$, whereas the reverse is true as x approaches 1 from the right.

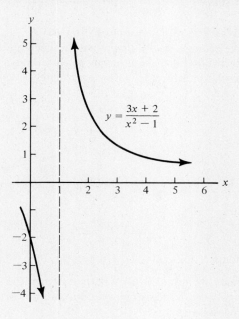

IX. If $y = f(x) = \dfrac{g(x)}{h(x)}$, and $\lim_{x \to a} g(x) = 0$, and $\lim_{x \to a} h(x) = 0$, then

$$\lim_{x \to a} \left[\frac{g(x)}{h(x)} \right] \qquad \text{may or may not exist}$$

When the situation described in technique IX arises, additional work is required, usually in the form of transforming the ratio $g(x)/h(x)$ into an equivalent form so that one of the techniques (I–VIII) can be applied. There is no way to predict in advance whether the limit does or does not exist, as Examples 12 and 13 illustrate.

Example 12 | If $y = f(x) = (3x - 6)/(x^2 + x - 6)$, find $\lim_{x \to 2} f(x)$.

Solution | Because $\lim_{x \to 2} (3x - 6) = 0$ and $\lim_{x \to 2} (x^2 + x - 6) = 0$, the ratio must be altered so that we can determine whether or not the limit exists. Factoring the numerator and denominator enables us to write

$$\frac{3x - 6}{x^2 + x - 6} = \frac{3(x - 2)}{(x + 3)(x - 2)} = \frac{3}{x + 3} \qquad x \neq 2$$

so we can now write

$$\lim_{x \to 2} \frac{3x - 6}{x^2 + x - 6} = \lim_{x \to 2} \frac{3}{x + 3} = \frac{3}{5}$$

NOTE: The replacement of $(3x - 6)/(x^2 + x - 6)$ by $3/(x + 3)$ in the preceding limit process is valid because the variable x approaches 2 but does not equal 2. The following partial graph of the function shows that as x approaches 2, the corresponding values of y approach $\frac{3}{5}$.

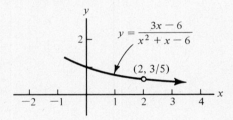

Example 13 | If $y = f(x) = (3x - 3)/(x^2 - 2x + 1)$, find $\lim_{x \to 1} f(x)$.

Solution | Because $\lim_{x \to 1} (3x - 3) = 0$ and $\lim_{x \to 1} (x^2 - 2x + 1) = 0$, we are again at an impasse; we cannot tell whether or not the limit exists. As in the previous example, the numerator and denominator are factorable, so that the ratio can be expressed as

$$\frac{3x - 3}{x^2 - 2x + 1} = \frac{3(x - 1)}{(x - 1)^2} = \frac{3}{x - 1} \qquad x \neq 1$$

In this case, we see that

$$\lim_{x \to 1} (x - 1) = 0 \qquad \text{but} \qquad \lim_{x \to 1} 3 = 3 \neq 0$$

According to rule VIII, we find that

$$\lim_{x \to 1} \frac{3x - 3}{x^2 - 2x + 1} \qquad \text{does not exist}$$

Reducing a fraction is only one of many techniques that can be used to determine whether $\lim_{x \to a} [g(x)/h(x)]$ exists when both $\lim_{x \to a} g(x)$ and $\lim_{x \to a} h(x)$ equal 0. It is not our intent to study a large number of these techniques, only to indicate that others beyond reduction of a rational expression exist. Often rationalizing either the numerator or the denominator enables us to determine whether or not the limit exists as the following example illustrates.

Example 14   If $y = f(x) = (x - 4)/(\sqrt{x} - 2)$, find $\lim\limits_{x \to 4} f(x)$.

Solution   Once more, we have $\lim\limits_{x \to 4} (x - 4) = 0$ and $\lim\limits_{x \to 4} (\sqrt{x} - 2) = 0$. To resolve this situation, the ratio can be written in an equivalent form by rationalizing the denominator as follows

$$\frac{x - 4}{\sqrt{x} - 2} = \frac{(x - 4)(\sqrt{x} + 2)}{(\sqrt{x} - 2)(\sqrt{x} + 2)} = \frac{(x - 4)(\sqrt{x} + 2)}{(x - 4)}$$

$$= \sqrt{x} + 2 \qquad x \neq 4$$

Next we proceed to find the limit

$$\lim_{x \to 4} \frac{x - 4}{\sqrt{x} - 2} = \lim_{x \to 4} (\sqrt{x} + 2) = 4$$

EXERCISE 2.2

Using one or more of the methods presented in this section, find each of the following limits when they exist.

1. $\lim\limits_{x \to 2} (7x^2 - 3x + 4)$

2. $\lim\limits_{x \to 9} \left(4\sqrt{x} + \dfrac{5}{x} \right)$

3. $\lim\limits_{x \to -1} \dfrac{8 + x}{4 - x^2}$

4. $\lim\limits_{x \to 2} \dfrac{4}{x^2 + 8}$

5. $\lim\limits_{x \to 1} \dfrac{14x - 14}{7x^2 - 7}$

6. $\lim\limits_{x \to 2} \dfrac{5x - 10}{x^2 - 4x + 4}$

7. $\lim\limits_{x \to 4} (\sqrt{x} + 1)(5x - 16)$

8. $\lim\limits_{x \to 2} \dfrac{x^3 - 8}{x^2 - x - 2}$

9. $\lim\limits_{x \to 3} \dfrac{2x^2 - 12x + 18}{x^2 - 9}$

10. $\lim\limits_{x \to 3} \dfrac{2x^2 - 7x - 3}{x^2 - 6x + 9}$

11. $\lim\limits_{x \to 0} \dfrac{2 - 5x}{3 - 6x}$

12. $\lim\limits_{x \to -2} \dfrac{x^3 + 2x^2 + 3x + 6}{3x + 6}$

13. $\lim\limits_{x \to 1} \dfrac{x - 1}{\sqrt{x} - 1}$

14. $\lim\limits_{x \to 2} \dfrac{3 - \sqrt{x + 7}}{x - 2}$

15. $\lim\limits_{x \to 1} \dfrac{x + 1}{\sqrt{x} - 1}$

16. $\lim\limits_{x \to 0} \dfrac{x}{(1 + x)^2 - 1}$

17. $\lim\limits_{x \to 0} \dfrac{x}{(1 + x)^3 - 1}$

18. $\lim\limits_{x \to 0} \dfrac{x}{(1 + x)^n - 1}$

➳ 2.3 Continuity

When we use the word *continuous* in referring to a function, we mean basically that its graph contains no breaks such as pinholes, gaps, or jumps. The curve is unbroken and can be sketched without lifting the pen from the paper as illustrated in Figure 2.7

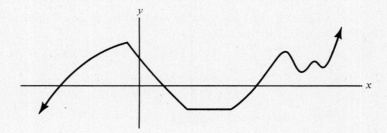

Figure 2.7

When a hole or gap appears in the graph of a function at $x = a$, the function is said to be discontinuous at $x = a$. The conditions that are described and illustrated in Figures 2.8–2.11 indicate some of the ways in which breaks can occur in the graph of a function at $x = a$.

A. $\lim\limits_{x \to a} f(x)$ exists, but $f(a)$ is *not defined*.

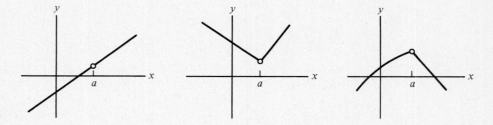

Figure 2.8

B. $f(a)$ is defined, but $\lim\limits_{x \to a} f(x)$ does *not* exist.

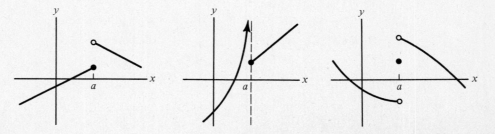

Figure 2.9

C. $f(a)$ is defined and $\lim\limits_{x \to a} f(x)$ exists, but $f(a) \neq \lim\limits_{x \to a} f(x)$.

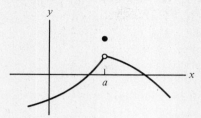

Figure 2.10

D. $f(a)$ is *not* defined and $\lim\limits_{x \to a} f(x)$ does *not* exist.

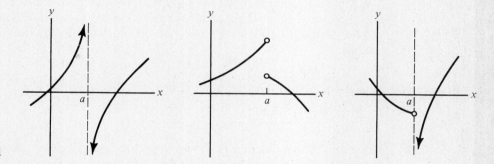

Figure 2.11

When any of the conditions A–D prevails at $x = a$, the function is said to be *discontinuous* at $x = a$. Because the behavior of the function at points such as these is somewhat unusual, additional time and effort is devoted to locating and analyzing the behavior of the function at points of discontinuity. In medicine and psychology, most professional time and effort is spent dealing with the unusual, namely those individuals who are physically or mentally ill; in the same way, the study of functions and their graphs deals extensively with the behavior of functions at points where they are discontinuous.

When a function is not exhibiting one of the bizarre features described by conditions A–D at $x = a$, the function is said to be *continuous* at $x = a$.

DEFINITION

A function $y = f(x)$ is said to be continuous at $x = a$ if

$$\lim_{x \to a} f(x) = f(a) \qquad\qquad (2.3.1)$$

---

that is, the limit of $f(x)$ as x approaches a equals $f(a)$, the value of the function at $x = a$.

Equation 2.3.1 implies that three conditions must be satisfied in order for a function to be continuous at $x = a$

1. $f(a)$ must be defined,
2. $\lim_{x \to a} f(x)$ must exist, and
3. $\lim_{x \to a} f(x) = f(a)$.

If one or more of these conditions is violated, then the function is discontinuous at $x = a$.

If a function is continuous at every point of an interval, it is said to be continuous over the interval; graphically, this means that the curve is unbroken in the sense that it can be drawn without holes or jumps as shown earlier in Figure 2.7. Almost all the functions we shall study will be continuous everywhere except for one or two values of x. Our attention in the remainder of this section will be centered on locating values of x, if any, where the function is discontinuous.

Example 1 | Find the points of discontinuity, if any, for the function

$$y = f(x) = -2x + 1$$

Solution | The graph of this function is shown in Figure 2.12. Because there are no holes or gaps, the function is continuous for all values of x. This intuitive result is in accord with what is obtained by applying Equation 2.3.1. As x approaches a, we find that

$$\lim_{x \to a} f(x) = \lim_{x \to a} (-2x + 1) = -2a + 1$$

The value of $f(x)$ at $x = a$ is given by

$$f(a) = -2a + 1$$

From this we see that

$$\lim_{x \to a} f(x) = f(a)$$

and we can conclude that the function is continuous for all values of x.

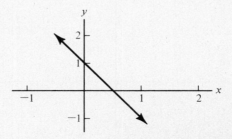

Figure 2.12

When they occur, discontinuities will generally be due to:

1. The function having the form

$$y = f(x) = \frac{g(x)}{h(x)}$$

and the discontinuities appear at those values of x for which $h(x) = 0$ as illustrated in Example 2, and/or,
2. Piecewise defined functions where there is a break in the graph due to a change in the mathematical form of the equation defining the function as in Example 3.

Example 2 | Find the values of x, if any, for which the function

$$y = f(x) = \frac{2}{x - 1}$$

is discontinuous.

Solution | An examination of the function indicates that it is not defined at $x = 1$. In addition, the limit of the function does not exist as $x \rightarrow 1$. So we can say that the function is not continuous at $x = 1$ for two reasons

1. $\lim\limits_{x \to 1} f(x) = \lim\limits_{x \to 1} \dfrac{2}{x - 1}$ does not exist, and

2. $f(1) = \frac{2}{0}$ is not defined

These conclusions are reinforced by the graph of the function shown in Figure 2.13.

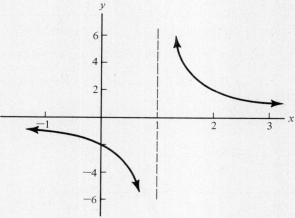

Figure 2.13

Example 3 | Find the points of discontinuity, if any, for the function

$$y = f(x) = \begin{cases} x + 1 & x \leq 1 \\ x^2 & x > 1 \end{cases}$$

Solution | As the graph of the function in Figure 2.14 shows, there is a jump in the function at $x = 1$. If we examine the situation a little more closely, we learn that the function is discontinuous at $x = 1$ because

$$\lim_{x \to 1} f(x) \quad \text{does not exist}$$

The function itself is defined at $x = 1$

$$f(1) = 1 + 1 = 2$$

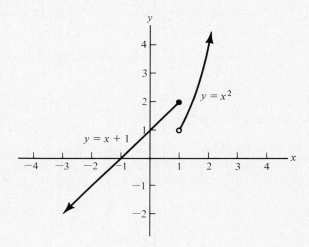

Figure 2.14

EXERCISE 2.3

1. For the functions whose graphs are shown in (i)–(vii), determine

   a. If the function is discontinuous anywhere over the interval shown, and
   b. The values of x, if any, where the function is discontinuous and what condition (A–D [p. 90–91]) best describes the discontinuity.

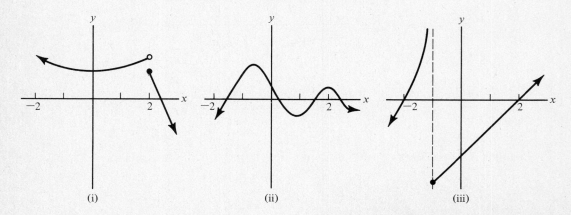

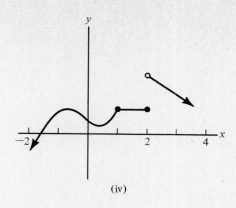

(iv)

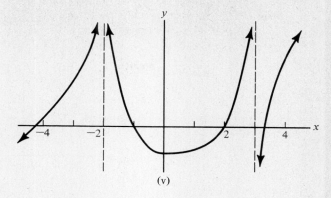

(v)

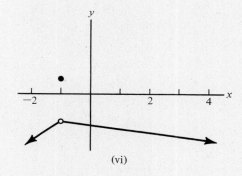

(vi)

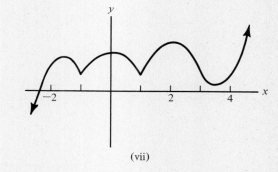

(vii)

2. Following are given a number of functions together with the values of x for which each function is discontinuous. Graph each function on the accompanying coordinate system and indicate which condition (A–D [p. 90–91]) holds at the discontinuity.

a. $y = f(x) = \dfrac{1}{x^2}$ Discontinuous at $x = 0$

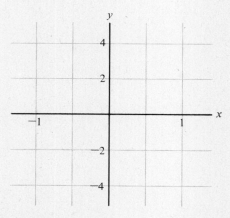

b. $y = f(x) = \dfrac{x^2 - 4}{4x - 8}$ Discontinuous at $x = 2$

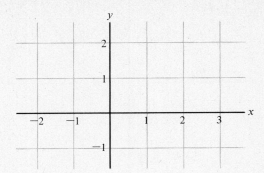

c. $y = f(x) = \begin{cases} \dfrac{x^2 - 4x + 4}{4x - 8} & (x \neq 2) \\ 1 & (x = 2) \end{cases}$ Discontinuous at $x = 2$

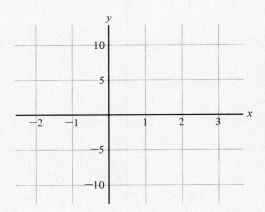

d. $y = f(x) = \begin{cases} 3x - 2 & x \leq 1 \\ x - 1 & x > 1 \end{cases}$ Discontinuous at $x = 1$

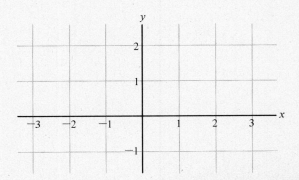

e. $y = f(x) = \begin{cases} x^3 & x < 0 \\ \dfrac{1}{x} & x > 0 \end{cases}$ Discontinuous at $x = 0$

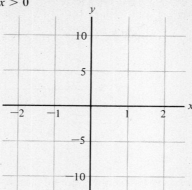

f. $y = f(x) = \begin{cases} -1 & x < 0 \\ 0 & x = 0 \\ 1 & x > 0 \end{cases}$ Discontinuous at $x = 0$

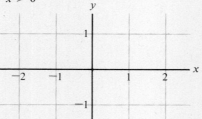

g. $y = f(x) = \begin{cases} \dfrac{1}{x + 1} & x \leq 0 \\ x^2 & x > 0 \end{cases}$ Discontinuous at $x = 0$ and $x = -1$

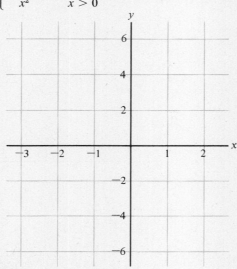

h. $y = f(x) = \dfrac{x^3 - 1}{4x - 4}$ Discontinuous at $x = 1$

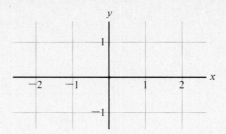

3. Determine whether there are any values of x for which the following functions are discontinuous; graph each function on the adjacent coordinate system.

a. $y = f(x) = \dfrac{x^2}{2} + 1$

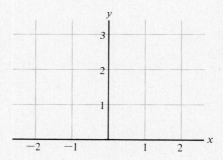

b. $y = f(x) = \dfrac{3}{x + 1}$

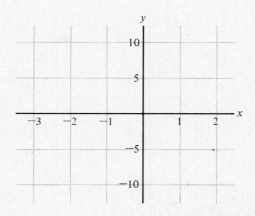

c. $y = f(x) = \begin{cases} x - 3 & x \le 0 \\ -3x - 1 & x > 0 \end{cases}$

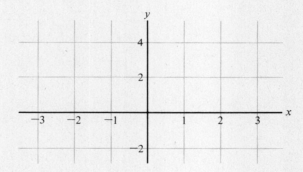

d. $y = f(x) = \dfrac{3x - 3}{x^3 - 1} \qquad x \ne 1$

$f(1) = 2$

3

First Derivative of a Function

This chapter will be devoted to defining and explaining the meaning and significance of the first derivative of a function. The first derivative is an important concept because it permits us to calculate the instantaneous rate of change of many quantities. For example, by using the first derivative, it is possible to determine how fast a firm's profits are growing, or how rapidly an oil slick is spreading out from an oil spill.

The concept of average rate of change of a function is introduced in Section 3.1, while the instantaneous rate of change, also known as the first derivative, is defined and its meaning explored in Section 3.2. Techniques for finding the first derivative quickly and efficiently are presented in Sections 3.3–3.5. The use of the first derivative in some basic applications is covered in Section 3.6. The last two sections deal with finding the second derivative and using a method called implicit differentiation to find the first derivative when the relationship between the variables x and y is not given in the form $y = f(x)$.

❧ 3.1 Average Rate of Change, Slope of Secant Line

Before discussing what is meant by the first derivative of a function, we would like to deal briefly with a preliminary concept, the *average rate of change* of a function. It should be noted that the emphasis is to be placed on the "rate of change" because change by itself is not as meaningful a concept. Rate of change is the terminology used to describe the change in one variable in terms of the changes occurring in a second variable.

To put the preceding remarks into perspective, consider the following situations:

1. A 4 percent rise in the consumer price index (change in one variable) could be a cause for concern if the increase occurred over a 1-month interval (change in second variable); however, hardly an eyebrow would be raised if the 4 percent increase were for a 12-month time period.
2. A 20°F change in temperature during a 1-hour period is far more dramatic and noteworthy than the same change occurring in a 12-hour interval.
3. A person who informs you that he paid $5000 more in federal income taxes this year than he paid last year is not likely to receive a great deal of sympathy because this increase has, in all likelihood, been accompanied by a much larger increase in his gross income.

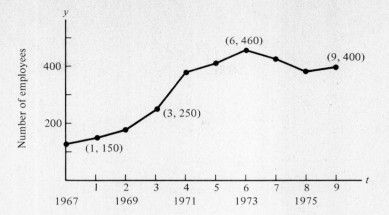

Figure 3.1

Suppose now that we look at Figure 3.1, which shows the number of employees in a small company from 1967 to 1976. From the graph, we can read the change in the number of employees from 1968 to 1970

Change in number of employees = 250 − 150 = 100

Change in years   = 1970 − 1968 = 2

This information enables us to determine the *average rate of change* in the number of employees from 1968 to 1970

$$\text{Average rate of change in the number of employees between 1968 and 1970} = \frac{\text{change in number of employees}}{\text{number of years}}$$

$$= \frac{250 - 150}{3 - 1}$$

$$= \frac{100}{2} = 50 \text{ employees/year}$$

Using this same approach, we can find the average rate of change in employment at this company between 1970 and 1973

$$\text{Average rate of change} = \frac{460 - 250}{3} = 70 \text{ employees/year}$$

while between 1973 and 1976

$$\text{Average rate of change} = \frac{400 - 460}{3} = -20 \text{ employees/year}$$

The process just described can be extended to any function $y = f(x)$. Referring to Figure 3.2 on the next page, we want to define the average rate of change of y with respect to x as x changes from x_1 to x_2. Defining

Change in $x = x_2 - x_1$

Change in $y = y_2 - y_1 = f(x_2) - f(x_1)$

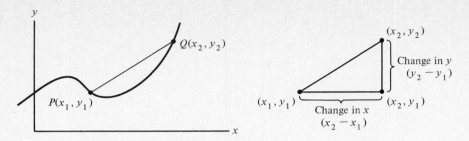

Figure 3.2

The average rate of change is defined as

$$\text{Average rate of change} = \frac{\text{change in } y}{\text{change in } x} = \frac{y_2 - y_1}{x_2 - x_1}$$

The line PQ, which connects the two points P and Q on the curve in Figure 3.2, is a *secant* line. Geometrically, the average rate of change can be interpreted as the slope of the secant line PQ. Denoting the slope of the secant line as m_S, we can write

$$m_S = \frac{y_2 - y_1}{x_2 - x_1} = \frac{f(x_2) - f(x_1)}{x_2 - x_1} \qquad \textbf{(3.1.1)}$$

Example 1 | Find the average rate of change of the function $y = f(x) = x^2/2$ as x changes

a. from 2 to 4   b. from 2 to 0   c. from 2 to -2

Solution | Using Equation 3.1.1, we get the following results

a. Average rate of change $= \dfrac{f(4) - f(2)}{4 - 2} = \dfrac{8 - 2}{2} = 3$

b. Average rate of change $= \dfrac{f(0) - f(2)}{0 - 2} = \dfrac{0 - 2}{-2} = 1$

c. Average rate of change $= \dfrac{f(-2) - f(2)}{(-2) - 2} = \dfrac{0}{-4} = 0$

The curve $y = f(x) = x^2/2$ is shown in Figure 3.3 together with the secant lines connecting (2, 2) with (4, 8), (0, 0), and $(-2, 2)$.

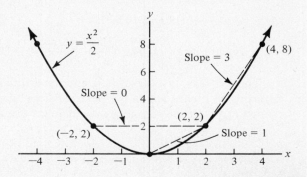

Figure 3.3

Example 2 If $y = f(x) = 1 - \sqrt{x}$, find the slope of the secant line connecting each of the following pairs of points:

a. $x_1 = 1$ and $x_2 = 4$

b. $x_1 = 1$ and $x_2 = \frac{1}{4}$

Solution Again, using Equation 3.1.1, we get the following results

a. $m_S = \dfrac{f(4) - f(1)}{4 - 1} = \dfrac{(-1) - 0}{3} = \dfrac{-1}{3}$

b. $m_S = \dfrac{f(1/4) - f(1)}{(1/4) - 1} = \dfrac{(1/2) - 0}{(-3/4)} = \dfrac{-2}{3}$

A sketch of the curve and the secant lines is shown in Figure 3.4.

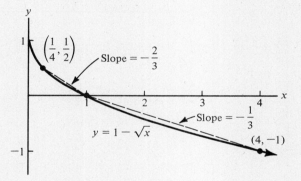

Figure 3.4

At the risk of becoming overly repetitious, let us again remind you that the average rate of change of the function $y = f(x)$ as x changes from x_1 to x_2 has the same meaning as the slope of the secant line connecting the points $[x_1, f(x_1)]$ and $[x_2, f(x_2)]$.

The procedures for describing the average rate of change of a function can be generalized in the following manner. First, let x represent the x coordinate of an arbitrary point on the curve and $x + h$ represent the x coordinate of a second point on the curve, where h represents the change in the variable x. The average rate of change of the function $y = f(x)$ in moving from $[x, f(x)]$ to $[x + h, f(x + h)]$ is given by the equation

$$m_S = \frac{f(x + h) - f(x)}{h} \qquad\qquad (3.1.2)$$

which is represented graphically in Figure 3.5 on the next page. The quantity $[f(x + h) - f(x)]/h$ is also called the *difference quotient*. Because this expression will be used subsequently in finding the first derivative of a function, it is important that you not only familiarize yourself with it, but in addition you should become proficient in using it for simple types of functions. To see what this quantity looks like in specific cases, consider the following examples in which the steps leading to the final form of the difference quotient are shown explicitly.

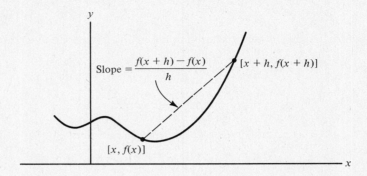

Figure 3.5

Example 3 | Find the difference quotient for the function $y = f(x) = 3x^2$.

Solution | STEP 1. Find $f(x + h)$

$$f(x + h) = 3(x + h)^2 = 3x^2 + 6xh + 3h^2$$

STEP 2. Evaluate $f(x + h) - f(x)$

$$f(x + h) - f(x) = 3x^2 + 6xh + 3h^2 - 3x^2 = 6xh + 3h^2$$

STEP 3. Divide the result found in step 2 by h

$$\frac{f(x + h) - f(x)}{h} = 6x + 3h$$

The result of this process is shown graphically in Figure 3.6.

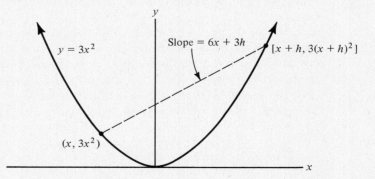

Figure 3.6

Example 4 | Find the difference quotient for the function

$$y = f(x) = \frac{1}{x}$$

Solution   STEP 1. $f(x + h) = \dfrac{1}{x + h}$

STEP 2. $f(x + h) - f(x) = \dfrac{1}{x + h} - \dfrac{1}{x} = \dfrac{x - (x - h)}{x(x + h)} = \dfrac{-h}{x(x + h)}$

NOTE: At this step, you should always simplify the difference $f(x + h) - f(x)$.

STEP 3. $\dfrac{f(x + h) - f(x)}{h} = \dfrac{-1}{x(x + h)}$

Again, the result is shown graphically in Figure 3.7.

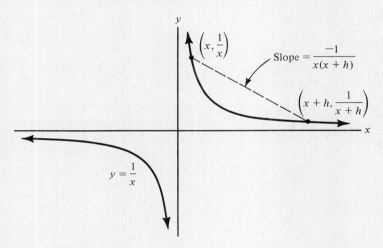

Figure 3.7

The difference quotient $[f(x + h) - f(x)]/h$ depends on both x and h. This means that values for both x and h must be provided in order to calculate the average rate of change or the slope of the secant line connecting two points. The difference quotient, once it is found, is a more efficient way of calculating the average rate of change, particularly if the average rates of change for a large number of pairs of points are to be tabulated.

Example 5   Given the curve $y = f(x) = 1/x$, find the slope of the secant line connecting the points whose x coordinates are $\tfrac{1}{2}$ and 3.

Solution   Using the result from Example 4, we have

$$m_S = \frac{-1}{x(x + h)}$$

Using $x = \tfrac{1}{2}$ and $h = 3 - \tfrac{1}{2} = \tfrac{5}{2}$, we get

$$m_S = -\tfrac{2}{3}$$

If we had set $x = 3$ and $h = -\tfrac{5}{2}$, we would have obtained the same result for m_S. The curve and the secant line are shown in Figure 3.8.

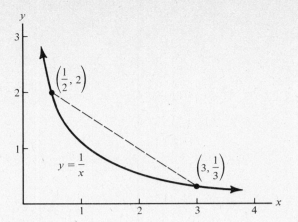

Figure 3.8

AVERAGE VELOCITY

When an object is in motion, the rate at which distance is traveled over a given interval of time yields the average velocity of the object. For example, if you traveled 160 miles in a period of 4 hours, the average velocity with which you were moving during this period is 40 miles/hour. When the distance s from some reference point is a known function of time t, the average velocity between time t and $t + h$ is defined as

$$\text{Average velocity} = \frac{s(t + h) - s(t)}{h}$$

Example 6   Suppose an object is moving along a straight line and the equation describing the distance s, in feet, of the object from some reference point is given by

$$s(t) = 5t^2 + 8$$

a. Find the expression for the average velocity between t and $t + h$.
b. Using the results from part a, find the average velocity between $t_1 = 2$ sec and $t_2 = 5$ sec.

Solution   a. STEP 1. $s(t + h) = 5(t + h)^2 + 8 = 5t^2 + 10th + 5h^2 + 8$

STEP 2. $s(t + h) - s(t) = 10th + 5h^2$

STEP 3. Average velocity $= \dfrac{10th + 5h^2}{h} = 10t + 5h$

b. Letting $t = 2$ and $h = 5 - 2 = 3$, we find that the average velocity from Step 3 in part a becomes

$$\text{Average velocity} = 10(2) + 5(3) = 35 \text{ ft/sec}$$

The same result can be obtained by setting $t = 5$ sec and $h = 2 - 5 = -3$ sec, that is,

$$\text{Average velocity} = 10(5) + 5(-3) = 35 \text{ ft/sec}$$

EXERCISE 3.1

1. One thousand dollars put into a savings account paying 6 percent per year grows in dollar value in the manner shown in the following diagram if no withdrawals or further deposits are made. If $t = 0$ designates the day of deposit, find the average rate of change of the amount in the account

a. From day of deposit to year 10   **b.** From year 5 to year 15
c. From year 5 to year 20

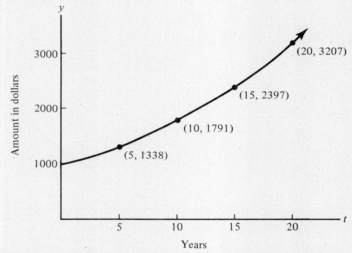

2. In addition to depreciating the value of an asset on a straight-line basis as discussed in Section 1.4, there are other methods for which the depreciation is larger in the early years than the corresponding depreciation calculated by the straight-line method. One of these methods, double-declining balance, will give the results shown on the accompanying graph for an asset that cost $20,000, has a useful life of 10 years, and has no salvage value at the end of the 10-year period. Using the following graph, determine the average rate of change of the value

a. Over the first 5 years   **b.** From year 2 to year 7
c. From year 2 to year 10

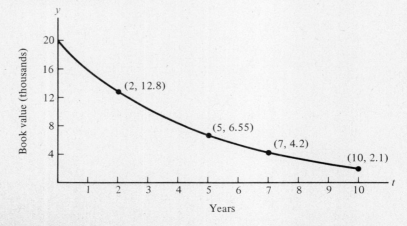

3. The amount of money an individual pays in federal income taxes depends on the amount of his or her net income. The following graph shows the relationship between annual taxes and net income for a single taxpayer. Using the graph, answer the following questions:

   a. What is the average rate of change of taxes paid as net income changes from $10,000 to $20,000?

   b. What is the average rate of change in taxes as net income changes from $20,000 to $40,000?

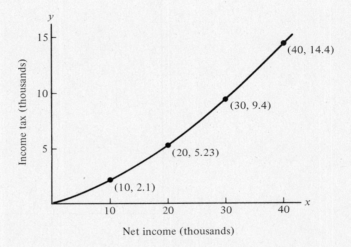

Net income (thousands)

For the functions in Exercises 4–9, find the slope of the secant line joining the points indicated; in addition, draw the secant line on the accompanying coordinate system.

4. $y = f(x) = \dfrac{x^2}{4}$

    a. $x_1 = -1$ $x_2 = 2$
    b. $x_1 = -1$ $x_2 = 0$
    c. $x_1 = -1.5$ $x_2 = 1$

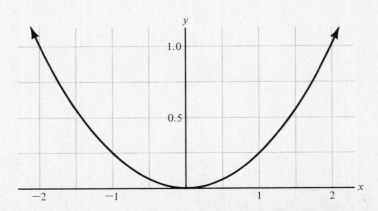

5. $y = f(x) = 2x + 1$

    a. $x_1 = -2$ $x_2 = 1$
    b. $x_1 = 0$ $x_2 = -3$

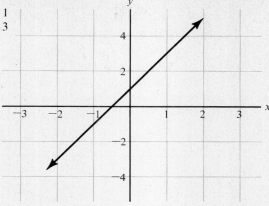

6. $y = f(x) = 1 - x^3$

    a. $x_1 = -2$ $x_2 = 0.5$
    b. $x_1 = -1$ $x_2 = 1$
    c. $x_1 = 0$ $x_2 = -1.5$

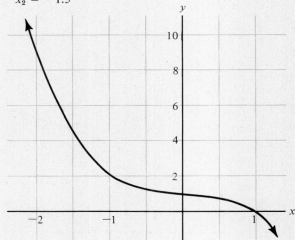

7. $y = f(x) = \dfrac{2}{x^2 + 1}$

    a. $x_1 = 0$ $x_2 = 1$
    b. $x_1 = -1$ $x_2 = 1$

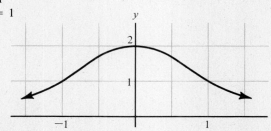

8. $y = f(x) = \sqrt[3]{x}$

   a. $x_1 = 0$ $x_2 = 8$

   b. $x_1 = -1$ $x_2 = +1$

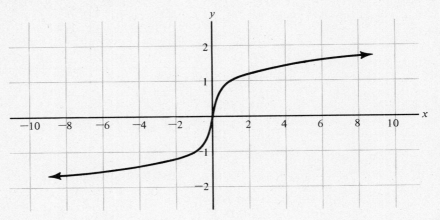

9. $y = f(x) = \sqrt{2x + 4}$

   a. $x_1 = 0$ $x_2 = -2$

   b. $x_1 = 6$ $x_2 = 0$

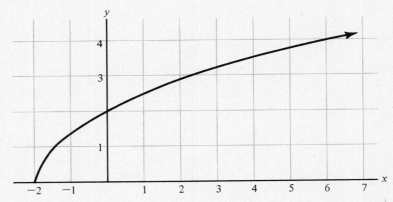

For the functions in Exercises 10–15, find the difference quotient $[f(x + h) - f(x)]/h$ in its simplest form.

10. $y = f(x) = 5 - 3x$

11. $y = f(x) = 2x - x^2$

12. $y = f(x) = 2x^3$

13. $y = f(x) = -\dfrac{1}{x^2}$

14. $y = f(x) = x^2 - 3x + 1$

15. $y = f(x) = x^4$

16. An object moves according to the equation

$$s = 4t^2 - 12t$$

where s is the distance in feet and t is the time in seconds.

a. Find the expression for the average velocity

$$\frac{s(t + h) - s(t)}{h}$$

Using the result of part a, find the average velocity
b. Between $t = 3$ sec and $t = 4$ sec
c. Between $t = 1$ sec and $t = 5$ sec

3.2 First Derivative of a Function

In determining either the average rate of change of a function or its equivalent, the slope of a secant line, in the previous section, we noted that the result depended on the coordinates of the two points selected on the curve under study. We would now like to turn our attention to the question of finding a way to determine the *slope of the line tangent* to a given curve at any *point* on the curve. While we are engaged in this endeavor, we will also find that we are determining what is called the *instantaneous* rate of change of the function, or, as it is more commonly called, the *first derivative* of the function.

Suppose we begin by asking: What is the slope of the line tangent to the curve $y = f(x) = x^2$ at $(1, 1)$? Although we could make a crude attempt to construct a line tangent to the curve at $(1, 1)$ as shown in Figure 3.9, we do not yet have an analytical method by which we can assign a definite value to the slope of the tangent line. However, it is possible to approximate the slope by using the expression for the difference quotient $[f(x + h) - f(x)]/h$ with $f(x) = x^2$, setting $x = 1$, and then assigning smaller and smaller values to h. A scenario of this process is shown in Figure 3.10 on the next page. As h approaches 0, the secant lines and the tangent line become more nearly parallel. On the basis of this observation, we define the *slope of the tangent line* m_T to the curve $y =$

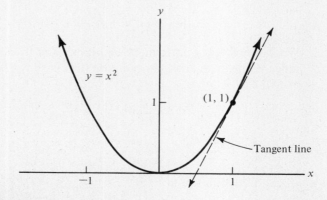

Figure 3.9

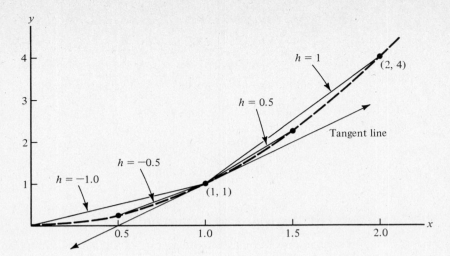

Figure 3.10

$f(x) = x^2$ at $x = 1$ to be the *limit* of the slopes of the secant lines as $h \to 0$. The difference quotient has the form

$$\frac{f(1 + h) - f(1)}{h} = \frac{(1 + h)^2 - 1^2}{h} = \frac{1 + 2h + h^2 - 1}{h}$$

$$= \frac{2h + h^2}{h} = 2 + h$$

The slope of the tangent line, then, at $x = 1$ is

$$m_T(x = 1) = \lim_{h \to 0} (2 + h) = 2$$

Thus, the slope of the line tangent to the curve $y = f(x) = x^2$ at $(1, 1)$ equals 2. If we want to find the slope of the line tangent to the curve at any other value, say $x = -2$, we go through the same process

$$m_T(x = -2) = \lim_{h \to 0} \left[\frac{f(-2 + h) - f(-2)}{h}\right] = \lim_{h \to 0}\left[\frac{(-2 + h)^2 - 4)}{h}\right]$$

$$= \lim_{h \to 0} \left[\frac{4 - 4h + h^2 - 4}{h}\right] = \lim_{h \to 0} \left[\frac{-4h + h^2}{h}\right]$$

$$= \lim_{h \to 0} (-4 + h) = -4$$

We can now get a little more ambitious and try to determine the slope of the tangent line at an arbitrary point on the curve $y = x^2$ and initially call the result $m_T(x)$; that is

$$m_T(x) = \lim_{h \to 0} \left[\frac{f(x + h) - f(x)}{h}\right] = \lim_{h \to 0} \left[\frac{(x + h)^2 - x^2}{h}\right]$$

$$= \lim_{h \to 0} \left[\frac{x^2 + 2xh + h^2 - x^2}{h}\right] = \lim_{h \to 0} \left[\frac{2xh + h^2}{h}\right]$$

$$= \lim_{h \to 0} (2x + h) = 2x$$

NOTE: In the limit process just described, h approached 0 while x, although arbitrary, remained constant.

For the function $y = f(x) = x^2$, we find that the slope m_T of a line tangent to the curve at any point (x, y) is given by $2x$.

Example 1   Find the slope of the line tangent to the curve $y = f(x) = x^2$ at $(3, 9)$.

Solution   We know that the slope m_T of a line tangent to the curve $y = x^2$ equals $2x$, so we can write

$$m_T(3) = 2(3) = 6$$

The tangent line at $(3, 9)$ is shown in Figure 3.11.

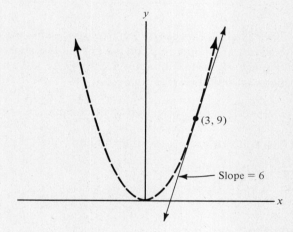

Figure 3.11

Before going on, it is only proper to notify you that although

$$\lim_{h \to 0} \frac{f(x + h) - f(x)}{h}$$

gives the slope of a line tangent to the curve $y = f(x)$ at any point, the quantity is ordinarily referred to as the *first derivative* of the function $y = f(x)$. In addition, the notation $m_T(x)$ is rarely used; the more common ways of denoting the first derivative of the function $y = f(x)$ are

$f'(x)$ read "f prime of x"

$y'(x)$ read "y prime of x"

$\dfrac{dy}{dx}$ read "dee y dee x"

$D_x y$ read "dee y sub x"

For the remainder of the text, we will generally use the notation $f'(x)$ or dy/dx when referring to the first derivative; however, it should be pointed out that this convention is not universal, so do not always expect it.

In summary, the first derivative is defined as follows.

DEFINITION:

The *first derivative* of the function $y = f(x)$ is denoted by $f'(x)$ or dy/dx and is defined as

$$f'(x) = \lim_{h \to 0} \left[\frac{f(x + h) - f(x)}{h} \right] \qquad (3.2.1)$$

provided that the limit exists.

The resulting expression for $f'(x)$ represents the slope of the line tangent to the curve at any point (x, y) on the curve. The steps used in finding the first derivative according to Equation 3.2.1 will be shown for a number of simple functions in the following examples.

Example 2   a. Find the first derivative of the function $y = f(x) = x^3$.

Solution   The following steps are used to find $f'(x)$.

STEP 1. Determine $f(x + h)$

$$f(x + h) = (x + h)^3$$
$$= x^3 + 3x^2h + 3xh^2 + h^3$$

STEP 2. Find the difference $[f(x + h) - f(x)]$

$$f(x + h) - f(x) = x^3 + 3x^2h + 3xh^2 + h^3 - x^3$$
$$= 3x^2h + 3xh^2 + h^3$$

STEP 3. Divide the result of step 2 by h to give $\dfrac{f(x + h) - f(x)}{h}$

$$\frac{f(x + h) - f(x)}{h} = \frac{3x^2h + 3xh^2 + h^3}{h}$$
$$= 3x^2 + 3xh + h^2$$

STEP 4. Find the limit as $h \to 0$ for the difference quotient found in step 3. The result will give you the first derivative $f'(x) = dy/dx$ of the function $y = f(x) = x^3$

$$\frac{dy}{dx} = f'(x) = \lim_{h \to 0} (3x^2 + 3xh + h^2) = 3x^2$$

Summarizing, we have found that the first derivative of the function $y = f(x) = x^3$ equals $3x^2$.

b. Find the slope of the line tangent to the curve $y = f(x) = x^3$ at the point $(-1, -1)$.

Solution   Because the slope of the line tangent to a curve at any point is given by $f'(x)$, the slope at $x = -1$ is given by $f'(-1)$

$$f'(-1) = 3(-1)^2 = 3$$

c. Find the equation of the line tangent to the curve $y = f(x) = x^3$ at $(-1, -1)$.

Solution   The equation of any line can be found if we know (1) the slope, and (2) the coordinates of one point on the line. From part b of this example, we know that the slope equals 3. In addition, the tangent line and the curve have one point in common, that is, the point of tangency, which in this case is $(-1, -1)$. Armed with this information, we apply the point-slope Formula 1.3.5

$$y - y_1 = m(x - x_1)$$

substituting $m = 3$, $x_1 = -1$, and $y_1 = -1$, getting

$$y + 1 = 3(x + 1)$$

Simplifying yields

$$y = 3x + 2$$

as the equation of the tangent line. The curve and the tangent line are shown in Figure 3.12. REMINDER: The variables x and y in the point-slope formula refer to the coordinates of points on the tangent line, *not* to the coordinates of points on the curve itself.

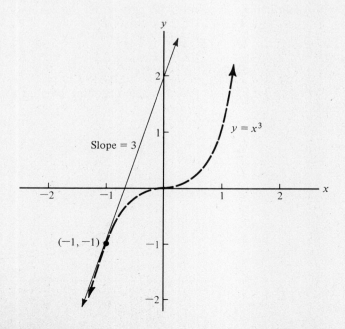

Figure 3.12

Problem 1 Find the slope and the equation of the line tangent to the curve $y = f(x) = x^3$ at $(\frac{1}{2}, \frac{1}{8})$; in addition, plot the tangent line on Figure 3.12.

Answer Slope $= \frac{3}{4}$; equation, $4y = 3x - 1$.

Example 3

a. Find the first derivative of the function $y = f(x) = 1/x$.

Solution

In Example 4 of Section 3.1, the difference quotient was found to be

$$\frac{f(x + h) - f(x)}{h} = \frac{-1}{x(x + h)}$$

From this, we can calculate $f'(x)$ directly

$$f'(x) = \lim_{h \to 0} \left[\frac{-1}{x(x + h)}\right] = \frac{-1}{x^2}$$

b. Find the slope of the line tangent to the curve at $(\frac{1}{2}, 2)$.

Solution

The slope of the tangent line equals $f'(\frac{1}{2})$, that is

$$f'(\tfrac{1}{2}) = \frac{-1}{(\frac{1}{2})^2} = -4$$

c. Find the equation of the line tangent to curve at $(\frac{1}{2}, 2)$.

Solution

Again using the result from part b and the point-slope formula, we can write

$$y - 2 = -4(x - \tfrac{1}{2})$$

giving

$$y = -4x + 4$$

Figure 3.13 shows the curve and the line tangent to the curve at $(\frac{1}{2}, 2)$.

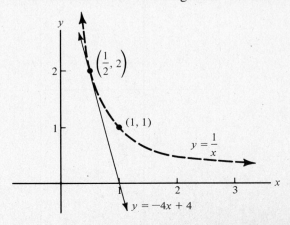

Figure 3.13

Problem 2 Find the slope and the equation of the line tangent to the curve $y = f(x) = 1/x$ at $(1, 1)$; plot the tangent line on Figure 3.13.

Answer Slope $= -1$; equation, $y = -x + 2$.

The process of finding the first derivative can be rather involved even for certain simple functions.

Example 4   a. Find the first derivative of the function

$$y = f(x) = \sqrt{x}$$

Solution As before, we illustrate the process step by step

STEP 1. $f(x + h) = \sqrt{x + h}$

STEP 2. $f(x + h) - f(x) = \sqrt{x + h} - \sqrt{x}$

STEP 3. $\dfrac{f(x + h) - f(x)}{h} = \dfrac{\sqrt{x + h} - \sqrt{x}}{h}$

STEP 4. $\lim\limits_{h \to 0} \dfrac{f(x + h) - f(x)}{h} = \lim\limits_{h \to 0} \dfrac{\sqrt{x + h} - \sqrt{x}}{h}$

If we attempt to substitute the limiting value of h, namely 0, into the right-hand side, the result is $\frac{0}{0}$, which is indeterminate. A situation similar to this one arose in Section 2.2, Example 14; however, in this case, we rationalize the numerator instead of the denominator

$$\frac{\sqrt{x + h} - \sqrt{x}}{h} = \left(\frac{\sqrt{x + h} - \sqrt{x}}{h}\right)\left(\frac{\sqrt{x + h} + \sqrt{x}}{\sqrt{x + h} + \sqrt{x}}\right)$$

$$= \frac{x + h - x}{h(\sqrt{x + h} + \sqrt{x})} = \frac{1}{\sqrt{x + h} + \sqrt{x}}$$

so now we can write

$$f'(x) = \lim\limits_{h \to 0}\left[\frac{\sqrt{x + h} - \sqrt{x}}{h}\right] = \lim\limits_{h \to 0}\left[\frac{1}{\sqrt{x + h} + \sqrt{x}}\right] = \frac{1}{2\sqrt{x}}$$

b. Find the slope of the line tangent to the curve $y = f(x) = \sqrt{x}$ at $(1, 1)$.

Solution From the result of part a, we know that $f'(x)$, which represents the slope of the tangent line, equals $1/(2\sqrt{x})$; at $x = 1$, we get

$$f'(1) = \frac{1}{2\sqrt{1}} = \tfrac{1}{2}$$

for the slope of the tangent line.

c. Find the equation of the line tangent to the curve $y = f(x) = \sqrt{x}$ at $(1, 1)$.

Solution   Again using the point-slope formula with $m = \frac{1}{2}$, $x_1 = 1$, and $y_1 = 1$, we get

$$y - 1 = \tfrac{1}{2}(x - 1)$$

simplifying gives

$$2y = x + 1$$

The tangent line is shown together with the curve in Figure 3.14.

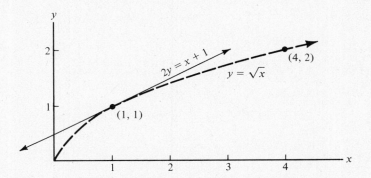

Figure 3.14

Problem 3 Find the slope and the equation of the line tangent to the curve $y = \sqrt{x}$ at $(4, 2)$. Sketch the graph of the tangent line on Figure 3.14.

Answer Slope $= \frac{1}{4}$; equation, $4y = x + 4$.

At this point, let us offer you some words of comfort. Rarely does one use the definition to find the first derivative of a function. A number of techniques exist that make the task of finding the first derivative a simple operation even for complex functions; these techniques will be presented in Sections 3.3–3.5.

The next topic relates to the problem of finding points on a curve where the slope of the tangent line has a given value. Up to this point, we have concentrated our attention on the problem of finding the slope and the equation of the line tangent to a curve at a given point. Now we want to examine the problem in reverse: *given the slope of a tangent line, find those coordinates of the point(s) on the curve where the first derivative equals the slope.* This problem is generally more difficult to solve than the problem of finding the slope at a given point because the equation that develops may be algebraically complex and therefore difficult to solve. For our initial exposure, the equations to be solved will be kept manageable.

Example 5   Find the point(s) on the curve $y = f(x) = x^2$ where the slope of the tangent line equals 3.

Solution   From Example 1 of this section, we know that the slope of the tangent line is given by $f'(x)$ where $f'(x) = 2x$. In this case, we are given $f'(x)$, that is,

$f'(x) = 3$, and we have to find the value(s) of x for which $f'(x) = 3$. Graphically, what we are attempting to do is to place the line shown in Figure 3.15 at some point on the curve where it will be tangent to the curve. To accomplish this, set $f'(x)$ equal to 3 and solve for x, that is, $3 = 2x$. Solving gives

$$x = \tfrac{3}{2} \qquad y = (\tfrac{3}{2})^2 = \tfrac{9}{4}$$

as the coordinates of the point where the slope of the tangent line equals 3.

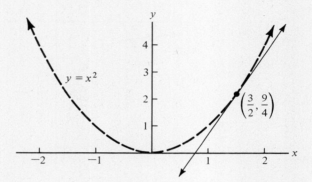

Figure 3.15

Problem 4 Find the point(s) on the curve $y = x^2$ where the slope of the tangent line equals (a) -2, and (b) 0. Sketch the tangent lines on Figure 3.15.

Answer a. $(-1, 1)$ b. $(0, 0)$

Example 6 Find the point(s) on the curve $y = f(x) = 1/x$ where the slope of the tangent line equals -4.

Solution From Example 3 of this section we know that

$$f'(x) = \frac{-1}{x^2}$$

Setting $f'(x) = -4$, we have

$$-4 = \frac{-1}{x^2}$$

Solving this equation for x yields two solutions

$$x_1 = \tfrac{1}{2} \quad \text{and} \quad x_2 = -\tfrac{1}{2}$$

Thus, there are two points $(\tfrac{1}{2}, 2)$ and $(-\tfrac{1}{2}, -2)$ where the tangent line to the curve has a slope equal to -4. Figure 3.16 on the next page contains the graph of the curve and the tangent lines.

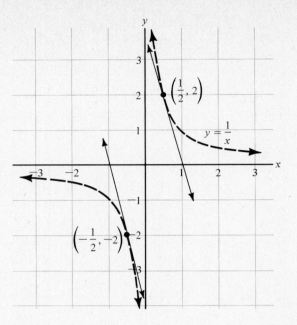

Figure 3.16

Problem 5 Find those points on the curve $y = 1/x$ where the slope of the curve equals $-\frac{1}{9}$. Sketch the tangent lines on Figure 3.16.

Answer $(3, \frac{1}{3})$ and $(-3, -\frac{1}{3})$

As a last item, you should be warned that the first derivative may not exist for some values of x, for a number of reasons. The first derivative of the function $y = f(x)$ will not exist at $x = a$ if

1. The function $y = f(x)$ is *not* continuous at $x = a$. The reader is advised to look over the material in Section 2.3 and see examples of this situation.
2. The function $y = f(x)$ is continuous at $x = a$, but
   a. The tangent line to the curve at $x = a$ is vertical and therefore its slope is not defined, as demonstrated in Example 7, or
   b. The direction of the curve changes abruptly as indicated by a sharp corner or a wedge-shaped appearance, as shown in Example 8.

Example 7 The graph of the curve $y = f(x) = \sqrt[3]{x - 1}$ is shown in Figure 3.17. Although $(1, 0)$ is a point on the curve, the slope of the tangent line is not defined. The first derivative*

$$f'(x) = \frac{1}{3\sqrt[3]{(x - 1)^2}}$$

is not defined at $x = 1$.

* Do not concern yourself about the derivation of $f'(x)$; it will be explained in Section 3.5.

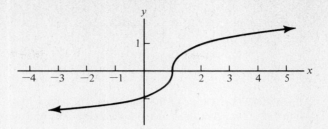

Figure 3.17

Example 8 | The function

$$y = f(x) = \begin{cases} 2x & x \leq 0 \\ -x & x > 0 \end{cases}$$

whose graph is shown in Figure 3.18, does not possess a first derivative at $x = 0$. To see why the derivative does not exist at $x = 0$, let us look at the difference quotient

$$\frac{f(0 + h) - f(0)}{h}$$

and try to see what happens as $h \to 0$. As $h \to 0$ *from the left*, the expression $2x$ is substituted for $f(x)$, while $-x$ is substituted for $f(x)$ as $h \to 0$ *from the right*.

From the left, $f(x) = 2x$

$$\frac{f(0 + h) - f(0)}{h} = \frac{2(0 + h) - 2(0)}{h} = 2$$

From the right, $f(x) = -x$

$$\frac{f(0 + h) - f(0)}{h} = \frac{-(0 + h) - 0}{h} = -1$$

Since $2 \neq -1$

$$\lim_{h \to 0} \frac{f(0 + h) - f(0)}{h}$$

does not exist and thus $f'(0)$ does not exist.

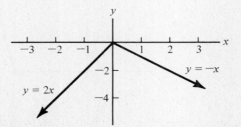

Figure 3.18

EXERCISE 3.2

Using the definition

$$\frac{dy}{dx} = f'(x) = \lim_{h \to 0} \frac{f(x + h) - f(x)}{h}$$

find the first derivative of the functions in Exercises 1–8. Also find the slopes and equations of the tangent lines at the indicated points and sketch the tangent lines on the accompanying coordinate systems.

1. $y = f(x) = \dfrac{x^2}{2}$

    a. $(2, 2)$
    b. $(-1, \frac{1}{2})$

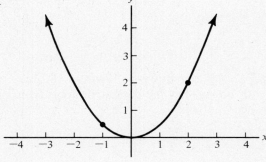

2. $y = f(x) = 1 - 2x$

    a. $(0, 1)$
    b. $(1, -1)$

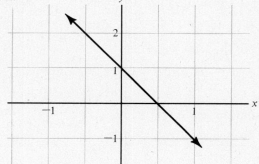

3. $y = f(x) = x^2 - 2x$

    a. $(1, -1)$
    b. $(2, 0)$

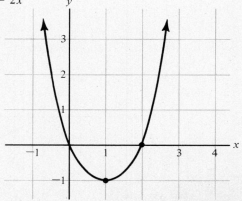

4. $y = f(x) = \dfrac{1}{x - 1}$

  a. $(0, -1)$
  b. $(3, \frac{1}{2})$

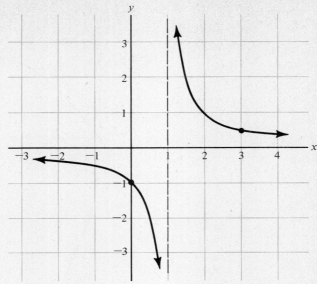

5. $y = f(x) = \sqrt{x - 1}$

  a. $(2, 1)$
  b. $(5, 2)$

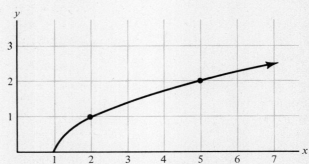

6. $y = f(x) = 1 - x^3$

  a. $(-1, 2)$
  b. $(1, 0)$

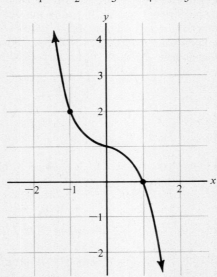

7. $y = f(x) = 3x - x^2$

    a. $(0, 0)$
    b. $(2, 2)$

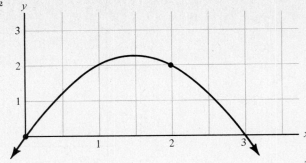

8. $y = f(x) = x^4$

    a. $(-1, 1)$
    b. $(1, 1)$

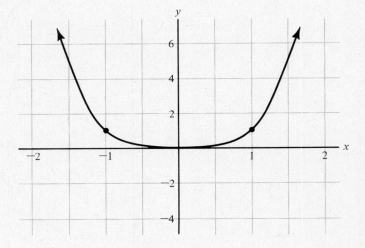

For Exercises 9–13, the first derivative $f'(x)$ will be needed. However, it is not necessary to begin from scratch by using the definition because the first derivatives of all the functions in this set of problems have been either given in an example in the text or been found earlier in Exercises 1–8. Find those point(s) on the curves for the following functions where the slope of the curve has the indicated value.

9. $y = f(x) = \dfrac{x^2}{2}$ **a.** $m_T = 3$, **b.** $m_T = -\dfrac{5}{2}$

10. $y = f(x) = 1 - x^3$ **a.** $m_T = -12$, **b.** $m_T = \dfrac{-1}{3}$

11. $y = f(x) = x^2 - 2x$ **a.** $m_T = 4$, **b.** $m_T = -3$

12. $y = f(x) = 3x - x^2$ **a.** $m_T = -5$, **b.** $m_T = \dfrac{3}{2}$

13. $y = f(x) = \dfrac{1}{x}$ **a.** $m_T = -9$, **b.** $m_T = -\dfrac{1}{16}$

14. Find those values of x on the following graph where the first derivative $f'(x)$ does not exist.

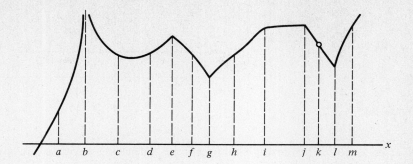

3.3 Simple Power Rule, Sum and Difference Rules

If it were necessary to use the definition

$$f'(x) = \lim_{h \to 0} \frac{f(x + h) - f(x)}{h}$$

each time you wanted to find the first derivative of a function, the process would, except for very simple functions, be difficult and very time consuming. Fortunately, other methods enable us to find the first derivative quickly and efficiently; in fact, the procedures are so straightforward that it is possible to find the first derivative without knowing or appreciating its meaning. In describing the techniques presented in Sections 3.3–3.5, our approach will be similar to that used in finding limits. The techniques will only be presented and illustrated in most cases. Derivations, unless they are very simple, are contained in Appendix C.

Before submerging ourselves in a description of these techniques, let us point out to you that, when we want to find the first derivative of an expression, the operation of finding the first derivative is also indicated by the notation

$$\frac{d}{dx} \left(\qquad \right)$$

which means that we are seeking the first derivative of the expression within the parentheses, for example

$$\frac{d}{dx} (15x^4 - 3x^2 + 5x + 2)$$

is a signal to find the first derivative of the function

$$y = f(x) = 15x^4 - 3x^2 + 5x + 2$$

I. THE FIRST DERIVATIVE OF A CONSTANT EQUALS 0

If $y = f(x) = c$, then

$$f'(x) = \frac{d(c)}{dx} = 0 \qquad\qquad\qquad \textbf{(3.3.1)}$$

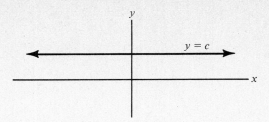

Figure 3.19

This result should not be surprising because the graph of the function $y = f(x) = c$ is a horizontal line (see Figure 3.19) whose slope equals 0 for all values of x. This result can also be obtained from the definition of the first derivative (3.2.1)

$$f(x) = c \qquad f(x + h) = c$$

$$f'(x) = \lim_{h \to 0} \frac{f(x + h) - f(x)}{h} = \lim_{h \to 0} \frac{c - c}{h}$$

$$= \lim_{h \to 0} \frac{0}{h} = 0$$

In Section 3.2, the following results emerged when we used the definition to find the first derivative of some simple power functions:

$$y = f(x) = \underline{x^2} \qquad \frac{dy}{dx} = f'(x) = \underline{2x}$$

$$y = f(x) = \underline{x^3} \qquad \frac{dy}{dx} = f'(x) = \underline{3x^2}$$

$$y = f(x) = \underline{x^4} \qquad \frac{dy}{dx} = f'(x) = \underline{4x^3}$$

These results can be obtained directly by a method known as the *simple power rule*. It is used whenever the function to be differentiated can be written as a power function, that is

$$y = f(x) = x^n \qquad \text{where } n \text{ is any real number}$$

II. SIMPLE POWER RULE

If $y = f(x) = x^n$, where n is a real number, then

$$f'(x) = \frac{dy}{dx} = nx^{n-1} \qquad\qquad\qquad \textbf{(3.3.2)}$$

The derivation of this result for the case where n is a positive integer is contained in Appendix C.

In Section 3.2 we found, using the definition, the first derivative of the following functions:

a. $y = f(x) = \dfrac{1}{x}$ $\dfrac{dy}{dx} = \dfrac{-1}{x^2}$

b. $y = f(x) = \sqrt{x}$ $\dfrac{dy}{dx} = \dfrac{1}{2\sqrt{x}}$

These results can be obtained more easily by using the simple power rule, for example

a. $y = f(x) = \dfrac{1}{x} = x^{-1}$

$$\frac{dy}{dx} = f'(x) = (-1)x^{-2} = \frac{-1}{x^2}$$

b. $y = f(x) = \sqrt{x} = x^{1/2}$

$$\frac{dy}{dx} = f'(x) = \tfrac{1}{2}x^{-1/2} = \frac{1}{2\sqrt{x}}$$

The next three examples illustrate the use of the simple power rule.

Example 1   Find the first derivative of the function

$$y = f(x) = x^8$$

Solution   Using the simple power rule with $n = 8$, we get

$$\frac{dy}{dx} = f'(x) = 8x^7$$

Example 2   Find the first derivative of the function

$$y = f(x) = \sqrt[4]{x}$$

Solution   First, write the expression to be differentiated as a power expression, that is, $\sqrt[4]{x} = x^{1/4}$. Then apply the simple power rule, yielding

$$\frac{dy}{dx} = f'(x) = \tfrac{1}{4}x^{-3/4} = \frac{1}{4x^{3/4}}$$

$$= \frac{1}{4\sqrt[4]{x^3}}$$

Example 3   Find the first derivative of the function

$$y = f(x) = \frac{1}{x^3}$$

Solution | Again, the expression to be differentiated is written as a power expression, that is, $1/x^3 = x^{-3}$. Applying the simple power rule gives

$$\frac{dy}{dx} = f'(x) = -3x^{-4} = \frac{-3}{x^4}$$

Problem 1 Find the first derivative of the following functions:

a. $y = f(x) = x^6$ b. $y = f(x) = \frac{1}{x^2}$ c. $y = f(x) = \sqrt[3]{x}$

Answer a. $6x^5$ b. $\dfrac{-2}{x^3}$ c. $\dfrac{1}{3\sqrt[3]{x^2}}$

> ## III. THE DERIVATIVE OF A CONSTANT TIMES A FUNCTION EQUALS THE CONSTANT TIMES THE DERIVATIVE OF THE FUNCTION
>
> If $y = f(x) = C \cdot F(x)$, where C is a constant and $F(x)$ is differentiable, then
>
> $$\frac{dy}{dx} = f'(x) = C \cdot F'(x) = C \cdot \frac{dF(x)}{dx} \qquad \textbf{(3.3.3)}$$

Example 4 | Find the first derivative of the function

$$y = f(x) = 3x^5$$

Solution | The function is 3 times x^5; thus we can write

$$f'(x) = 3\,\frac{d(x^5)}{dx} = 3(5x^4) = 15x^4$$

Rule III can be obtained directly from the definition of the first derivative as follows:

$$f(x) = C \cdot F(x) \qquad f(x + h) = C \cdot F(x + h)$$

$$f'(x) = \lim_{h \to 0} \frac{f(x + h) - f(x)}{h}$$

$$= \lim_{h \to 0} \frac{C \cdot F(x + h) - C \cdot F(x)}{h}$$

$$= C \cdot \left[\lim_{h \to 0} \frac{F(x + h) - F(x)}{h} \right]$$

$$= C \cdot F'(x)$$

By incorporating rule III with rule II, the simple power rule can be expressed as follows:

SIMPLE POWER RULE

If $y = f(x) = C \cdot x^n$, where n is a real number, then

$$\frac{dy}{dx} = f'(x) = C \cdot n \cdot x^{n-1} \qquad\qquad (3.3.4)$$

The next technique tells us how to differentiate a function containing two or more terms.

IV. THE FIRST DERIVATIVE OF A SUM OR DIFFERENCE OF TWO FUNCTIONS EQUALS THE SUM OR DIFFERENCE OF THE FIRST DERIVATIVES

If $y = f(x) = G(x) \pm H(x)$, where $G(x)$ and $H(x)$ are differentiable functions, then

$$\frac{dy}{dx} = f'(x) = \frac{dG}{dx} \pm \frac{dH}{dx} = G'(x) \pm H'(x) \qquad\qquad (3.3.5)$$

If the expression to be differentiated contains two or more terms, then rule IV allows us to differentiate term by term.

Example 5 | Find the first derivative of the function

$$y = f(x) = 7x^3 + 8x + 5$$

Solution | Because the expression to be differentiated contains three terms, and each term is differentiable, that is

$$\frac{d}{dx}(7x^3) = 21x^2 \qquad \frac{d}{dx}(8x) = 8 \qquad \text{and} \qquad \frac{d}{dx}(5) = 0$$

we apply the sum rule

$$\frac{d}{dx}(7x^3 + 8x + 5) = \frac{d}{dx}(7x^3) + \frac{d}{dx}(8x) + \frac{d}{dx}(5)$$

$$= 21x^2 + 8$$

Example 6 | Find the first derivative of the function

$$y = f(x) = 4\sqrt{x} + \frac{3}{x}$$

Solution | Each of the terms is differentiable

$$\frac{d}{dx}(4\sqrt{x}) = \frac{d}{dx}(4x^{1/2}) = 4(\tfrac{1}{2})x^{-1/2} = \frac{2}{\sqrt{x}}$$

and

$$\frac{d}{dx}(3/x) = \frac{d}{dx}(3x^{-1}) = 3(-1)x^{-2} = -\frac{3}{x^2}$$

so we get

$$\frac{dy}{dx} = f'(x) = \frac{2}{\sqrt{x}} - \frac{3}{x^2}$$

Problem 2 Find the first derivative of each of the following:

a. $y = f(x) = x^4 - 8x^3 + 5x^2 - 2x - 6$

b. $y = f(x) = 7x^4 - \dfrac{1}{x^2}$

c. $y = f(x) = 2\sqrt[3]{x} + 5x^3$

Answer a. $4x^3 - 24x^2 + 10x - 2$

b. $28x^3 + \dfrac{2}{x^3}$

c. $\dfrac{2}{3\sqrt[3]{x^2}} + 15x^2$

Rule IV can also be obtained directly from the definition of the first derivative (3.2.1). Our analysis will be restricted to the case in which we have the sum of two functions.

$$f(x) = G(x) + H(x) \qquad f(x + h) = G(x + h) + H(x + h)$$

$$f'(x) = \lim_{h \to 0} \left[\frac{G(x + h) + H(x + h) - G(x) - H(x)}{h} \right]$$

$$= \lim_{h \to 0} \left[\frac{G(x + h) - G(x)}{h} + \frac{H(x + h) - H(x)}{h} \right]$$

$$= \lim_{h \to 0} \left[\frac{G(x + h) - G(x)}{h} \right] + \lim_{h \to 0} \left[\frac{H(x + h) - H(x)}{h} \right]$$

The last step follows from property V of limits, that is, the limit of a sum equals the sum of the limits. Noting that each term on the right-hand side denotes $G'(x)$ and $H'(x)$, respectively, we can write, finally

$$f'(x) = G'(x) + H'(x)$$

EXERCISE 3.3

Find the first derivative of the functions in Exercises 1–18.

1. $y = f(x) = x^4 - 3x^2 + 5$ **2.** $y = f(x) = 6x^5 - 8x^4 + 7$

3. $y = f(x) = \sqrt{8}$ **4.** $y = f(x) = x^7 + 6x^4 - 9x^3 - 1$

5. $y = f(x) = 2x^{15} + 11x^9 - 3$ **6.** $y = f(x) = 8 - 4x^{10}$

7. $y = f(x) = \dfrac{5}{x^2}$ **8.** $y = f(x) = \dfrac{-2}{x^4}$

9. $y = f(x) = 7x^3 - \dfrac{4}{x^3}$ **10.** $y = f(x) = \dfrac{3}{x} - \dfrac{7}{x^3} + 9$

11. $y = f(x) = \dfrac{6}{x^{13}}$ **12.** $y = f(x) = 8x^4 - \dfrac{9}{x^5} + 2$

13. $y = f(x) = \sqrt[3]{x^4}$ **14.** $y = f(x) = \dfrac{7}{\sqrt{x}}$

15. $y = f(x) = 5x - \dfrac{4}{x^2}$ **16.** $y = f(x) = 6\sqrt{x} + \dfrac{2}{x} + 1$

17. $y = f(x) = \sqrt{x^5} + 3x^4 - \dfrac{1}{x}$ **18.** $y = f(x) = \dfrac{8}{x} - \dfrac{3}{\sqrt{x}} + 7$

Find the slope and the equation of the line tangent to the curves in Exercises 19–21. Sketch the tangent line on the accompanying coordinate systems.

19. $y = f(x) = x^3 - 6x + 1$ at $(-1, 6)$

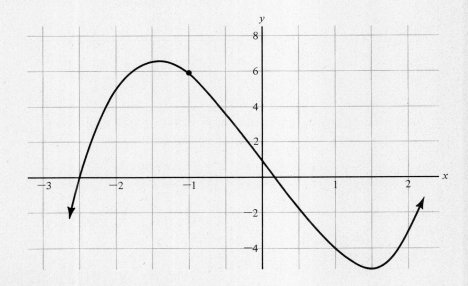

20. $y = f(x) = 4\sqrt{x} - x$ at $(1, 3)$

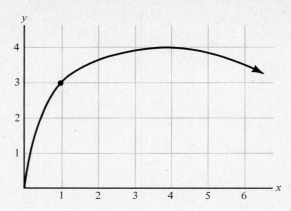

21. $y = f(x) = x^2 - x^3 + 2$ at $(-1, 4)$

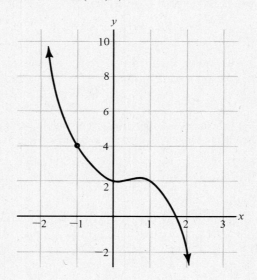

In Exercises 22–26, find those points on each of the curves where the slope of the tangent line has the value given.

22. $y = f(x) = x^3 + 3x^2 - x + 1$ $m_T = -1$

23. $y = f(x) = 7x^2 - 8x + 3$ $m_T = 6$

24. $y = f(x) = x^3 + 3x^2 + 4x + 5$ $m_T = 1$

25. $y = f(x) = \dfrac{x}{2} - \dfrac{1}{x}$ $m_T = \frac{3}{4}$

26. $y = f(x) = 2\sqrt{x} - \dfrac{x}{3} - 1$ $m_T = \frac{2}{3}$

3.4 Product and Quotient Rules

When a function is written as a product of two factors, for example

$$y = f(x) = (x^6 + 3x^2 - 9)(4x^5 - 8x^3 + 6x)$$

its first derivative can be found by using a technique known as the *product rule*.

V. PRODUCT RULE

If $u(x)$ and $v(x)$ are two differentiable functions, the first derivative of the function

$$y = f(x) = u(x) \cdot v(x)$$

is given by the formula

$$\frac{dy}{dx} = f'(x) = u(x) \cdot v'(x) + v(x) \cdot u'(x) \qquad \textbf{(3.4.1)}$$

In words, the rule states: *The first derivative of a product equals the first factor times the derivative of the second plus the second factor times the derivative of the first.* The derivation of this rule is contained in Appendix C.

Example 1 | Use the product rule to find the first derivative of the function

$$y = f(x) = (x^3)(x^4)$$

Solution | First calling one factor $u(x)$ and the second $v(x)$

$$u(x) = x^3 \qquad v(x) = x^4$$

$u'(x)$ and $v'(x)$ are

$$u'(x) = 3x^2 \qquad v'(x) = 4x^3$$

Multiplying the expressions indicated by the arrows and then adding gives

$$f'(x) = (x^3)(4x^3) + (x^4)(3x^2) = 7x^6$$

which agrees with the result obtained when you differentiate the same function $y = f(x) = x^7$ using the simple power rule. This example also serves to illustrate that **the derivative of a product does not equal the product of the derivatives,** that is, if

$$y = f(x) = (x^3)(x^4)$$

then

$$f'(x) \neq (3x^2)(4x^3) = 12x^5$$

Example 2 | Find the first derivative of the function

$$y = f(x) = (x^2 + 5x - 3)(x^3 - 2x^2 + 7x)$$

Solution | To utilize the product rule, we label one factor $u(x)$ and the other $v(x)$. Letting

$$u(x) = x^2 + 5x - 3 \qquad v(x) = x^3 - 2x^2 + 7x$$

Differentiating each gives

$$u'(x) = 2x + 5 \qquad v'(x) = 3x^2 - 4x + 7$$

Now combining the expressions indicated by the arrows, we get

$$\frac{dy}{dx} = f'(x) = (x^2 + 5x - 3)(3x^2 - 4x + 7)$$
$$+ (x^3 - 2x^2 + 7x)(2x + 5)$$

or, after simplifying, the first derivative becomes

$$\frac{dy}{dx} = f'(x) = 5x^4 + 12x^3 - 18x^2 + 82x - 21$$

You may have observed that it would have been easier to multiply the two polynomials immediately and then to differentiate term by term. Although the product rule is the lengthier of the two approaches in this example, problems will arise in which the product rule will be the only avenue open to you in obtaining the derivative, so it is in your best interest to master it while the expressions to be differentiated are not especially lengthy or complex.

Example 3 | Find the first derivative of the function

$$y = f(x) = \sqrt{x}(2 + x^2)$$

Solution | Proceeding as we did in Example 2, we let

$$u(x) = \sqrt{x} \qquad v(x) = 2 + x^2$$

and differentiating

$$u'(x) = \frac{1}{2\sqrt{x}} \qquad v'(x) = 2x$$

Applying the product rule yields

$$\frac{dy}{dx} = f'(x) = (\sqrt{x})(2x) + (2 + x^2)\left(\frac{1}{2\sqrt{x}}\right)$$

Simplifying gives

$$f'(x) = \frac{2\sqrt{x}(\sqrt{x})(2x)}{2\sqrt{x}} + \frac{2 + x^2}{2\sqrt{x}} = \frac{4x^2 + 2 + x^2}{2\sqrt{x}} = \frac{5x^2 + 2}{2\sqrt{x}}$$

Problem 1 Using the product rule, find the first derivative of each of the following

a. $y = f(x) = (3x^5 - x^3 + 1)(6x^2 + 3)$

Answer $f'(x) = (3x^5 - x^3 + 1)(12x)$
$$+ (6x^2 + 3)(15x^4 - 3x^2)$$

b. $y = f(x) = 5x\left(3x^2 + \dfrac{2}{x^3}\right)$

Answer $f'(x) = 5x\left(6x - \dfrac{6}{x^4}\right) + \left(3x^2 + \dfrac{2}{x^3}\right)5 = 45x^2 - \dfrac{20}{x^3}$

The next method, known as the *quotient rule*, is used to find the first derivative of a ratio of two functions.

VI. QUOTIENT RULE

If $u(x)$ and $v(x)$ are two differentiable functions, the first derivative of the function

$$y = f(x) = \frac{u(x)}{v(x)} \qquad v(x) \neq 0$$

is given by the formula

$$\frac{dy}{dx} = f'(x) = \frac{v(x)u'(x) - u(x)v'(x)}{[v(x)]^2} \qquad\qquad \textbf{(3.4.2)}$$

The derivation of this rule is contained in Appendix C. The use of this rule is illustrated in Examples 4 and 5.

Example 4 | Use the quotient rule to find the first derivative of the function

$$y = f(x) = \frac{x^7}{x^4}$$

Solution | Setting

$$u(x) = x^7 \qquad v(x) = x^4$$

we get for $u'(x)$ and $v'(x)$

$$u'(x) = 7x^6 \qquad v'(x) = 4x^3$$

Applying the quotient rule (Equation 3.4.2) gives

$$f'(x) = \frac{(x^4)(7x^6) - (x^7)(4x^3)}{(x^4)^2} = \frac{7x^{10} - 4x^{10}}{x^8}$$

$$= \frac{3x^{10}}{x^8} = 3x^2 \qquad x \neq 0$$

This result could also have been obtained by applying the simple power rule to the function $y = f(x) = x^3$. This example also illustrates that **the derivative of a quotient does not equal the quotient of the derivatives,** that is, if

$$y = f(x) = \frac{x^7}{x^4}$$

then

$$f'(x) \neq \frac{7x^6}{4x^3} = \frac{7x^3}{4}$$

Example 5 | Find the first derivative of the function

$$y = f(x) = \frac{x}{2x + 1}$$

Solution | Because $f(x)$ is written as the ratio of two functions, we apply the quotient rule with

$$u(x) = x \qquad v(x) = 2x + 1$$

Differentiating each gives

$$u'(x) = 1 \qquad v'(x) = 2$$

Applying the quotient rule, Equation 3.4.2, gives

$$\frac{dy}{dx} = f'(x) = \frac{(2x + 1)(1) - (x)(2)}{(2x + 1)^2} = \frac{1}{(2x + 1)^2}$$

Problem 2 Using the quotient rule, find the first derivative of each of the following functions:

a. $y = f(x) = \dfrac{x}{x + 1}$

Answer $f'(x) = \dfrac{(x + 1)(1) - (x)(1)}{(x + 1)^2} = \dfrac{1}{(x + 1)^2}$

b. $y = f(x) = \dfrac{3x^2}{1 - 2x}$

Answer $f'(x) = \dfrac{(1 - 2x)(6x) - (3x^2)(-2)}{(1 - 2x)^2} = \dfrac{6x - 6x^2}{(1 - 2x)^2}$

EXERCISE 3.4

Use either the product rule or the quotient rule to find the first derivative of the functions in Exercises 1–15.

1. $y = f(x) = (3x + 2)(x - 5)$

2. $y = f(x) = (x^2 + 4)(2x - 7)$

3. $y = f(x) = (x^3 - x + 3)(6x - 4)$

4. $y = f(x) = \left(x + \dfrac{2}{x}\right)(x^2 - 1)$

5. $y = f(x) = (x^4 + x^2 + 1)(x^6 - 1)$

6. $y = f(x) = (\sqrt{x} + 2)(x^3 - 7x^2 + 1)$

7. $y = f(x) = \dfrac{2x}{x - 5}$

8. $y = f(x) = \dfrac{3x}{x^2 + 2}$

9. $y = f(x) = \dfrac{x^2}{3x + 4}$

10. $y = f(x) = \dfrac{x^3}{x^2 + 5}$

11. $y = f(x) = \dfrac{2x + 6}{x - 4}$

12. $y = f(x) = \dfrac{3x - 8}{6x + 1}$

13. $y = f(x) = \dfrac{x + 1}{1 - x}$

14. $y = f(x) = \dfrac{4x^2 + 7x - 2}{3x^2 - 5x + 1}$

15. $y = f(x) = \dfrac{\sqrt{x}}{x + 1}$

In Exercises 16–18, find the slope and the equation of the line tangent to each of the curves at the point indicated. In addition, sketch the tangent line on the accompanying coordinate system.

16. $y = f(x) = \frac{1}{8}x^2(x^2 - 8)$ at $(-2, -2)$

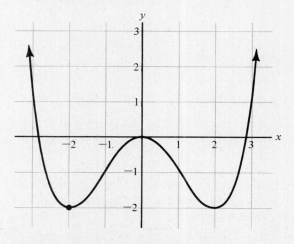

17. $y = f(x) = \dfrac{x}{x - 1}$ at $(2, 2)$

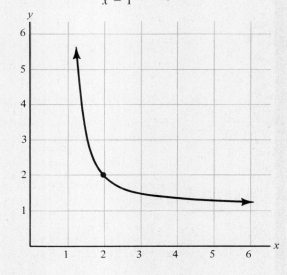

18. $y = f(x) = \sqrt{x}\,(x - 6)$ at $(1, -5)$

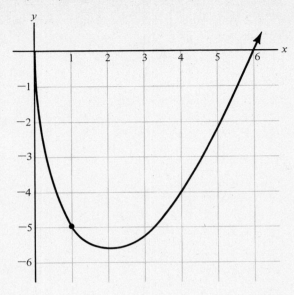

In Exercises 19–24, find the point(s) on each curve where the slope of the tangent line equals the given value.

19. $y = f(x) = x^2(3x - 4)$ $m_T = 1$

20. $y = f(x) = (x^2 + 2)(1 - x)$ $m_T = -3$

21. $y = f(x) = \sqrt{x}\,(2x + 6)$ $m_T = 6$

22. $y = f(x) = \dfrac{x^2}{x - 1}$ $m_T = 0$

23. $y = f(x) = \dfrac{x - 2}{x - 3}$ $m_T = -1$

24. $y = f(x) = \dfrac{x^2 + 1}{x - 1}$ $m_T = \dfrac{1}{2}$

3.5 General Power Rule, Chain Rule

The *general power rule* is used when we want to find the first derivative of functions having the form

$$y = f(x) = C \cdot [u(x)]^n \tag{3.5.1}$$

where $u(x)$ is a function of x, and C and n are real numbers. The right-hand side of Equation 3.5.1 is a power expression whose base is a function of x. Examples of functions that fall into this category are

$$y = f(x) = 2(x^2 + 3x - 1)^4$$

where

$$u(x) = x^2 + 3x - 1 \qquad C = 2 \qquad \text{and} \qquad n = 4$$

or

$$y = f(x) = \sqrt{5 - x}$$

where

$$u(x) = 5 - x \qquad C = 1 \qquad \text{and} \qquad n = \tfrac{1}{2}$$

To give you some feeling for the origin of the rule, suppose you were asked to find the first derivative of the function

$$y = f(x) = [u(x)]^2$$

At this point, the only alternative open to you is to write $[u(x)]^2$ as $u(x) \cdot u(x)$ and then to apply the product rule

$$f(x) = u(x) \cdot u(x)$$

$$f'(x) = u(x) \cdot u'(x) + u(x) \cdot u'(x) = 2u(x) \cdot u'(x)$$

This result looks like something you would expect from the simple power rule except for the additional factor $u'(x)$, the derivative of the base. This additional factor is the basic difference between the simple and general power rules.

VII. GENERAL POWER RULE

If $u(x)$ is differentiable, the first derivative of the function

$$y = f(x) = C[u(x)]^n$$

where C and n are real numbers, is given by

$$\frac{dy}{dx} = f'(x) = Cn[u(x)]^{n-1}u'(x) \tag{3.5.2}$$

The derivation of the general power rule when n is a positive integer is contained in Appendix C.

Example 1 | Find the first derivative of the function

$$y = f(x) = (3x^2 + 8x - 1)^4$$

Solution | Applying the general power rule with

$$u(x) = 3x^2 + 8x - 1 \qquad n = 4 \qquad C = 1 \qquad u'(x) = 6x + 8$$

we get

$$\frac{dy}{dx} = f'(x) = 4(3x^2 + 8x - 1)^3(6x + 8)$$

$$\phantom{\frac{dy}{dx} = f'(x) = }\; \underset{n}{\uparrow} \quad \underset{u(x)}{\uparrow} \quad \underset{n-1}{\uparrow} \quad \underset{u'(x)}{\uparrow}$$

Simplifying then yields

$$\frac{dy}{dx} = f'(x) = (24x + 32)(3x^2 + 8x - 1)^3$$

Example 2 | Find the first derivative of the function

$$y = f(x) = 5\sqrt{3x^4 + 10x} = 5(3x^4 + 10x)^{1/2}$$

Solution | Again applying the general power rule with

$$u(x) = 3x^4 + 10x \qquad C = 5 \qquad n = \tfrac{1}{2} \qquad u'(x) = 12x^3 + 10$$

we can write

$$\frac{dy}{dx} = f'(x) = 5(\tfrac{1}{2})(3x^4 + 10x)^{-1/2}(12x^3 + 10)$$
$$\qquad\qquad\quad C\;n \qquad u(x) \quad n - 1u'(x)$$

Simplifying then yields

$$\frac{dy}{dx} = f'(x) = \frac{30x^3 + 25}{\sqrt{3x^4 + 10x}}$$

Problem 1 Using the general power rule, find the first derivative of each of the following functions:

   a. $y = f(x) = (x^5 + 7x^2 - 3)^6$

   Answer $(30x^4 + 84x)(x^5 + 7x^2 - 3)^5$

   b. $y = f(x) = \sqrt{8x^2 + 5x - 1}$

   Answer $\dfrac{16x + 5}{2\sqrt{8x^2 + 5x - 1}}$

The general power rule is a special case of a more powerful technique known as the *chain rule*. Suppose we have the function $y = f(x) = (x^2 - 3x + 5)^4$ and we introduce a new variable u defined as $u = x^2 - 3x + 5$. Roughly speaking, u can be regarded as a "middleman" in expressing the functional relationship between x and y, that is, u is a function of x and y is a function of u as defined by the following equations:

$$u = x^2 - 3x + 5 \qquad \text{and} \qquad y = u^4$$

When the functional relationship between x and y can be expressed via an intermediate variable such as u, the first derivative of y with respect to x can be found by the following equation, which defines the chain rule

$$\frac{dy}{dx} = \frac{dy}{du} \cdot \frac{du}{dx}$$

In words, it says that dy/dx, the first derivative of y with respect to x, is equal to dy/du, the first derivative of y with respect to u, times du/dx, the first derivative of u with respect to x. For the preceding example, we have

$$\frac{dy}{du} = 4u^3 \qquad \frac{du}{dx} = 2x - 3$$

so that

$$\frac{dy}{dx} = (4u^3)(2x - 3) = 4(x^2 - 3x + 5)^3(2x - 3)$$

The proof of the chain rule is beyond the scope of this text, so no further development will be attempted.

VIII. CHAIN RULE

If $y = g(u)$ and $u = h(x)$ define the function $y = f(x) = g[h(x)]$ and both du/dx and dy/du exist, then

$$\frac{dy}{dx} = \frac{dy}{du} \cdot \frac{du}{dx} \qquad\qquad (3.5.3)$$

Example 3 Use the chain rule to find the first derivative dy/dx for the function

$$y = f(x) = (2x^5 - 3)^8$$

Solution Letting $u = h(x) = 2x^5 - 3$, we get $y = g(u) = u^8$

$$\frac{du}{dx} = 10x^4 \qquad \frac{dy}{du} = 8u^7$$

According to the chain rule

$$\frac{dy}{dx} = \frac{dy}{du} \cdot \frac{du}{dx} = (8u^7)(10x^4) = 80x^4u^7$$

Substituting $(2x^5 - 3)$ for u gives

$$\frac{dy}{dx} = 80x^4(2x^5 - 3)^7$$

When $y = g(u) = Cu^n$, $dy/du = Cnu^{n-1}$, and the chain rule yields the general power rule (Equation 3.5.2)

$$\frac{dy}{dx} = \frac{dy}{du} \cdot \frac{du}{dx} = Cnu^{n-1}u'(x)$$

It is only fair to notify you that many problems will require the use of two or more methods before the first derivative can be found. Because of this, it is important to determine which technique should be applied first, which second, and so forth. To determine which technique to apply first, ask yourself which algebraic operation is carried out *last* in constructing the expression to be differentiated. The technique to apply first is determined as shown in the accompanying table.

LAST OPERATION	TECHNIQUE APPLIED
Addition-subtraction	Sum-difference rule
Multiplication	Product rule
Division	Quotient rule
Raising to a power	Power rule

Although these procedures may seem self-evident, many students encounter difficulties when taking the first derivative of seemingly complex functions because they are not sure where to begin. The following examples illustrate situations calling for the application of two or three techniques within each problem.

Example 4   Find the first derivative of the function

$$y = f(x) = 3\sqrt{x^2 + 6x} + 8x$$

Solution   The last operation to be performed is addition; therefore, the sum rule is applied first by differentiating each term, that is

$$3(x^2 + 6x)^{1/2} \qquad \text{and} \qquad 8x$$

In differentiating $3(x^2 + 6x)^{1/2}$, the general power rule must be applied. We get

$$\frac{d}{dx}\left[3(x^2 + 6x)^{1/2}\right] = 3(\tfrac{1}{2})(x^2 + 6x)^{-1/2}(2x + 6) = \frac{3x + 9}{(x^2 + 6x)^{1/2}} = \frac{3x + 9}{\sqrt{x^2 + 6x}}$$

while

$$\frac{d}{dx}(8x) = 8$$

Now adding the derivatives, we get

$$\frac{d}{dx}(3\sqrt{x^2 + 6x} + 8x) = \frac{3x + 9}{\sqrt{x^2 + 6x}} + 8$$

Example 5   Find the first derivative of the function

$$y = f(x) = \frac{(x^2 + 1)^5}{x^3 + 7x}$$

Solution   Because the last operation performed is *division,* we must apply the *quotient* rule first. Letting

$$u(x) = (x^2 + 1)^5 \qquad v(x) = x^3 + 7x$$

we now differentiate each of these expressions. Note that we have to apply the general power rule to find $u'(x)$ and the sum rule to find $v'(x)$. Carrying out these operations gives us

$$u'(x) = 10x(x^2 + 1)^4 \qquad v'(x) = 3x^2 + 7$$

Now putting the pieces together, we get

$$\frac{dy}{dx} = f'(x) = \frac{(x^3 + 7x)10x(x^2 + 1)^4 - (x^2 + 1)^5(3x^2 + 7)}{(x^3 + 7x)^2}$$

which, after simplification, reduces to

$$f'(x) = \frac{(x^2 + 1)^4(7x^4 + 60x^2 - 7)}{(x^3 + 7x)^2}$$

Problem 2 Find the first derivative of each of the following functions:

a. $y = f(x) = (x^4 - 2)^3 - \sqrt{7x + 3}$

Answer $12x^3(x^4 - 2)^2 - \dfrac{7}{2\sqrt{7x + 3}}$

b. $y = f(x) = (x^2 + 1)\sqrt{2x^3 + 5}$

Answer $\dfrac{3x^2(x^2 + 1)}{\sqrt{2x^3 + 5}} + 2x\sqrt{2x^3 + 5}$

One last item: in mastering the techniques presented in this section, make an extra effort to avoid algebraic mistakes. As you familiarize yourself with the methods of differentiation, errors are bound to occur; detecting and analyzing these errors will be made more difficult if the algebra is not carried out flawlessly. There is yet another reason for exercising care when carrying out the algebraic operations. In many cases, successful application of the techniques represents only a fraction of the energy and time required to reach the final form of the derivative; the bulk of the effort consists in algebraically manipulating the expressions to achieve a simple form for the derivative.

EXERCISE 3.5

1. For each of the following functions, indicate which rule among those listed below you would apply *first* in finding the first derivative of the function as written. NOTE: We are not asking you to find the first derivative at this point.

    I. Sum-difference rule   IV. Simple power rule
    II. Product rule   V. General power rule
   III. Quotient rule

    a. $y = f(x) = \dfrac{x(3 - x^2)}{x - 2}$ **b.** $y = f(x) = \sqrt{x + 3} + 5x - 3$

c. $y = f(x) = \dfrac{x + 1}{3 - x} + \dfrac{2x}{x - 1}$ **d.** $y = f(x) = x(7x^3 - 4x^2 + 3x - 9)$

e. $y = f(x) = \sqrt{\dfrac{x + 1}{2 - x}}$ **f.** $y = f(x) = (x + 5)\sqrt{x - 6}$

Find the first derivative of the functions in Exercises 2–19.

2. $y = f(x) = 8(x^3 + 6)^5$ **3.** $y = f(x) = 3(x^2 - 7x + 1)^6$

4. $y = f(x) = 6(x^{10} + 4)^{12}$ **5.** $y = f(x) = \left(2 + \dfrac{1}{x}\right)^8$

6. $y = f(x) = \sqrt{x^3 + 5x - 1}$ **7.** $y = f(x) = \sqrt[3]{6x - 10}$

8. $y = f(x) = \dfrac{4}{(3 - 2x - x^2)^3}$ **9.** $y = f(x) = \dfrac{5}{\sqrt{7 - 2x^6}}$

10. $y = f(x) = 3x(5 - x^6)^2$ **11.** $y = f(x) = (7x + 2)(x^4 + 5)^8$

12. $y = f(x) = (x^2 + 1)^4(x^3 + 2)^5$ **13.** $y = f(x) = \dfrac{(6x + 4)}{(x - 3)^2}$

14. $y = \dfrac{(x^2 + 1)^3}{2x - 5}$ **15.** $y = f(x) = \dfrac{\sqrt{7x + 2}}{x^2}$

16. $y = f(x) = x^3\sqrt{x^2 + 1}$ **17.** $y = f(x) = (x^4 + 2)\sqrt{3 - x^2}$

18. $y = f(x) = \sqrt{x}\left(3 + \dfrac{2}{x}\right)^5$

19. $y = f(x) = (x^3 + 7x + 2)^3(2x^4 - x^3 + 5)^2$

Find the slopes and the equations of the tangent lines at the indicated points for the curves in Exercises 20–21. Sketch each tangent line on the accompanying coordinate system.

20. $y = f(x) = (2 - x)^4$ at (1, 1)

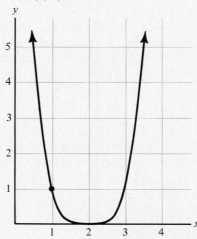

21. $y = f(x) = \sqrt{2x + 4}$ at $(0, 2)$

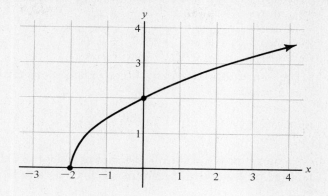

In Exercises 22–24, find those points on the curves where the slope of the tangent line has the value given.

22. $y = f(x) = \frac{1}{12}(4x + 6)^3$ $m_T = 4$

23. $y = f(x) = \sqrt{3x - 4}$ $m_T = \frac{1}{2}$

24. $y = f(x) = (x^2 - 6x + 8)^4$ $m_T = 0$

❧ 3.6 Applications of the Derivative

So far, the geometrical interpretation of the first derivative as the slope of a line tangent to a curve has received the bulk of our attention. However, applications using the first derivative stress the meaning of the derivative as the *instantaneous rate of change* of the dependent variable with respect to the independent variable. In addition, each area of study has developed its own terminology or jargon for the first derivative. This multiplicity of meanings can be confusing when first encountered. For these reasons, the applications will be restricted to those that can be grasped without possessing an extensive knowledge of the areas under study.

VELOCITY

The initial developments in the area of calculus were motivated by a desire to describe mathematically characteristics such as velocity and acceleration for moving objects. In Section 3.1, the average velocity of a moving object between $t = t_1$ and $t = t_2$ was given by the expression

$$\text{Average velocity} = \frac{s(t_2) - s(t_1)}{t_2 - t_1}$$

where s represents the distance of the object from some reference point. The *average velocity* between two arbitrary instants of time, t and $t + h$, is given by the difference quotient

$$\text{Average velocity} = \frac{s(t + h) - s(t)}{h}$$

The *instantaneous velocity* $V(t)$ is defined to be the limit of the average velocity as h approaches 0

$$V(t) = s'(t) = \frac{ds}{dt} = \lim_{h \to 0} \frac{s(t + h) - s(t)}{h} \qquad \text{(3.6.1)}$$

Example 1 An object is dropped from the roof of a 400-ft building. The distance s (in feet) of the object above the ground is given as a function of t (in seconds) by the equation

$$s(t) = -16t^2 + 400$$

The position of the object for various values of t is shown in the accompanying figure.

    a. Find the velocity as a function of t.
    b. How fast is the object moving when $t = 3$ sec?

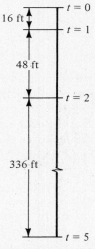

Solution a. The velocity $V(t)$ at any time is found by taking the first derivative of s with respect to t. The result is

$$V(t) = \frac{ds}{dt} = -32t$$

b. The velocity V when $t = 3$ sec equals

$$V(3) = -32(3) = -96 \text{ ft/sec}$$

The negative sign indicates that the object is falling; a positive sign indicates that the object is rising.

The velocity function $V(t)$ is rarely constant. The rate of change of the velocity is called the *acceleration* $a(t)$ and is defined as

$$a(t) = V'(t) = \frac{dV}{dt} \tag{3.6.2}$$

For Example 1, the acceleration $a(t)$ is given by

$$a(t) = \frac{dV}{dt} = -32 \text{ ft/sec}^2$$

Problem 1   The displacement s (in feet) of an object from a given point is given by the equation

$$s(t) = 15t^2 - 40t + 10$$

a. Find the velocity $V(t)$ and the acceleration $a(t)$.
b. Find the velocity and the acceleration when $t = 5$.

Answers   a. $V(t) = 30t - 40$ $a(t) = 30$
   b. $V(5) = 110$ ft/sec $a(5) = 30$ ft/sec^2

MARGINAL INTERPRETATION

When a company produces or sells a product, the revenue, cost and, ultimately, the profit functions will depend on the number of items produced or sold. Managers and accountants are interested not only in determining the value of these functions at a given level of production or sales, but also in determining how sensitive such functions are to changes in production or sales. The word *marginal* is used to describe the rate of change of any of these functions with respect to the number of items produced or sold. If we denote the number of items sold or produced as x, marginal revenue, cost, and profit are defined as follows:

1. Given a revenue function $R(x)$

$$\text{Marginal revenue} = R'(x) = \frac{dR}{dx} \tag{3.6.3}$$

2. Given a cost function $C(x)$

$$\text{Marginal cost} = C'(x) = \frac{dC}{dx} \tag{3.6.4}$$

3. Given the profit function $P(x)$

$$\text{Marginal profit} = P'(x) = \frac{dP}{dx} \tag{3.6.5}$$

Because $R(x)$, $C(x)$, and $P(x)$ are related, that is

$$P(x) = R(x) - C(x)$$

we have the following relationship for the marginal profit, revenue, and cost functions:

$$P'(x) = R'(x) - C'(x) \qquad\qquad\qquad \textbf{(3.6.6)}$$

Example 2   The weekly revenue and cost functions for a firm selling x units of their most popular motorcycle model, the Macho, are given by the equations

$$R(x) = 2000x - 25x^2 \qquad C(x) = 1000 + 1300x - 20x^2$$

Find the marginal revenue, marginal cost, and marginal profit functions.

Solution   The marginal revenue and cost functions can be found by taking the first derivative of $R(x)$ and $C(x)$, respectively. The results are

$$R'(x) = 2000 - 50x \qquad C'(x) = 1300 - 40x$$

The marginal profit function $P'(x)$ is found by taking the difference $[R'(x) - C'(x)]$, yielding

$$P'(x) = 700 - 10x$$

The marginal concept is a useful one to employ if one wants to determine the change in a function when an additional item is produced or sold. Suppose the firm selling motorcycles in the previous example is selling five motorcycles per week. The weekly revenue function $R(5)$ can be found easily

$$R(5) = 2000(5) - 25(5)^2 = 9375 \text{ dollars/week}$$

If sales were to increase to six motorcycles per week, the weekly revenue would increase to

$$R(6) = 2000(6) - 25(6)^2 = 11,100 \text{ dollars/week}$$

The change in weekly revenue because of the additional sale is

$$R(6) - R(5) = 1725 \text{ dollars/week}$$

If instead, we had calculated the marginal revenue $R'(x)$ at $x = 5$, we would have found

$$R'(5) = 2000 - 50(5) = 1750 \text{ dollars/week}$$

Although $[R(6) - R(5)]$ and $R'(5)$ are not equal, they generally differ by such a small amount that their difference is ignored and one writes

$$R(6) - R(5) \cong R'(5)$$

where $\cong$ means *approximately equal*.

Generalizing this approach, we can write

$$R(x + 1) - R(x) \cong R'(x) \tag{3.6.7}$$

The rationale behind this result can be seen by referring to Figure 3.20, where the graph of an arbitrary revenue function is plotted, together with the secant line connecting $[x, R(x)]$ and $[x + 1, R(x + 1)]$ and the tangent line to the curve at $[x, R(x)]$. Because the points $[x, R(x)]$ and $[x + 1, R(x + 1)]$ are close together, the slope of the secant line

$$\frac{R(x + 1) - R(x)}{(x + 1) - x} = R(x + 1) - R(x)$$

will not differ significantly from the slope $R'(x)$ of the line tangent to the curve at $[x, R(x)]$. More will be said about this technique when the differential is presented in Section 7.1.

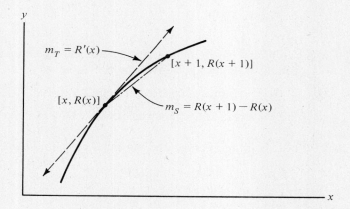

Figure 3.20

Example 3 | For the cost function

$$C(x) = 1000 + 1300x - 20x^2$$

find the change in cost when x changes from 5 to 6

a. Using the first derivative $C'(x)$ at $x = 5$, and

b. Using the relation $C(x + 1) - C(x)$ with $x = 5$

Solution | a. $C'(x)$ can be written as

$$C'(x) = 1300 - 40x$$

$$C'(5) = 1100 \text{ dollars/unit}$$

b. $C(x + 1) - C(x) = C(6) - C(5)$

$$= 1080 \text{ dollars/unit}$$

Problem 2 Find the change in the profit function

$$P(x) = 700x - 5x^2 - 1000$$

when x goes from 6 to 7

a. Using $P'(x)$ when $x = 6$
b. Using $P(x + 1) - P(x)$ with $x = 6$

Answers a. $P'(6) = 640$ b. $P(7) - P(6) = 635$

The concept of *marginal* should not be confused with that of *average*. When given a total cost function $C(x)$, the average cost associated with producing x items is denoted by $\bar{C}(x)$ and is defined as

$$\bar{C}(x) = \frac{C(x)}{x}$$

Example 4 | A company has found that the relationship between total cost C and the number of items produced x is given by the equation

$$C(x) = 2000 + 200x + x^2 \qquad 0 \le x \le 100$$

Find the average and marginal costs when 25 items are produced.

Solution | First, find the average and marginal cost functions, $\bar{C}(x)$ and $C'(x)$

$$\bar{C}(x) = \frac{2000}{x} + x + 200$$

$$C'(x) = 200 + 2x$$

Evaluating these functions at $x = 25$ gives

$$\bar{C}(25) = 305 \text{ dollars/unit}$$

$$C'(25) = 250 \text{ dollars/unit}$$

The average and marginal cost functions are shown in Figure 3.21.

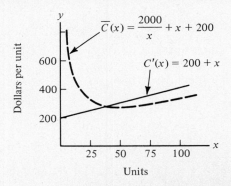

Figure 3.21

It should be pointed out that the average cost function $\bar{C}(x)$ represents the cost per item for the entire lot (0 through x), whereas the marginal cost $C'(x)$ is approximately the unit cost for an additional item to be produced or sold. Roughly speaking, the historical orientations of $\bar{C}(x)$ and $C'(x)$ are different. $\bar{C}(x)$ takes into account all past cost data, whereas $C'(x)$ presents unit costs within a more narrow time frame, that is, the immediate future.

TIME RATE OF CHANGE

The *time rate of change* of any function $y = f(t)$ with respect to the variable t is given by

$$\frac{dy}{dt} = f'(t)$$

Example 5   The number of citizens receiving medical assistance payments as a function of time is given by the equation

$$N(t) = 5 + 0.5t - 0.01t^2$$

where t represents the time in years since 1960 and $N(t)$ equals the number of citizens in millions. Find the rate of increase in the number of people receiving medical assistance payments during the tenth year of the program.

Solution   The rate of increase is given by $N'(t)$, calculated at $t = 10$. First, we find $N'(t)$

$$N'(t) = 0.5 - 0.02t$$

Evaluating this when $t = 10$ gives

$$N'(10) = 0.3 \text{ million/year}$$

or approximately 300,000 people are added during the tenth year to the roles of those receiving medical assistance payments.

EXERCISE 3.6

For Exercises 1–5, find the velocity $V(t)$ and acceleration $a(t)$. In addition, calculate the velocity and acceleration at the times indicated.

1. $s(t) = 5t^3 - 6t^2 + 9t + 100 \qquad t = 2$

2. $s(t) = \dfrac{2}{1 + t^2} \qquad t = 1$

3. $s(t) = 7t^2 + \sqrt{t + 2} \qquad t = 7$

4. $s(t) = 800 + 50t - 16t^2 \qquad t = 3$

5. $s(t) = \dfrac{2t}{t + 3} \qquad t = 0$

6. A small rocket is launched from ground level with a speed of 960 ft/sec. The vertical height s of the rocket above the ground is given by the equation

$$s(t) = 960t - 16t^2$$

    a. Find the velocity and acceleration functions.
    b. Using one of the results from part a, determine how long it takes for the rocket to reach its maximum height.
    c. How long is the rocket airborne?

7. The annual revenue function for an appliance store selling portable color TV sets is given by the equation.

$$R(x) = 200x - x^2$$

where x is the number of units sold.

    a. Find the marginal revenue function $R'(x)$.
    b. On the coordinate system below, plot both the revenue function $R(x)$ and $R'(x)$.
    c. Find the marginal revenue when $x = 30$.
    d. Find the change in revenue when x changes from 30 to 31 and compare with the result from part c.

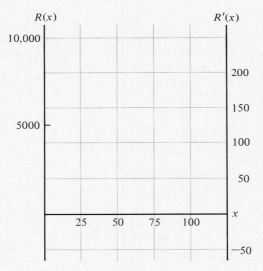

8. The total monthly cost function for a small manufacturer of swimwear is given by the equation

$$C(x) = 10,000 + 4x - 0.001x^2$$

where x represents the number of items produced.

    a. Find the marginal cost function $C'(x)$.
    b. Find the marginal cost when 300 units are produced.
    c. Find the cost of producing the 301st item and compare with answer to part b.
    d. Find the average cost function $\bar{C}(x)$.
    e. What is the average cost when monthly production is 300 suits?

f. On the accompanying coordinate system, plot the functions $C'(x)$ and $\overline{C}(x)$.

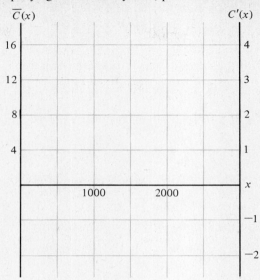

9. The level of sales S for an item in a very competitive area is a function of the amount of money x spent on advertising and is given by the relation

$$S(x) = 3 + 0.01x - 0.001x^2 \qquad 0 \le x \le 50$$

where S is expressed in millions of dollars and x is expressed in thousands of dollars. Find the rate of change of sales with respect to the amount of money spent on advertising.

10. Suppose the concentration C of alcohol in the bloodstream after consuming 1 oz of vodka is given by the equation

$$C(t) = \frac{1000}{30 + t}$$

where t is the time in minutes

a. Find the rate at which the alcohol concentration is changing when $t = 0$, $t = 30$, and $t = 60$.

3.7 Higher-order Derivatives

Although all our efforts have been directed toward defining and developing techniques for finding the first derivative, it would be misleading to imply that the process could not be carried further. The first derivative, $f'(x)$ or dy/dx, is itself a function and therefore one can attempt to find its first derivative. The result, not too surprisingly, is called the second derivative of the original function. The notation generally used to denote the second derivative takes the form of either

$$f''(x) \qquad \text{or} \qquad \frac{d^2y}{dx^2}$$

that is

$$\frac{d}{dx}[f'(x)] = f''(x) \qquad \frac{d}{dx}\left(\frac{dy}{dx}\right) = \frac{d^2y}{dx^2}$$

Example 1 | Find the second derivative of the function

$$y = f(x) = 9x^4 - 3x^3 + x^2 + 5x - 1$$

Solution | We first find $f'(x)$. The result is

$$f'(x) = \frac{dy}{dx} = 36x^3 - 9x^2 + 2x + 5$$

To find the second derivative, we differentiate the function $f'(x)$. The result is

$$f''(x) = 108x^2 - 18x + 2$$

Example 2 | Find the second derivative of the function

$$y = f(x) = \frac{2}{x} = 2x^{-1}$$

Solution | Proceeding step by step, we get for $f'(x)$ and $f''(x)$

$$f'(x) = -2x^{-2}$$

$$f''(x) = 4x^{-3} = \frac{4}{x^3}$$

Just as the first derivative $f'(x)$ can be differentiated to produce the second derivative $f''(x)$, so too can the second derivative be differentiated to produce the third, $f'''(x)$, the third to generate the fourth $f^{(4)}(x)$, and so on. The third, fourth, and higher-order derivatives are denoted by

$$f'''(x) \qquad \text{or} \qquad \frac{d^3y}{dx^3}$$

$$f^{(4)}(x) \qquad \text{or} \qquad \frac{d^4y}{dx^4}$$

$$f^{(5)}(x) \qquad \text{or} \qquad \frac{d^5y}{dx^5}$$

Example 3 | Find the third and fourth derivatives of the function given in Example 1 of this section.

Solution | Because we know $f''(x)$, we can generate $f'''(x)$ immediately

$$f'''(x) = \frac{d^3y}{dx^3} = 216x - 18$$

$$f^{(4)}(x) = \frac{d^4y}{dx^4} = 216$$

Example 4 | Find the third and fourth derivatives of the function given in Example 2 of this section.

Solution | We have already found $f''(x)$, so we differentiate to yield

$$f'''(x) = \frac{d^3y}{dx^3} = -12x^{-4} = -\frac{12}{x^4}$$

$$f^{(4)}(x) = \frac{d^4y}{dx^4} = 48x^{-5} = \frac{48}{x^5}$$

This might be an appropriate point to mention that your efforts in finding the second and higher-order derivatives will be made easier if you simplify algebraically all expressions as much as possible before attempting to differentiate. The time you spend on a problem and the chances for error will be reduced considerably if you make an effort to simplify as much as possible. In addition, keep in mind that all the techniques developed in Sections 3.3–3.5 are applicable in finding higher-order derivatives.

Example 5 | Find the second derivative of the function

$$y = f(x) = (x^2 + 5)^4$$

Solution | The general power rule (3.5.1) is used to find $f'(x)$, giving

$$f'(x) = 4(x^2 + 5)^3(2x) = 8x(x^2 + 5)^3$$

Next the product rule (3.4.1), with $u(x) = 8x$ and $v(x) = (x^2 + 5)^3$, is used to find $f''(x)$; the result is

$$f''(x) = 8x[3(x^2 + 5)^2(2x)] + (x^2 + 5)^3(8)$$

$$= 48x^2(x^2 + 5)^2 + 8(x^2 + 5)^3$$

A discussion of the meaning and significance of the second derivative will not be taken up at this point, but will be postponed until the next chapter. Applications in which the third and higher-order derivatives play an important role are beyond the scope of this text, and, for this reason, will not be encountered again in the text.

EXERCISE 3.7

Find the second derivative of each of the following functions.

1. $y = f(x) = 4x + 7$

2. $y = f(x) = 3x^2 - 5x + 2$

3. $y = f(x) = 6x^3 - 9x^2 + 11$

4. $y = f(x) = x^7 - 5x^4 + 3x^2 - 9$

5. $y = f(x) = \dfrac{8}{x^2}$

6. $y = f(x) = \dfrac{-3}{x^4}$

7. $y = f(x) = 5\sqrt{x}$ **8.** $y = f(x) = \sqrt{5x}$

9. $y = f(x) = (1 - 6x)^3$ **10.** $y = f(x) = 4(x^2 - 3)^5$

11. $y = f(x) = \sqrt{1 - x}$ **12.** $y = f(x) = \sqrt{1 - x^2}$

13. $y = f(x) = \dfrac{2x}{1 + x}$ **14.** $y = f(x) = \dfrac{x^2 + 4}{3x + 4}$

15. $y = f(x) = (x^4 + 1)^3 - \dfrac{1}{x}$ **16.** $y = f(x) = x^2(4 - x)^3$

17. $y = f(x) = 2x\sqrt{x^2 + 1}$ **18.** $y = f(x) = \dfrac{x + 1}{\sqrt{x}}$

19. $y = f(x) = \sqrt[3]{x} + \dfrac{1}{x}$ **20.** $y = f(x) = (x^2 + 5)(x + 1)^2$

3.8 Implicit Differentiation

All the functions presented to this point have been expressed in the form $y = f(x)$. That is, the dependent variable y appears alone on one side of the equation, while the mathematical rule that generates the value of y for a given value of x appears on the other side. When this occurs, y is called an *explicit* function of x.

However, in many cases the equation defining the relationship between x and y does not assume the form $y = f(x)$, and, in addition, may represent one or more functions of x. When this occurs, the function or functions are defined *implicitly* by the equation.

In the following, y is defined implicitly in terms of x:

$$2xy + x = 3 \tag{3.8.1}$$

$$y^2 - x + 1 = 0 \tag{3.8.2}$$

$$\sqrt{y + 2} + xy^4 = 3x^2y^2 \tag{3.8.3}$$

Occasionally, the equation can be rewritten so that the function(s) can be written explicitly. Equation 3.8.1 can be written as

$$y = f(x) = \frac{3 - x}{2x}$$

Equation 3.8.2 can be solved for y by first writing

$$y^2 = x - 1$$

Next, taking the square root of both sides reveals two functions

$$y = f_1(x) = +\sqrt{x - 1} \qquad y = f_2(x) = -\sqrt{x - 1}$$

whose graphs are shown in Figure 3.22

In most cases, however, rewriting the equation to express y explicitly in terms of one or more functions of x is either extremely difficult or impossible.

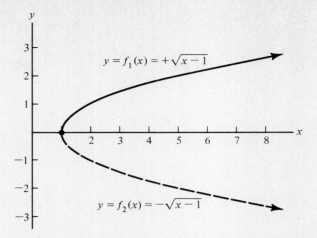

Figure 3.22

Equation 3.8.3 represents an example of this situation. The failure or inability to express y explicitly as a function of x does not prevent us from finding the first derivative when y is differentiable. A technique called *implicit differentiation* is used, in which we differentiate term by term, noting that

$$\frac{d}{dx}(y) = \frac{dy}{dx}$$

and that, according to the general power rule,

$$\frac{d}{dx}(y^n) = ny^{n-1} \cdot \frac{dy}{dx}$$

When the formal differentiation has been completed, all terms containing the factor dy/dx are gathered on one side of the equation, and the resulting equation is then solved for dy/dx in terms of x and y.

Example 1   Find the first derivative of the function $y = f(x)$ defined implicitly by the equation

$$3xy - x = 2$$

Solution   Differentiating term by term gives

$$\frac{d}{dx}(3xy) - \frac{d}{dx}(x) = \frac{d}{dx}(2)$$

The product rule must be applied to the first term on the left-hand side yielding

$$\frac{d}{dx}(3xy) = 3x \cdot \frac{dy}{dx} + y \cdot \frac{d}{dx}(3x) = 3x \cdot \frac{dy}{dx} + 3y$$

The other two terms can be differentiated quickly

$$\frac{d}{dx}(x) = 1 \qquad \frac{d}{dx}(2) = 0$$

so we now have

$$3x\frac{dy}{dx} + 3y - 1 = 0$$

Solving now for dy/dx gives

$$\frac{dy}{dx} = \frac{1 - 3y}{3x}$$

Because the original equation has a simple algebraic structure, we could have solved for y explicitly in terms of x, getting

$$y = f(x) = \frac{2 + x}{3x} = \frac{2}{3x} + \frac{1}{3}$$

If we differentiate term by term, we obtain the result

$$\frac{dy}{dx} = f'(x) = \frac{-2}{3x^2}$$

On the surface, the results obtained by the two approaches do not appear to be consistent. You should not be misled by this superficial discrepancy; the two results are in agreement as the following rewriting indicates

$$\frac{1 - 3y}{3x} = \frac{1 - 3[(2 + x)/3x]}{3x} = \frac{1 - [(2 + x)/x]}{3x} = \frac{x - (2 + x)}{3x^2}$$
$$= \frac{-2}{3x^2}$$

In general, you will find that the results obtained by implicit and explicit differentiation, although equivalent, will assume different algebraic forms. Some algebraic manipulation will bring the two forms into agreement.

Example 2   Find the first derivative of the functions defined implicitly by the equation

$$x^2 + y^3 = 2xy^2$$

Solution   Differentiating term by term, we obtain

$$\frac{d}{dx}(x^2) = 2x \qquad \frac{d}{dx}(y^3) = 3y^2\left(\frac{dy}{dx}\right)$$

$$\frac{d}{dx}(2xy^2) = 2x \cdot \frac{d}{dx}(y^2) + y^2 \cdot \frac{d}{dx}(2x) = 4xy\left(\frac{dy}{dx}\right) + 2y^2$$

where the product rule was used to find the first derivative of $2xy^2$. Putting the pieces together gives

$$2x + 3y^2 \left(\frac{dy}{dx}\right) = 2y^2 + 4xy \left(\frac{dy}{dx}\right)$$

Rewriting the equation with all terms containing the factor dy/dx on one side and the rest of the terms on the other side yields

$$3y^2 \left(\frac{dy}{dx}\right) - 4xy \left(\frac{dy}{dx}\right) = 2y^2 - 2x$$

Solving for dy/dx gives

$$\frac{dy}{dx} = \frac{2y^2 - 2x}{3y^2 - 4xy}$$

Example 3

Find the equation of the line tangent to the curve

$$y^2 - 4x = 0$$

at the point $(1, -2)$

Solution

Recalling that dy/dx represents the slope of the line tangent to curve, we first find the first derivative by implicit differentiation, obtaining

$$2y \left(\frac{dy}{dx}\right) - 4 = 0$$

Solving for dy/dx gives

$$\frac{dy}{dx} = \frac{2}{y}$$

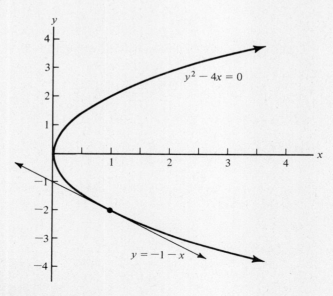

Figure 3.23

so that the slope of the tangent line at $(1, -2)$ becomes

$$\frac{dy}{dx} = \frac{2}{-2} = -1$$

The equation of the tangent line can now be found from the point-slope formula $y - y_1 = m(x - x_1)$. The result is

$$y + 2 = -1(x - 1)$$

or

$$y = -x - 1$$

The curve and the tangent line are shown in Figure 3.23

It is often necessary to specify both the x and the y coordinates of the point on the curve where the slope of the tangent line is sought because generally the equations do *not* represent functions so that two or more values of y may be paired with each value of x.

EXERCISE 3.8

For Problems 1–6 find the first derivative by each of two methods:

a. Solve for y explicitly in terms of x and then differentiate.
b. Implicit differentiation of both sides of the equation as written.

1. $5x + 2y = 3$ **2.** $xy + 5 = x^4$

3. $x^2 + 3y = x^3 - 2xy$ **4.** $x^2 - 2y^3 = 7$

5. $x^2 + y^2 = 4$ **6.** $\dfrac{1}{x} + \dfrac{1}{y} = 2$

Find dy/dx by implicit differentiation for Problems 7–14.

7. $2x^2 - y^2 = x + 4y$ **8.** $xy^2 - 5x = 4y$

9. $y^5 + 3x^2y^2 = 10$ **10.** $(3 + y)^2 = 3x^4 - 5$

11. $(x + y)^3 - y^2 = 7x^2$ **12.** $\sqrt{x} + \sqrt{y} = 2$

13. $\dfrac{x + y}{y + 1} = 6x$ **14.** $(2x - y^2)^3 = y + 4$

Find the slope and equation of the line tangent to each of the curves described in Problems 15–17.

15. $x^2 - y^2 = 5$ at $(3, 2)$

16. $x^3 + y^3 = 9$ at $x = 1$

17. $x^2 + y^2 = 25$ at $(3, 4)$

4

Curve Sketching and Optimization

CHAPTER 4 &⤳ INTRODUCTION

Sketching the graph of a function can be carried out more quickly if we can determine the shape of the curve in the vicinity of two classes of points: *critical points* and *inflection points*. Much of this chapter will be devoted to studying and illustrating how these points are found and how the shape of the curve in the vicinity of each is determined.

The ability to locate critical points can be used to "optimize," that is, maximize or minimize some quantity. For example, a farmer wants to know the amount of fertilizer that will maximize (optimize) the yield from his crops. On the other hand, the owner of a fleet of trucks wants to know the number of miles between tune-ups that minimizes the combined annual costs of gasoline plus tune-ups.

&⤳ 4.1 Critical Points, Maxima and Minima, First Derivative Test

When discussing the geometrical significance of the slope of a straight line, we found the following relationships between the slope m and the behavior of the line:

1. When $m > 0$, the line is rising.
2. When $m < 0$, the line is falling.
3. When $m = 0$, the line is horizontal.

Recalling that the first derivative $f'(x)$ represents the slope of the tangent line, we can draw similar conclusions about the behavior of a nonlinear curve in the vicinity of a point where the first derivative is known. Denoting the point as $[a, f(a)]$, the relationship between $f'(a)$ and the behavior of the curve can be described as follows:

1. If $f'(a)$ is *positive,* that is, $f'(a) > 0$, then the curve is *rising* or the function is *increasing* in the vicinity of $[a, f(a)]$.
2. If $f'(a)$ is *negative,* that is, $f'(a) < 0$, then the curve is *falling* or the function is *decreasing* in the vicinity of $[a, f(a)]$.
3. If $f'(a)$ *equals* 0, that is, $f'(a) = 0$, then the *tangent line* is *horizontal* at the point $[a, f(a)]$.

These relationships are illustrated in Figure 4.1 where the graph of the function $y = f(x) = x^3 - 12x + 4$ is shown. In addition, the sign of the first derivative

$$f'(x) = 3x^2 - 12 = 3(x + 2)(x - 2)$$

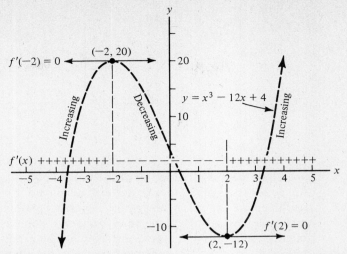

Figure 4.1

is displayed for various segments of the curve. For example, $f'(x)$ is positive for all $x < -2$ where the function is increasing, is negative for all x in the interval $-2 < x < 2$ where the function is decreasing, and is positive for all $x > 2$ where the function is increasing. The first derivative $f'(x)$ equals 0 at $(-2, 20)$ and $(2, -12)$ where the tangent lines are horizontal.

Points such as $(-2, 20)$ where the curve "peaks" and $(2, -12)$ where the curve "bottoms out" are important in the curve-sketching process because they indicate where the direction of the curve may change from rising to falling and vice versa. They belong to a class of points known as *critical points*. A point on a curve is called a critical point if either of the following conditions holds:

I. $f'(x) = 0$ or   II. $f'(x)$ Does not exist   **(4.1.1)**

To see why critical points are extremely useful, suppose we have a situation such as that shown in Figure 4.2, where all the critical points and the shape of the curve in the vicinity of each are shown for a function that is continuous for all values of x. The first derivative $f'(x)$ equals 0 at four of the five points, that is

$$f'(A) = f'(B) = f'(C) = f'(E) = 0$$

while $f'(D)$ does not exist.

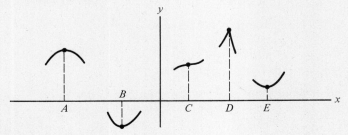

Figure 4.2

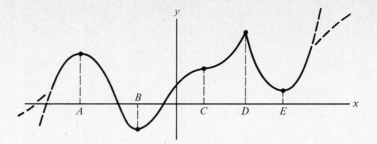

Figure 4.3

Suppose now that you were asked to complete the graph by making a rough sketch of the curve. Since all of the critical points are shown, it is not possible for peaks or valleys other than those displayed in Figure 4.2 to exist. By connecting adjacent tails with smooth segments and extending the outer tails at A and E down and up, respectively, a curve such as that shown in Figure 4.3 might result. In most cases, the sketch would be a fairly accurate representation of the graph. Some features, such as the shape of the curve as x becomes very large or small, cannot be determined from a knowledge of the critical points; that is the reason for drawing multiple tails on both ends of the curve. The subject of curve sketching will be discussed in detail in Section 4.4 where we will find that locating all the critical points and determining the shape of the curve in the vicinity of each one is an essential step in obtaining the graph.

The peaks that occur on the curve at $x = A$ and $x = D$ are called *relative maxima*. The word "relative" signifies that the y coordinate associated with such a point is greater than the y coordinate of any other point in its vicinity. In the same way, the valleys or troughs that appear at $x = B$ and $x = E$ are called *relative minima*. A critical point such as that which appears at $x = C$ is neither a maximum nor a minimum.

It is now time to learn how critical points are found and how the shape of the curve in the vicinity of each is determined. The critical points are found, according to Equation 4.1.1, by locating those points where $f'(x) = 0$ or where $f'(x)$ does not exist. Example 1 illustrates this process.

Example 1 | Find the critical points for the function

$$y = f(x) = x^3 - 3x^2 - 9x + 10$$

Solution | First, we find $f'(x)$, getting

$$f'(x) = 3x^2 - 6x - 9$$

Next, we set $f'(x)$ equal to 0 and solve for x

$$0 = 3x^2 - 6x - 9$$

$$0 = 3(x - 3)(x + 1)$$

from which we get the solutions

$$x_1 = -1 \qquad y_1 = 15$$

$$x_2 = 3 \qquad y_2 = -17$$

Because $f'(x)$ is defined for all values of x, condition II in Equation 4.1.1 does not apply. Thus $(-1, 15)$ and $(3, -17)$ are the only critical points on the curve; they are shown on the following graph together with the horizontal tangent line at each point.

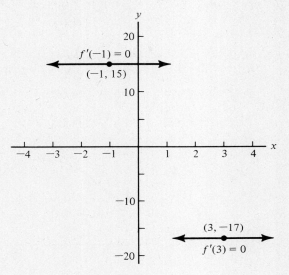

Problem 1 Find the critical points for the function

$$y = f(x) = 24x - 15x^2 + 2x^3$$

Answer $(1, 11)$ and $(4, -16)$

After locating the critical point(s), the next step in the analysis is to determine the shape of the curve in the vicinity of each. One method, the *first derivative test,* utilizes the sign of the first derivative at two neighboring points, one located to the left, the other to the right of each critical point to determine whether the curve is rising or falling at each of these points. It is then possible to determine the shape of the curve in the vicinity of each critical point. The first derivative test together with its graphical interpretation is described next.

FIRST DERIVATIVE TEST

Suppose that $[a, f(a)]$ is a critical point and that $f'(x)$ exists for all values of x close to $[a, f(a)]$. Select two values of x, one to the left of $x = a$ called x_L, the other to the right called x_R; calculate $f'(x_L)$ and $f'(x_R)$. Based on the results of these calculations, we can determine the shape of the curve in the vicinity of $[a, f(a)]$ as follows:

A. If $f'(x_L) < 0$ and $f'(x_R) > 0$, then $[a, f(a)]$ is a relative *minimum*.

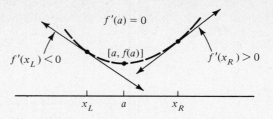

Figure 4.4

B. If $f'(x_L) > 0$ and $f'(x_R) < 0$, then $[a, f(a)]$ is a relative *maximum*.

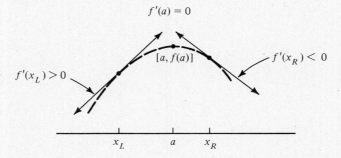

Figure 4.5

C. If $f'(x_L) < 0$ and $f'(x_R) < 0$, then $[a, f(a)]$ is *neither* a *maximum nor* a *minimum*.

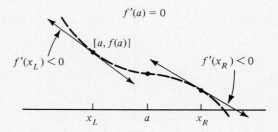

Figure 4.6

D. If $f'(x_L) > 0$ and $f'(x_R) > 0$, then $[a, f(a)]$ is *neither* a *maximum nor* a *minimum*.

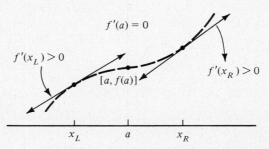

Figure 4.7

The first derivative test is summarized in Table 4.1.

Table 4.1

$f'(x_L)$	$f'(x_R)$	$[a, f(a)]$
−	+	Minimum
+	−	Maximum
− +	−} +}	Neither

We will now apply the first derivative test to various types of functions in the following examples.

Example 2 | Find the critical points and then use the first derivative test to determine the shape of the curve in the vicinity of each critical point for the function

$$y = f(x) = x^2 - 6x + 10$$

Solution | STEP 1. The critical points are found by setting $f'(x)$ equal to 0 and solving the resulting equation for x.

$$f'(x) = 2x - 6$$

$$0 = 2x - 6$$

Solution:

$$x = 3 \qquad y = 1$$

STEP 2. Apply the first derivative test.
   a. Select a value of x to the left of the critical point, say $x_L = 2$, and calculate $f'(2)$. The result is

$$f'(x_L) = f'(2) = -2 < 0 \qquad \text{Curve is falling}$$

   b. Select a value of x to the right, say $x_R = 4$, and calculate $f'(x_R)$. The result is

$$f'(x_R) = f'(4) = +2 > 0 \qquad \text{Curve is rising}$$

We can conclude that $(3, 1)$ is a relative minimum. A rough sketch of the curve in the vicinity of the critical point is shown on the following graph.

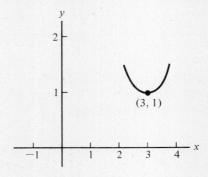

The first derivative test can be simplified for those functions that have two or more critical points and that are continuous for all values of x. Noting that changes in the sign of the first derivative can occur only at critical points, we can use the x coordinates of the critical points as reference points in order to subdivide the x axis into intervals on which the function is either increasing or decreasing. Calculating the first derivative at an arbitrary point on each interval will provide us with enough information to determine the shape of the curve in the vicinity of each critical point. The procedure is illustrated in the next example.

Example 3 Use the first derivative test to determine the shape of the curve $y = f(x) = x^3 - 3x^2 - 9x + 10$ (see Example 1) in the vicinity of its critical points.

Solution We found that the function has two critical points $(-1, 15)$ and $(3, -17)$. Using $x = -1$ and $x = 3$ as reference points, we subdivide the x axis into the three intervals, $x < -1$, $-1 < x < 3$, and $x > 3$, shown in the following graph.

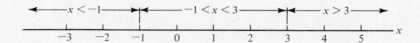

Next, the first derivative $f'(x) = 3x^2 - 6x - 9$ is calculated at an arbitrary point on each interval to determine whether the curve is rising or falling over each. This process is summarized in the following table and graph.

INTERVAL	VALUE OF x	$f'(x)$	BEHAVIOR OF FUNCTION
$x < -1$	-2	$+15$	Increasing
$-1 < x < 3$	0	-9	Decreasing
$x > 3$	4	$+15$	Increasing

On the basis of this analysis, we can conclude that $(-1, 15)$ is a relative maximum and $(3, -17)$ is a relative minimum as shown in the rough sketch at the top of the following page.

The next example shows that all critical points are not necessarily maxima or minima.

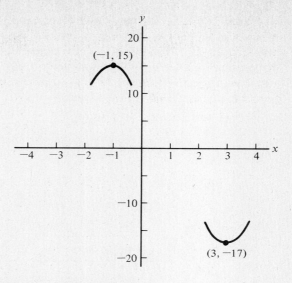

Example 4   Find the critical points for the function

$$y = f(x) = 2x^3 - 6x^2 + 6x + 1$$

and apply the first derivative test to determine the behavior of the function in the vicinity of each critical point.

Solution   First we find the critical points

$$f'(x) = 6x^2 - 12x + 6$$

$$0 = 6(x^2 - 2x + 1) = 6(x - 1)^2$$

Solving for x gives

$$x = 1 \qquad y = 3$$

   Next, we apply the first derivative test. Because $f'(x) = 6(x - 1)^2 > 0$ for all values of x except $x = 1$, the curve is rising on both sides of $(1, 3)$. From this, we can conclude that $(1, 3)$ is neither maximum nor minimum and that the curve has the shape shown in the next figure.

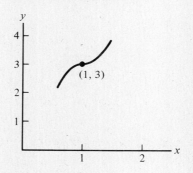

When using the first derivative test, it is also important to find those values of x where the function is discontinuous because the sign of the first derivative may change at a discontinuity. For example, the function $y = f(x) = 1/x^2$ is discontinuous at $x = 0$. The first derivative $f'(x) = -2/x^3$ is positive when $x < 0$ and negative when $x > 0$, as shown in the accompanying graph of the function where the sign of the first derivative is shown on each branch. For this

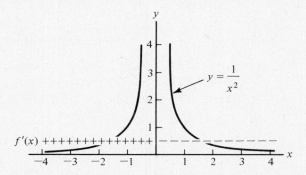

reason, any analysis based on the first derivative test should include not only the critical values of x but also values of x where the function is discontinuous. These values together with the critical values of x are then used to subdivide the x axis into distinct intervals over which the function is either increasing or decreasing. The next example illustrates the procedure.

Example 5 Find the critical points for the function

$$y = f(x) = \frac{x^2}{x - 1}$$

and apply the first derivative test to determine the shape of the curve in the vicinity of each critical point.

Solution First we find $f'(x)$ using the quotient rule. We get

$$f'(x) = \frac{2x(x - 1) - x^2(1)}{(x - 1)^2} = \frac{x(x - 2)}{(x - 1)^2}$$

Next, we set $f'(x)$ equal to 0 and solve the equation for x

$$0 = \frac{x(x - 2)}{(x - 1)^2}$$

Solutions:

$x_1 = 0 \qquad y_1 = 0$

$x_2 = 2 \qquad y_2 = 4$

Before applying the first derivative test, we note that the function is discontinuous at $x = 1$. Using $x = 0$, 1, and 2 as reference points, the x axis is subdivided into four separate intervals, as shown on the following graph. The sign of $f'(x)$ remains the same for all x in each interval.

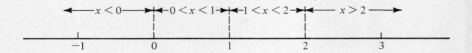

Next, the first derivative $f'(x)$ is calculated at an arbitrary point on each of the intervals to determine whether the curve is rising or falling over each. The procedure is shown in the next table and graph.

INTERVAL	x	$f'(x)$	BEHAVIOR OF FUNCTION
$x < 0$	-1	$+\frac{3}{4}$	Increasing
$0 < x < 1$	$\frac{1}{2}$	-3	Decreasing
$1 < x < 2$	$\frac{3}{2}$	-3	Decreasing
$x > 2$	3	$+\frac{3}{4}$	Increasing

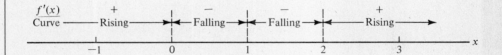

Based on the results of this analysis, we can conclude that $(0, 0)$ is a relative maximum and $(2, 4)$ a relative minimum. A rough sketch of the curve in the vicinity of each critical point is shown in the following graph.

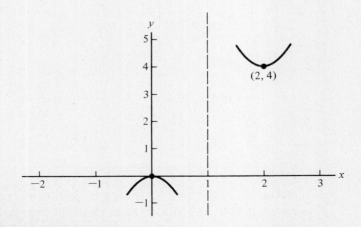

The next example illustrates a case where the critical point is a point where $f'(x)$ does not exist.

Example 6 | Find the critical points for the function

$$y = f(x) = \sqrt[3]{x^2} = x^{2/3}$$

and apply the first derivative test to determine the behavior of the function in the vicinity of each critical point.

Solution | STEP 1.   First, we find $f'(x)$

$$f'(x) = (\tfrac{2}{3})x^{-1/3} = \frac{2}{3\sqrt[3]{x}}$$

STEP 2.   Set $f'(x)$ equal to 0 and solve for x, if possible,

$$0 = \frac{2}{3\sqrt[3]{x}}$$

There are no values of x that satisfy this equation, so there are no critical points where $f'(x)$ equals 0. However, there is a point on the curve, namely, $(0, 0)$, where $f'(x)$ is not defined.

STEP 3.   Next, we apply the first derivative test by selecting values of x to the left and right of $(0, 0)$, say $x_L = -1$ and $x_R = 1$

$$f'(x_L) = f'(-1) = -\tfrac{2}{3} \qquad \text{Decreasing on left}$$

$$f'(x_R) = f'(1) = \tfrac{2}{3} \qquad \text{Increasing on right}$$

These results indicate that the point $(0, 0)$ is a minimum. To get a feeling for what is happening to the slope in the vicinity of $(0, 0)$, let us look at the following table, which is similar to those used in Chapter 2 when limits were introduced.

	FROM THE LEFT		FROM THE RIGHT
x	$f'(x) = \dfrac{2}{3\sqrt[3]{x}}$	x	$f'(x) = \dfrac{2}{3\sqrt[3]{x}}$
-1.000000	$-\tfrac{2}{3}$	1.000000	$\tfrac{2}{3}$
-0.001000	$-\tfrac{20}{3}$	0.001000	$\tfrac{20}{3}$
-0.000001	$-\tfrac{200}{3}$	0.000001	$\tfrac{200}{3}$
.	.	.	.
.	.	.	.
.	.	.	.

From the table, we see that the tangent line is becoming steeper or more nearly vertical as $x \to 0$. On the basis of this, we can make a rough sketch of the curve in the vicinity of $(0, 0)$ as shown in the next figure.

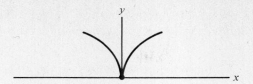

Situations will arise in which we will want to find either the largest or the smallest value of a continuous function whose domain is a closed interval, $a \leq x \leq b$, that is, the interval contains its endpoints. The largest (*absolute maximum*) or smallest (*absolute minimum*) value may or may not be located at a critical point. One or both may be located at one of the endpoints of the interval. Some of the possibilities are illustrated in Figures 4.8–4.11.

A. Absolute maximum and minimum occur at a critical point.

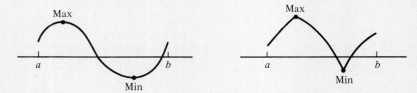

Figure 4.8

B. Absolute maximum is located at a critical point; absolute minimum at an endpoint.

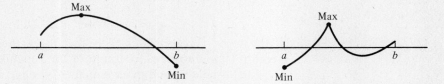

Figure 4.9

C. Absolute maximum located at endpoint; absolute minimum at a critical point.

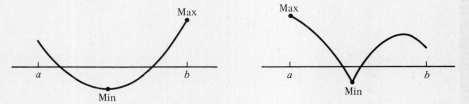

Figure 4.10

D. Absolute maximum and minimum are each located at an endpoint.

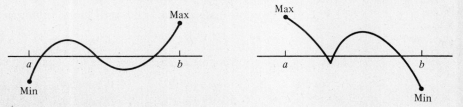

Figure 4.11

When attempting to locate the absolute maximum and/or minimum of a continuous function on a closed interval $a \le x \le b$, the following procedure is used:

1. Find the critical points on the closed interval.
2. Find the y coordinates of all points found in step 1.
3. Calculate $f(a)$ and $f(b)$, the y coordinates of the endpoint.
4. The largest and smallest values obtained from steps 2 and 3 represent the absolute maximum and minimum, respectively.

This procedure is illustrated in the following examples.

Example 7 | Find the absolute maximum and minimum of the function

$$y = f(x) = x^2 - 6x + 2$$

on the interval $0 \le x \le 5$.

Solution | STEP 1. The critical points are found by solving the equation

$$f'(x) = 2x - 6 = 0$$

for x, giving

$$x = 3$$

STEP 2. $y = f(3) = -7$

STEP 3. $f(0) = 2 \qquad f(5) = -3$

STEP 4. Of the three y coordinates, 2 is the largest and -7 the smallest; therefore, the absolute maximum of the function is 2, occurring at $x = 0$, and the absolute minimum is -7, occurring at $x = 3$.

Example 8 | Find the absolute maximum and minimum of the function

$$y = f(x) = x^3 + 3x^2 + 4$$

on the interval $-4 \le x \le 2$.

Solution | STEP 1. The critical points are found by solving the equation

$$f'(x) = 3x^2 + 6x = 0$$

for x yielding

$$x_1 = -2 \qquad x_2 = 0$$

STEP 2. The y coordinates associated with these values are

$$y_1 = f(-2) = 8 \qquad y_2 = f(0) = 4$$

STEP 3. The y coordinates of the endpoints are obtained next

$$f(-4) = -12 \qquad f(2) = 24$$

STEP 4. Comparison of the four values obtained in steps 2 and 3 indicates the absolute maximum is 24, appearing at $x = 2$, and the absolute minimum is -12, appearing at $x = -4$.

Example 9 | Find the absolute maximum and minimum of the function

$$y = f(x) = 3x^4 + 4x^3 - 36x^2 + 1$$

on the interval $-1 \leq x \leq 1$

Solution | STEP 1. The critical points are found by solving the equation

$$f'(x) = 12x^3 + 12x^2 - 72x = 0$$

for x. Writing the equation in factored form gives

$$12x(x + 3)(x - 2) = 0$$

The solutions are

$$x_1 = 0 \qquad x_2 = -3 \qquad x_3 = 2$$

Only one of the critical values, $x_1 = 0$, is in the interval, $-1 \leq x \leq 1$. The other two will be ignored.

STEP 2. $y_1 = f(0) = 1$

STEP 3. The y coordinates of the endpoints are found next and we get

$$f(-1) = -36 \qquad f(1) = -28$$

STEP 4. Of the three values found in steps 2 and 3, the largest is 1, located at $x = 0$, and the smallest is -36, located at $x = -1$.

EXERCISE 4.1

Find the critical points for each function in Problems 1–15. Use the first derivative test to determine the shape of the curve in the vicinity of each critical point. Finally, make a rough sketch of the curve in the vicinity of each critical point.

1. $y = f(x) = 2x^2 - 4x - 3$

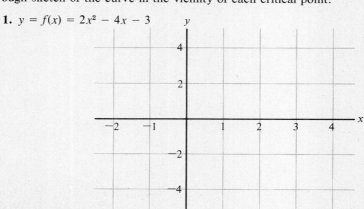

2. $y = f(x) = 8x - 2x^2$

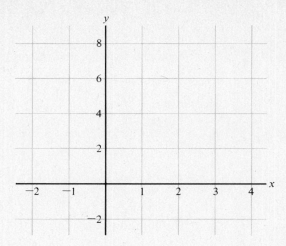

3. $y = f(x) = x^2 + 5x - 1$

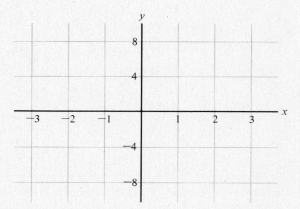

4. $y = f(x) = \frac{1}{2}x^2 + x - 2$

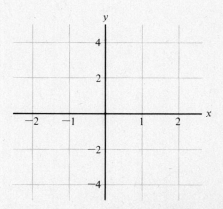

5. $y = f(x) = x^3 + 3x^2 - 1$

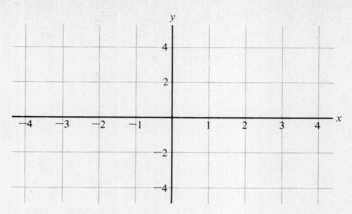

6. $y = f(x) = x^4 - 4x + 2$

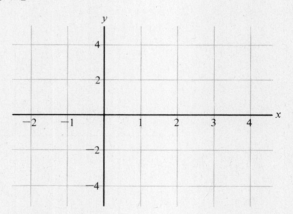

7. $y = f(x) = 2x^3 - 6x^2 - 18x + 14$

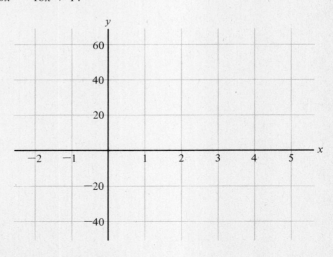

8. $y = f(x) = x^3 + 6x^2 - 4$

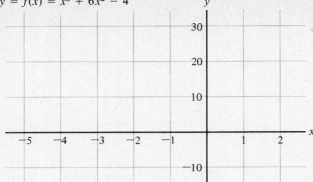

9. $y = f(x) = x^4 - 2x^2 + 1$

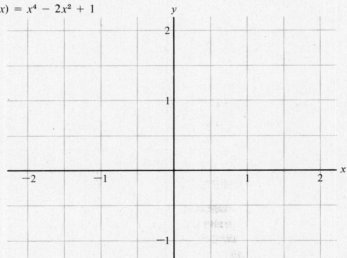

10. $y = f(x) = x^3 + 2$

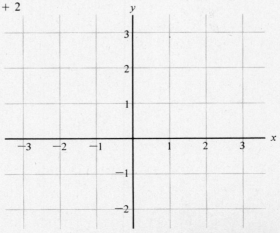

11. $y = f(x) = (x - 1)^2(x - 4)$

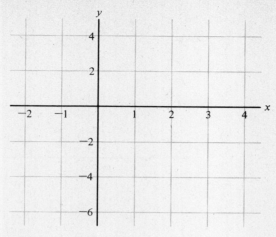

12. $y = f(x) = \sqrt[3]{x^4}$

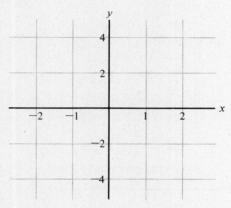

13. $y = f(x) = \dfrac{x^2}{x - 2}$

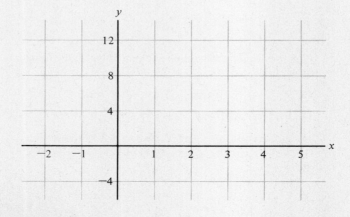

14. $y = f(x) = x + \dfrac{1}{x}$

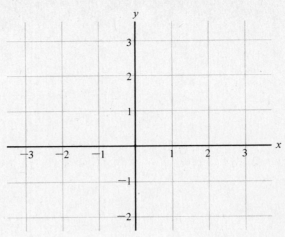

15. $y = f(x) = 3x^5 - 5x^3 + 1$

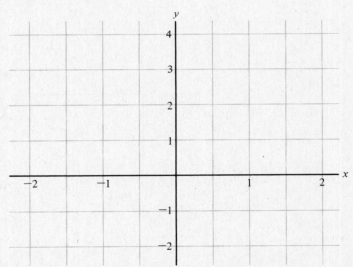

Find the absolute maximum and minimum for the functions in Problems 16–23 on the interval indicated.

16. $y = f(x) = 7 - 4x - x^2 \qquad -1 \le x \le 3$

17. $y = f(x) = x^3 + 1 \qquad -1 \le x \le 1$

18. $y = f(x) = 2x^3 - 9x^2 - 24x \qquad -2 \le x \le 0$

19. $y = f(x) = \dfrac{1}{x} \qquad 1 \le x \le 3$

20. $y = f(x) = x^3 - 3x^2 + 5 \qquad 1 \le x \le 4$

21. $y = f(x) = 4\sqrt{x} - x \qquad 1 \le x \le 16$

22. $y = f(x) = \dfrac{x^5}{5} - \dfrac{3x^4}{4} + 1 \qquad -1 \le x \le 2$

23. $y = f(x) = 2x - \dfrac{1}{x} \qquad 1 \le x \le 5$

24. Find constants A, B, and C for which the function

$$y = f(x) = Ax^2 + Bx + C$$

has a relative maximum at $(1, -3)$ and a y intercept equal to -5.

4.2 Applications of Maxima and Minima

In many situations, it is desirable to maximize or minimize some quantity; problems of this sort are categorized as *optimization* problems. For example, a manufacturer will try to set the selling price of an item so as to *maximize* company profits; the price at which this occurs is called the *optimum* price. On the other hand, the company will try to schedule production so as to *minimize* total manufacturing costs. On a personal level, each person tries to achieve or maintain an *optimum* weight, that is, a weight for which his or her physical well-being is *maximized*.

The following examples are designed to illustrate the basic principles used in formulating an optimization problem and determining the optimum value of the independent variable. It is important to point out at the beginning that most of your time and efforts will be concentrated on "translating" the problem from English to mathematics. To expedite the process, you should take enough time to read each problem carefully until it is well understood.

Example 1   The recreation department in a small town has decided to set aside a rectangular, 10,000 ft² parcel of land for a playground. The area is to be enclosed on all four sides by a fence. What are the dimensions of the rectangular plot that minimize the number of linear feet of fencing required to enclose the playground?

Solution   Before trying to solve a problem mathematically, it is sometimes preferable to visualize it by drawing a diagram and considering a number of different configurations. Some extreme cases might even be quite illuminating; one of the configurations that will produce a 10,000 ft² plot, shown in the next diagram, is a field measuring 1 ft by 10,000 ft. Aside from the fact that it is highly unsuitable for a playground, it has the dubious distinction of requiring 20,002 ft of fencing to enclose it, almost 4 miles of fence, a situation that would undoubtedly please the fencing supplier. The configuration can be improved slightly by doubling

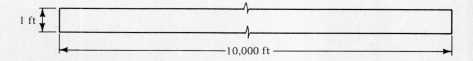

1 ft

10,000 ft

the width to 2 ft and reducing the length to 5,000 ft; the number of linear feet needed to enclose the field has been reduced to 10,004 ft. Note that the area is constant in both cases.

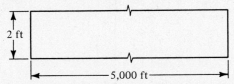

The optimum configuration can be obtained by generalizing this process. If the length is designated x and the width y, the perimeter P, the quantity to be minimized, can be expressed as

$$P = 2x + 2y \qquad x, y > 0 \tag{4.2.1}$$

The expression for P contains the two variables x and y. However, they are not independent variables because the area must be kept constant at 10,000 ft^2, that is

$$10,000 = xy \tag{4.2.2}$$

This relationship enables us to express P in terms of one variable alone. Expressing P in terms of x alone yields the function

$$P(x) = 2x + 2\left(\frac{10,000}{x}\right) = 2x + \frac{20,000}{x} \tag{4.2.3}$$

At this point, we now ask the question: What value of x minimizes $P(x)$? This question can be answered by finding $P'(x)$, setting $P'(x)$ equal to 0 and solving the resulting equation for x

$$P'(x) = 2 - \frac{20,000}{x^2}$$

$$0 = 2 - \frac{20,000}{x^2} = \frac{2x^2 - 20,000}{x^2}$$

$$0 = 2x^2 - 20,000 = 2(x + 100)(x - 100)$$

Solutions:

$$x_1 = 100 \text{ ft} \qquad x_2 = -100 \text{ ft}$$

The only realistic solution is x_1 because the length and the width must be positive numbers. The corresponding value of y is obtained by substituting $x = 100$ into Equation 4.2.2, yielding

$$y = 100 \text{ ft}$$

The corresponding value of P is found from Equation 4.2.3 to be

$$P(x_1) = 400 \text{ ft}$$

To determine if $P(x_1)$ represents a minimum, the first derivative test can be applied; let

$$x_L = 90 \qquad P'(90) = 2 - \frac{20,000}{(90)^2} < 0 \qquad \text{Decreasing on left}$$

$$x_R = 110 \qquad P'(110) = 2 - \frac{20,000}{(110)^2} > 0 \qquad \text{Increasing on right}$$

Therefore, we conclude that 100,400 is a minimum. Following is a graph of $P(x)$.

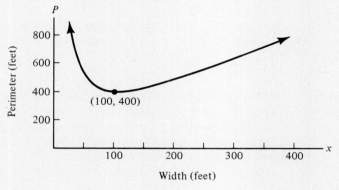

Example 2 Suppose the recreation department decides to add a wading pool to the playground and to fence it off from the rest of the equipment by a fence running across the width of the field as shown in the following diagram. What are the dimensions of the rectangle that will minimize the length of fencing required if the total area enclosed remains 10,000 ft²?

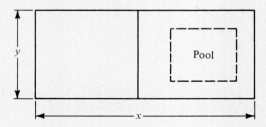

Solution As before, x designates the length and y the width of the playground and they are related via the equation

$$xy = 10,000$$

L, the number of feet of fencing, can be written

$$L = 2x + 3y$$

Proceeding as in Example 1, we express L in terms of x alone, getting

$$L(x) = 2x + \frac{30,000}{x}$$

$$L'(x) = 2 - \frac{30,000}{x^2}$$

$$0 = 2x^2 - 30,000$$

$$x^2 = 15,000$$

yielding

$$x = \sqrt{15,000} = 122.47 \text{ ft} \qquad y = 81.65 \text{ ft}$$

as the values of x and y that minimize the amount of fencing required. The minimum amount of fencing can be calculated easily, giving

$$L(122.47) = 489.9 \text{ ft}$$

Example 3 | A flat piece of sheet metal measuring 30 in. by 30 in. is to be used for making an open box to hold rivets that did not pass inspection in a quality control process. The box is to be made by cutting identical square pieces from the four corners and folding up the flaps to complete the box as shown in the accompanying figure, where x denotes the length and width of the cut-out pieces. For what values of x is the volume of the box maximized?

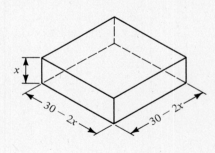

Solution | To see how the volume V of the box depends on the size of the cut-out pieces, a specific case or two will be considered before generalizing. Integral values of x will be used, only because the arithmetic is easier

$x = 1$ Base of box measures 28 in. by 28 in., height = 1 in.
$\qquad V = $ (area of base) $\cdot$ (height) $= 28 \cdot 28 \cdot 1 = 784$ in.^{3}

$x = 2$ Base of box measures 26 in. by 26 in., height = 2 in.
$\qquad V = 26 \cdot 26 \cdot 2 = 1354$ in.^{3}

x is arbitrary   Base measures $(30 - 2x)$ by $(30 - 2x)$, height $= x$

$$V = (30 - 2x)^2 \cdot x$$

$\cdot$

$\cdot$

$\cdot$

$x = 14$ Base measures 2 in. by 2 in., height $= 14$ $V = 56$ in.³

The expression for $V(x)$ then is

$$V(x) = x(30 - 2x)^2 \qquad 0 \le x \le 15$$

where the inequality indicates the range of values that can be assigned to the variable x. This is an important part of the problem because the equation $V'(x) = 0$ may have solutions outside this range. Explicitly denoting the domain will reduce the probability of accepting as a bona fide solution a value of x that is not realistic.

Going through the customary procedure, $V'(x)$ is found using in this case the product rule

$$V'(x) = -4x(30 - 2x) + (30 - 2x)^2 = (30 - 2x)(30 - 6x)$$

Setting $V'(x)$ equal to 0 and solving gives

$$0 = 30 - 2x \qquad \text{and} \qquad 0 = 30 - 6x$$

Solutions:

$$x_1 = 15 \text{ in.} \qquad x_2 = 5 \text{ in.}$$

One of these solutions, $x_1 = 15$, obviously does not correspond to a maximum volume since

$$V(x = 15) = 0 \qquad \text{while} \qquad V(x = 5) = 2000$$

Because the solution $x = 5$ is the only critical value for which V is positive, and because $V(x)$ is 0 at the endpoints $x = 0$ and $x = 15$, it represents the width and length of the cut-out piece that maximizes the volume.

The flexibility available in solving problems of this type can be demonstrated by denoting the length and width of the box as x as shown in the next figure. The volume $V(x)$ can then be written as

$$V(x) = x^2 \left(\frac{30 - x}{2}\right) = 15x^2 - \frac{x^3}{2}$$

$$V'(x) = 30x - \frac{3x^2}{2}$$

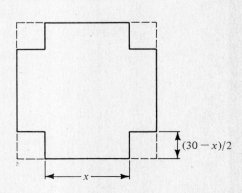

Next, $V'(x)$ is set equal to 0 and resulting equation is solved for x

$$0 = 30x - \frac{3x^2}{2}$$

$$= \frac{60x - 3x^2}{2}$$

$$= 3x(20 - x)$$

Solutions:

$$x_1 = 0 \qquad x_2 = 20$$

$$V(0) = 0$$

$$V(20) = 2000 \text{ in.}^3$$

These techniques can also be applied to profit maximization problems, as the next example illustrates.

Example 4   The owners of the Economy Motel have observed that all 50 rooms can be rented if they charge $12 or less per day per unit. Whenever they attempt to increase the daily rate, occupancy declines. They estimate that for each dollar increase in the daily rate, two additional units go unoccupied. It costs them $2 per day to clean and maintain each occupied unit. What rate should they charge to maximize daily profit?

Solution   Contrary to what one might expect, maximum profit is not achieved at full occupancy, as some simple calculations will reveal.

RATE	UNITS OCCUPIED	REVENUE	COST	PROFIT
12	50	600	100	500
13	48	624	96	528
14	46	644	92	552
.	.	.	.	.
.	.	.	.	.
.	.	.	.	.

The problem can be solved in a number of ways

I. Let x represent the daily rate
   The number of rooms occupied on any day depends on the daily rate:

   $$\text{Number of rooms occupied} = 50 - 2(x - 12) = 74 - 2x$$

   Using this we can find the revenue $R(x)$

   $$R(x) = (\text{Daily rate}) \cdot (\text{number of rooms occupied})$$
   $$= x(74 - 2x) = 74x - 2x^2$$

The cost of cleaning and maintenance C is given by the equation

$$C(x) = 2(74 - 2x) = 148 - 4x$$

The daily profit P can be found next

$$P(x) = R(x) - C(x)$$
$$= 78x - 2x^2 - 148 \qquad 12 \le x \le 37$$

Now the value of x that maximizes $P(x)$ can be found as follows:

$$P'(x) = 78 - 4x$$
$$0 = 78 - 4x$$

Solution:

$$x = 19.50 \qquad P = 612.50$$

The profits at the endpoints, $x = 12$ and $x = 37$, are less than that at $x = 19.50$, so that the point $(19.50, 612.50)$ represents an *absolute maximum*. The number of rooms occupied equals $74 - 2(19.50) = \$35$.

II. The problem can also be solved by letting the variable x represent the number of rooms that are vacant; then

$$50 - x \qquad \text{Represents the number of rooms occupied.}$$

The "daily room rate" is related to the number of vacant rooms by the equation

$$\text{Daily rate} = 12 + \frac{x}{2}$$

The daily revenue $R(x)$ can be written as

$$R = \left(12 + \frac{x}{2}\right)(50 - x) = 600 + 13x - \frac{x^2}{2}$$

while the cost C can be written as

$$C = 2(50 - x) = 100 - 2x$$

From these expressions for R and C, we can write the profit P as

$$P(x) = R(x) - C(x)$$
$$= 600 + 13x - \frac{x^2}{2} - (100 - 2x)$$
$$= 500 + 15x - \frac{x^2}{2}$$

The value of x that maximizes P can now be found

$$P'(x) = 15 - x$$
$$0 = 15 - x$$

Solution:

$$x = 15 \qquad P(15) = \$612.50$$

which agrees with the result obtained in part I.

These optimization techniques can also be applied to the problem of minimizing the total costs associated with producing and storing a product as illustrated in the next example.

Example 5

The Rain-and-Snow Tire Company has a factory in New Hampshire that produces and distributes, among other items, steel-belted radial tires for the New England area. Demand for these tires is constant throughout the year and amounts to 2,700,000 tires annually. Because of the numerous items produced at this plant, production of the tires is not a continuous process; they are produced in large batches and then stored in a warehouse nearby from which they are then shipped as orders are filled. When the inventory gets very low, another batch is produced and brought to the warehouse until the number runs low again. To visualize the inventory as a function of time, suppose that the tires are produced in batches of 900,000. This indicates that three production runs per year are required and that the inventory level over the 12-month period looks like that shown in the accompanying figure. The vertical segments on the

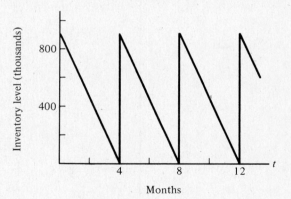

Months

graph represent the replenishment of the inventory via production, while the declining segments represent depletion of the inventory due to filling orders from wholesalers throughout New England. Management wants to determine the number of batches per year that will minimize total annual costs associated with this process based on the following estimates:

Machine set-up costs per run equal \$18,000
Material and labor costs are \$5 per tire
Storage cost is \$3 per tire per year

The total annual cost of this operation consists of three items

$$\begin{pmatrix} \text{Total annual} \\ \text{cost} \end{pmatrix} = \begin{pmatrix} \text{annual set-up} \\ \text{costs} \end{pmatrix} + \begin{pmatrix} \text{annual material} \\ \text{and labor costs} \end{pmatrix} + \begin{pmatrix} \text{annual storage} \\ \text{costs} \end{pmatrix}$$

Designating the batch size, that is, the number of items produced during each run as x, the three types of costs are found as follows:

1. Annual set-up costs = (cost per set-up) $\begin{pmatrix} \text{number of set-ups} \\ \text{per year} \end{pmatrix}$

$$= 18,000 \left(\frac{2,700,000}{x} \right) = \frac{(486)(10^8)}{x}$$

2. $\begin{pmatrix} \text{Annual material} \\ \text{and labor costs} \end{pmatrix} = \begin{pmatrix} \text{number of tires} \\ \text{produced annually} \end{pmatrix} \begin{pmatrix} \text{material and labor} \\ \text{cost per tire} \end{pmatrix}$

$$= (2,700,000)(5) = 13,500,000$$

3. Annual storage or inventory costs. This is probably one of the more difficult items to get a grip on because the inventory does not remain constant. However, if the average number of items in storage during the year were known, then the annual inventory costs could be found by finding the product

$$\begin{pmatrix} \text{Annual storage} \\ \text{cost per tire} \end{pmatrix} \begin{pmatrix} \text{average number of} \\ \text{tires in inventory} \end{pmatrix}$$

Because the inventory level declines uniformly from a maximum of x units to 0 units during each cycle, the average number of items in inventory during the year is given by $x/2$. Therefore, annual storage costs are $3(x/2)$. Putting all this together gives the following for the annual cost function $C(x)$:

$$C(x) = \frac{(486)(10^8)}{x} + \frac{3x}{2} + 13,500,000 \qquad \textbf{(4.2.4)}$$

The value of x that minimizes $C(x)$ is found in the usual manner

$$C'(x) = -\frac{(486)(10^8)}{x^2} + \frac{3}{2}$$

$$0 = -\frac{(486)(10^8)}{x^2} + \frac{3}{2}$$

Multiplying both sides by $2x^2$ gives

$$0 = -(972)(10^8) + 3x^2$$

Rewriting gives

$$0 = -324(10^8) + x^2$$

Solutions:

$$x_1 = 180{,}000 \qquad x_2 = -180{,}000$$

The only viable solution is x_1, so each production run should be scheduled to produce 180,000 tires. To see if this value of x corresponds to a minimum, we use the first derivative test:

let

$$x_L = 100{,}000 = 10^5$$

$$C'(x_L) = -\frac{(486)(10^8)}{(10^{10})} + \frac{3}{2} = -3.66 < 0 \qquad \text{Decreasing}$$

let

$$x_R = 200{,}000 = (2)(10^5)$$

$$C'(x_R) = -\frac{(486)(10^8)}{4(10^{10})} + \frac{3}{2} = +0.29 > 0 \qquad \text{Increasing}$$

We can conclude that costs are minimized when 180,000 tires are produced on each run, or, equivalently, when $2{,}700{,}000/180{,}000 = 15$ batches per year are scheduled. The cost for this production schedule can be found from Equation 4.2.4; the result is

$$C(180{,}000) = \$14{,}040{,}000$$

EXERCISE 4.2

1. A piece of wire 12 ft long is to be bent into a rectangle. What is the largest possible area that can be enclosed by the wire?

2. From the set of all pairs of numbers whose sum is 80, find that pair whose product is a maximum.

3. From the set of all pairs of numbers whose sum is 55, find that pair whose product is a maximum.

4. From the set of all pairs of numbers whose difference is 14, find that pair whose product is a minimum.

5. What positive number exceeds its square by the largest amount?

6. One hundred feet of fencing are available for enclosing a rectangular plot to be used for a small vegetable garden. What are the dimensions of the plot that will maximize the area of the garden?

7. The Lee-Kee Boat Rental Co. wants to install a fence around a rectangular plot of land where its canoes and rowboats can be stored. If 1800 ft² of land are to be enclosed and if fencing is needed on only three sides, what are the dimensions of the plot that will require the minimum length of fence?

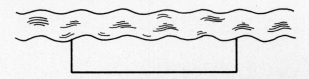

8. A field located next to a main highway is to be fenced in for storage of the state highway department's heavy equipment. The field is to have an area of 20,000 ft². What are the dimensions of the rectangular field that will minimize the cost of fencing material if material for the side running along the highway costs $3 per linear foot while material for the other three sides costs $1 per linear foot?

9. A closed rectangular box is to be made with a square base and to have a volume of 324 in.³. If the material for the sides and top cost 1¢/in.², while the material for the base costs 2¢/in.², what are the dimensions of the box that will minimize the cost of material?

10. An open rectangular box is to be made with a square base and to have a volume of 6000 in.³. If the material for the sides costs 2¢/in.² and that for the base costs 3¢/in.², what are the dimensions of the box that minimize the cost of material?

11. Identical square pieces are cut out from the four corners of a thin piece of cardboard measuring 25 in. by 25 in.; the flaps are then turned up to make an open rectangular box. What are the dimensions of the cut-out pieces that will maximize the volume?

12. Identical square pieces are cut out from the four corners of a thin piece of cardboard measuring 8 in. by 15 in.; the flaps are then turned up to make an open rectangular box. What are the dimensions of the cut-out pieces that will maximize the volume? What is the maximum volume?

13. A restaurant finds that if it charges $5.95 for its prime rib dinner, it will serve 100 dinners each night. For every 25¢ increase in the price, five fewer dinners are ordered. If the cost of preparing and serving each dinner is $2.25, what price should the restaurant charge if it wants to maximize its daily profit from the sale of prime rib dinners?

14. A retailer who sells TV recorders finds that she can sell 50 recorders per month when the price is $1000 per unit. For each $10 decrease in price, one additional recorder can be sold. If she purchases each unit for $600 and pays $20 to a technician to adjust and fine tune each unit when it is delivered, what level of sales will maximize monthly profits?

15. The publisher of *Skin*, a professional magazine for dermatologists, has learned that the demand equation for the magazine is given by the equation

$$p = 3 - 0.01x$$

where x represents the number of magazines (in hundreds) and p is the price (in dollars) of each magazine. The cost of printing, distributing, and advertising is given by the equation

$$C(x) = 10 + x + 0.03x^2$$

where $C(x)$ is expressed in hundreds of dollars. At what level of sales will the company's profit be maximized? At what price should the magazine be selling in order to maximize profits?

16. The daily output per person in a small machine shop is constant at 500 units when the number of stations is 18 or less. For each station over 18, the daily output per person declines by 25 due to overcrowding and the inability to service breakdowns quickly. For what number of stations is daily output for the entire shop maximized?

17. The owner of an apple orchard finds that the annual yield per tree is constant at 320 lb when the number of trees per acre is 50 or less. For each additional tree over 50,

the annual yield per tree decreases by 4 lb due to overcrowding. How many trees should be planted on each acre to maximize the annual yield from an acre?

18. The state legislature is considering a bill that would impose a meals tax of t cents per dollar spent on food in restaurants, cafeterias, and the like. Annual spending in restaurants is currently $216 million. Economists estimate that the tax will cause a decline in A, the amount of money spent by the public on food. The relationship between A and t is described by the equation

$$A(t) = 216 - 2t^2 \qquad 0 \le t \le 8$$

Find the equation that describes the government's revenue as a function of t. What value of t maximizes the government's revenue?

19. The publishers of a popular paperback book find that demand is steady at 400,000 copies per year. Set-up costs for printing a batch of these books is $2000. The cost of labor and material is $0.75 per book, while the cost to hold one book in inventory for one year is $0.25. How many production runs should the company schedule each year to minimize total costs?

20. Counterfeit Charlie is the West Coast printer and distributor of bogus 10-dollar bills for organized crime. Charlie has a contract to print 50,000 bills annually. Set-up costs for each printing run are $1250, material and labor costs are $0.25 for each bill, and storage costs for each bill amount to $0.20 per year. The high cost of storage is due to the requirement that the bills not be stored at one location. How many bills should be printed during each run if Charlie wants to minimize the cost of printing and storing the bogus money? Assume that demand is uniform during the year.

21. A loan company is restricted by state law to charge no more than 25 percent interest on loans to customers. Even at that exorbitant rate of interest, the company can lend out all the money it has available. To raise money for loans, the company offers saving accounts to individuals with surplus cash. The amount of money placed by depositors is proportional to the interest rate that is paid on saving accounts by the company. What interest rate should be paid on saving accounts to maximize the profit earned by the company?

22. The moonshine produced and bottled in a still hidden in a wooded area is to be transported for storage across a river to a barn located three miles down the river from point A (see the figure). The moonshiners want to transport the liquor to the barn in the shortest possible time to avoid the legal authorities. Travel by boat is at the rate of 6 mph, while travel over land is at the rate of 10 mph. How far from point A should the boat land in order to minimize transport time?

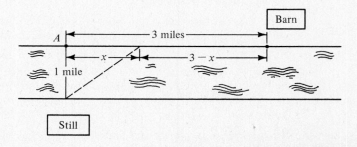

23. A 36-ft piece of wire is to be cut into two parts. One part is to be bent into a square and the second into a circle. At what point should the wire be cut so that the sum of the areas is a minimum?

4.3 Second Derivative Test, Inflection Points

Although the second derivative $f''(x)$ has received little attention so far, it would be a mistake to regard it as an unimportant quantity. The second derivative is very useful in determining the behavior of a function because it describes the rate and direction of bending on a curve.

Before studying the second derivative in more detail, it is necessary to introduce you to the concept of *concavity;* the concept is illustrated in the two curves shown in Figure 4.12. When a curve opens *upward*, such as that shown in Figure 4.12(a), it is said to be *concave up;* a curve like the one shown in Figure 4.12(b) is said to be *concave down.* The two types of concavity can be distinguished by the relative positions of the curve and its tangent lines in the two examples shown in Figure 4.12.

1. When the *tangent line* lies *below* the curve, the curve is said to be *concave up* in the vicinity of the point of tangency.
2. When the *tangent line* lies *above* the curve, the curve is said to be *concave down* in the vicinity of the point of tangency.

Figure 4.12   (a)   (b)

An understanding of the geometrical meaning of $f''(x)$ requires a knowledge of the relationship between $f''(x)$ and $f'(x)$. Noting that the second derivative $f''(x)$ bears the same relationship to the first derivative $f'(x)$ that $f'(x)$ bears to the function $f(x)$, the relationship can be stated as follows:

1. If $f''(a)$ is *positive,* that is, $f''(a) > 0$, then the *first derivative $f'(x)$* is *increasing* in the vicinity of $[a, f(a)]$.
2. If $f''(a)$ is *negative,* that is, $f''(a) < 0$, then the *first derivative $f'(x)$* is *decreasing* in the vicinity of $[a, f(a)]$.

To visualize what an increasing or decreasing first derivative means, let us examine the curves in Figure 4.13 on the next page. As we move from left to right along the curve shown in Figure 4.13(a), the first derivative becomes larger or is increasing; on the other hand, the first derivative is decreasing for the curve in Figure 4.13(b). These curves indicate that an increasing first derivative causes the curve to bend upward, while a decreasing first derivative causes the curve to

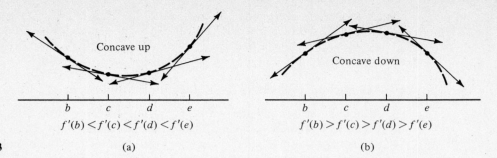

Figure 4.13   (a)   (b)

bend down. From this analysis, we can conclude the following important relationship between the second derivative $f''(a)$ and the shape of the curve at $[a, f(a)]$:

> 1. If $f''(a) > 0$, the curve is *concave up* in the vicinity
>   of $[a, f(a)]$.
> 2. If $f''(a) < 0$, the curve is *concave down* in the vicinity
>   of $[a, f(a)]$.
>
> **(4.3.1)**

The relationship between $f''(x)$ and the shape of a curve is illustrated in Figure 4.14, which contains the graph of the function $y = f(x) = x^4 - 6x^2 + 12$; in addition, the sign of $f''(x) = 12x^2 - 12$ is displayed on the x axis. The figure indicates that when $f''(x) > 0$, the curve is concave up and when $f''(x) < 0$, the curve is concave down.

It is now possible to combine our knowledge of the signs of the first and second derivatives at a point to determine the shape of a curve in the vicinity of the point. The results are shown in Table 4.2. Situations 5 and 6 in Table 4.2 form the basis of the second derivative test to determine whether a critical point represents a maximum or a minimum.

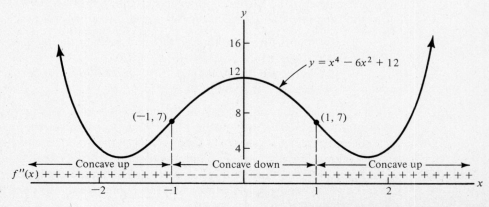

Figure 4.14

Table 4.2

$f'(a)$	$f''(a)$	CURVE	GRAPH
1. +	+	Rising—concave up	
2. +	−	Rising—concave down	
3. −	+	Falling—concave up	
4. −	−	Falling—concave down	
5. 0	+	Horizontal tangent line—concave up	
6. 0	−	Horizontal tangent line—concave down	

SECOND DERIVATIVE TEST

If the first derivative $f'(x)$ equals 0 at $x = a$, then

$[a, f(a)]$

I. $[a, f(a)]$ is a relative maximum if $f''(a) < 0$

II. $[a, f(a)]$ is a relative minimum if $f''(a) > 0$

$[a, f(a)]$

III. Test fails when $f''(a) = 0$; the critical point can be a relative maximum, relative minimum, or neither. Use the first derivative test to determine the shape of the curve.

An explanation for the reasons behind the failure of the test when $f''(a) = 0$ will have to wait until the topic of inflection points has been presented later in this section. The following example illustrates the use of the second derivative test at a critical point.

Example 1 Find the critical points for the function

$$y = f(x) = 2x^3 + 3x^2 - 12x - 8$$

and apply the second derivative test to determine whether each is a maximum or minimum.

Solution First, we find $f'(x)$, set it equal to 0, and solve for x

$$f'(x) = 6x^2 + 6x - 12$$

$$0 = 6(x^2 + x - 2) = 6(x + 2)(x - 1)$$

Solutions:

$$x_1 = -2 \qquad y_1 = +12$$
$$x_2 = +1 \qquad y_2 = -15$$

Next, we evaluate the second derivative $f''(x)$ at each critical point

$$f''(x) = 12x + 6$$

$$f''(-2) = -18 < 0 \qquad \text{Concave down}$$

Therefore $(-2, 12)$ is a relative maximum.

$$f''(1) = +18 > 0 \qquad \text{Concave up}$$

so $(1, -15)$ is a relative minimum. The two points and the curve in the vicinity of each are shown in the following graph.

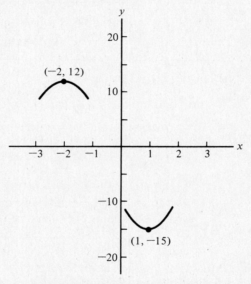

Although easy to apply, the second derivative test is sometimes less desirable than an alternate test such as the first derivative test for the following reasons:

1. If the function is complex, finding the second derivative may require much more time and effort than that required by the first derivative test.
2. The second derivative test fails when both $f'(a)$ and $f''(a)$ equal 0, thus forcing us to fall back on the reliable first derivative test to determine the shape of the curve in the vicinity of the critical point.

INFLECTION POINTS

Our study thus far has avoided discussing any situation where $f''(x) = 0$. Before investigating this case, let us again look at the graph of the function $y = f(x) = x^4 - 6x^2 + 12$ shown in the next figure. Points such as $(-1, 7)$ and

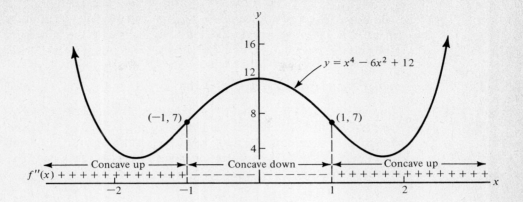

(1, 7), which separate a segment of the curve that is concave downward from one that is concave upward, are called *inflection points*. Closer inspection of the curve in the vicinity of these points (Figure 4.15) indicates that the tangent line at an inflection point intersects the curve so that it is above the curve on one side of the inflection point and below the curve on the other.

Figure 4.15

Having described the graphical characteristics of an inflection point, we now have to address the question: How do we analytically locate inflection points on a curve? Finding the inflection points on a curve requires two steps. First, because an inflection point separates a region of positive concavity from one of negative concavity, the second derivative $f''(x)$ can be neither positive nor negative at such a point. Therefore, if $[a, f(a)]$ is an inflection point, we can conclude that either (1) $f''(a) = 0$ or (2) $f''(a)$ does not exist. Condition 2 occurs infrequently so we shall ignore it in future considerations. Thus, if we are given a function $y = f(x)$ and we want to locate points of inflection, we look for those points where

$$f''(x) = 0 \qquad \qquad \text{(4.3.2)}$$

After finding those points, if any, where $f''(x) = 0$, the second step involves a test to determine whether the points found are indeed inflection points. This step is required because the second derivative can equal 0 at points that are *not* inflection points; such a situation will be illustrated in Example 3. The test itself consists in determining whether the concavity to the left of each point where $f''(x) = 0$ differs from the concavity to the right. If the concavity is different, the point is an inflection point; otherwise, it is not. The test is described next.

TEST FOR INFLECTION POINTS

Suppose $f''(a) = 0$ and that $f''(x)$ exists for all values of x close to $[a, f(a)]$. Select two values of x, one to the left of $[a, f(a)]$ called x_L, the other to the right called x_R, and calculate $f''(x_L)$ and $f''(x_R)$. If the sign of $f''(x_L)$ differs from that of $f''(x_R)$, $[a, f(a)]$ is an inflection point; otherwise it is not. The test is summarized in Table 4.3.

Table 4.3

		$f''(a) = 0$
$f''(x_L)$	$f''(x_R)$	$[a, f(a)]$
+	−	Inflection point
−	+	Inflection point
+	+⎫	
−	−⎭	Not an inflection point

The following example is intended not only to demonstrate the process of locating and testing for inflection points, but also to show how a rough sketch of the curve in the vicinity of an inflection point can be drawn.

Example 2 Find those points, if any, on the curve

$$y = f(x) = x^3 - 3x^2 + 4$$

where $f''(x)$ equals 0 and determine whether the points found are inflection points. In addition, make a rough sketch of the curve in the vicinity of these points.

Solution STEP 1. Find the second derivative $f''(x)$

$$f'(x) = 3x^2 - 6x$$

$$f''(x) = 6x - 6$$

STEP 2. Find those points where $f''(x)$ equals 0

$$0 = 6x - 6$$

Solving for x gives

$$x = 1 \qquad y = f(1) = 2$$

STEP 3. Carry out the test to see if $(1, 2)$ is an inflection point.

a. Select a value for x_L, say $x_L = 0$; evaluate $f''(x_L)$

$$f''(x_L) = f''(0) = -6 \qquad \text{Concave down}$$

b. Select a value for x_R, say $x_R = 2$; evaluate $f''(x_R)$

$$f''(x_R) = f''(2) = +6 \qquad \text{Concave up}$$

STEP 4. Since the curve is concave down to the left and concave up to the right of the point (1, 2), we can conclude that (1, 2) is an inflection point.

STEP 5. A rough sketch of the curve in the vicinity of (1, 2) can be drawn once we find the value of the first derivative $f'(x)$ at (1, 2).

$$f'(1) = -3$$

which means that the curve is falling in the vicinity of (1, 2). Combining this information with that contained in step 3 enables us to make the following sketch.

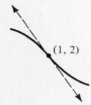

The next example illustrates a case in which a point where $f''(x) = 0$ is not an inflection point.

Example 3 Find those points on the curve

$$y = f(x) = x^4 - 4x^3 + 6x^2 + 2$$

where $f''(x)$ equals 0. Determine whether the points are inflection points and make a rough sketch of the curve in the vicinity of each point.

Solution STEP 1. Find the second derivative $f''(x)$. The result is

$$f''(x) = 12x^2 - 24x + 12$$

STEP 2. As before, we set $f''(x)$ equal to 0 and solve for x

$$0 = 12x^2 - 24x + 12$$

$$0 = 12(x - 1)^2$$

Solving for x gives

$$x = 1 \qquad y = f(1) = 5$$

STEP 3. Test to determine if (1, 5) is an inflection point. Because $f''(x) = 12(x - 1)^2 > 0$ for all values of x except $x = 1$, the curve is concave up on both sides of (1, 5). Therefore (1, 5) is not an inflection point.

STEP 4. A rough sketch of the curve in the vicinity of $(1, 5)$ can be made once we know the sign of $f'(1)$.

$$f'(x) = 4x^3 - 12x^2 + 12x$$

$$f'(1) = 4 > 0 \qquad \text{Curve is rising}$$

Knowing that the curve is rising and is concave up in the vicinity of $(1, 5)$ enables us to draw a rough sketch such as that shown in the following figure.

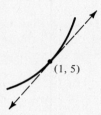

Having found that the concavity may or may not change as we move along a curve through a point where $f''(x) = 0$, we can begin to see why the second derivative test applied at a critical point fails when $f''(a) = 0$. Figure 4.16 contains the graphs of two curves, $y = f(x) = x^3 + 2$ and $y = g(x) = x^4 + 2$, both of which have critical points at $(0, 2)$. In addition, $f''(x) = g''(x) = 0$ at $(0, 2)$. For the function $f(x) = x^3 + 2$, the critical point $(0, 2)$ is also an inflection point. For the function $g(x) = x^4 + 2$, the critical point $(0, 2)$ represents a minimum.

It might appear at first glance that inflection points are merely geometrical curiosities and have little or no importance in the world of business and eco-

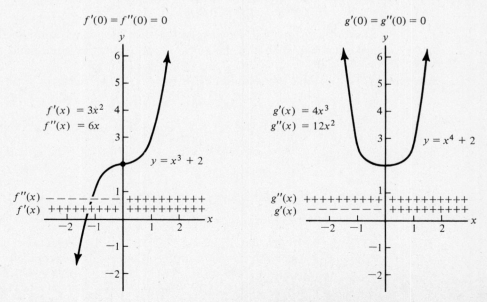

Figure 4.16

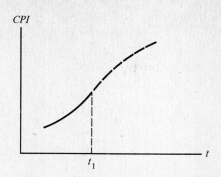

Figure 4.17

nomics. To remove that impression, let us consider the situation shown in Figure 4.17 where the consumer price index (CPI) is plotted as a function of time during a runaway inflationary period such as that which occurred during 1973–1974 or 1979 (shown by the solid curve). The curve is concave upward, indicating that the rate of increase in CPI, that is, $\frac{d}{dt}$ (CPI), is increasing. Suppose at $t = t_1$, the government takes action to brake the accelerating inflationary trend by instituting wage and price controls, sharply reducing the money supply, or raising taxes. The results of these actions, if effective, would cause a decrease in the rate at which the CPI is increasing, that is, the curve would become concave downward (shown by the dashed portion). The point at which the change from an accelerating to a decelerating rate of increase in the CPI occurs represents an inflection point. However, unlike the situations we will encounter, one cannot predict when or even if the economic actions proposed will have the desired effects.

EXERCISE 4.3

Find the critical points, if any, for the functions in Problems 1–10. Use the second derivative test to determine whether a critical point is a maximum or a minimum. If the test fails, use the first derivative test to determine the shape of the curve in the vicinity of each critical point.

1. $y = f(x) = x^2 + 2x + 3$ **2.** $y = f(x) = 4x - x^2 - 1$

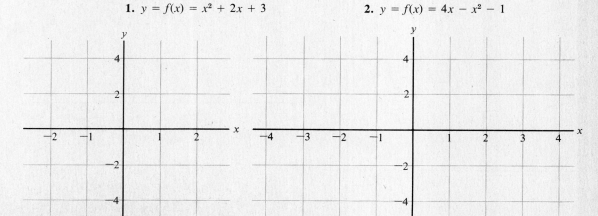

3. $y = f(x) = \dfrac{x^3}{3} - \dfrac{x^2}{2} + 1$

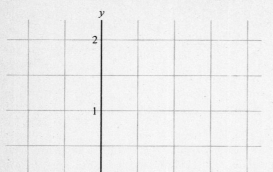

4. $y = f(x) = 4 + 12x - x^3$

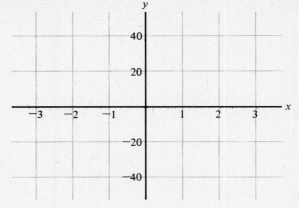

5. $y = f(x) = \dfrac{x^3}{3} - \dfrac{x^2}{2} - 2x + 1$

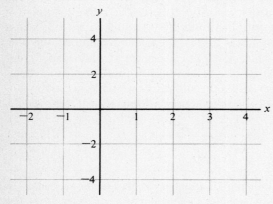

6. $y = f(x) = x^3 + 3x - 1$

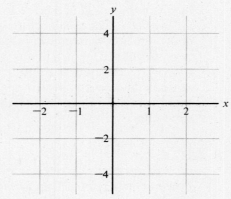

7. $y = f(x) = \dfrac{x^4}{4} - \dfrac{x^2}{2} + 2$

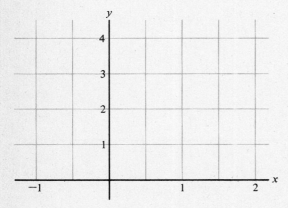

8. $y = f(x) = x + \dfrac{4}{x}$

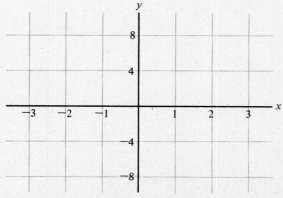

9. $y = f(x) = 2x^4 - 4x^2 + 3$

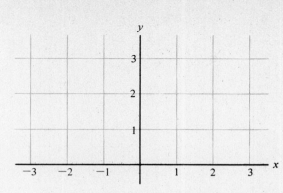

10. $y = f(x) = \dfrac{x^2}{x + 2}$

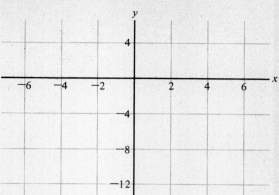

For the functions in Problems 11–17

a. Locate those points, if any, where $f''(x) = 0$.
b. Determine whether the points found in part a are inflection points.
c. Make a rough sketch of the curve in the vicinity of the points found in part a.

11. $y = f(x) = x^3 - 9x^2 + 15x + 6$

12. $y = f(x) = \dfrac{x^4}{12} - 2x^2 + 3x + 7$

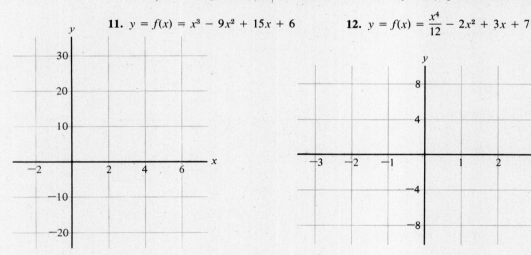

13. $y = f(x) = x^4 - 10x + 4$

14. $y = f(x) = x^3 + 3x - 2$

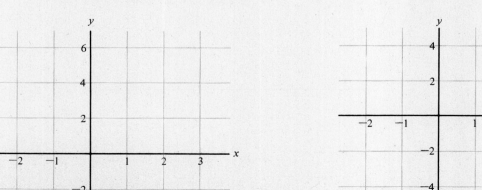

15. $y = f(x) = \dfrac{x^5}{20} - \dfrac{x^4}{6} + x + 3$

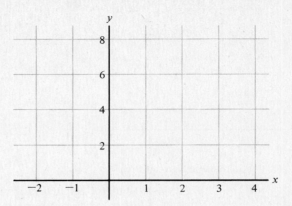

16. $y = f(x) = x^4 - 2x^3 - 12x^2 - 10x + 20$

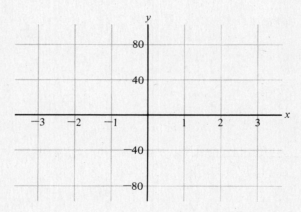

17. $y = f(x) = \dfrac{3}{x} - \dfrac{1}{x^2} - 1$

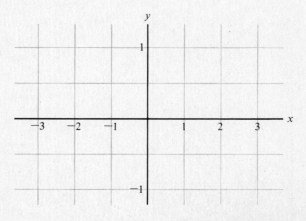

18. In recent years, a well-known manufacturer of ball-point pens has seen its sales decline because of heavy competition from a foreign manufacturer. The sales curve is shown in the following diagram. If the company begins an intensive advertising campaign when $t = t_1$, what would you expect the shape of the curve to look like shortly after the advertising campaign begins, if it should be successful?

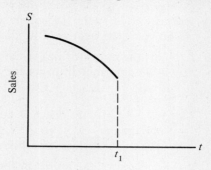

₴ 4.4 Curve Sketching

Our attention in this chapter has been limited to determining the behavior of a function in the vicinity of its critical and inflection points. For polynomial functions, this information is sufficient to enable us to make a rough sketch of the entire curve, as shown in Examples 1 and 2.

Example 1 | Make a rough sketch of the function

$$y = f(x) = 2x^3 - 3x^2 - 12x + 15$$

Solution | STEP 1. First, we locate the critical points

$$f'(x) = 6x^2 - 6x - 12$$

$$0 = 6(x^2 - x - 2) = 6(x - 2)(x + 1)$$

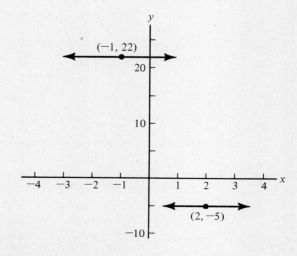

Solutions:

$$x_1 = -1 \qquad y_1 = 22$$
$$x_2 = 2 \qquad y_2 = -5$$

NOTE: There are no points on the curve where $f'(x)$ does not exist. The critical points together with the horizontal tangent lines are shown in the figure on the preceding page.

STEP 2. Next, the behavior of the curve in the vicinity of each critical point will be determined, using the second derivative test

$$f''(x) = 12x - 6$$

a. $f''(-1) = -18$, curve is concave down; so $(-1, 22)$ is a relative maximum.
b. $f''(2) = +18$, curve is concave up; so $(2, -5)$ is a relative minimum.

The shape of the curve in the vicinity of each critical point is displayed on the next figure.

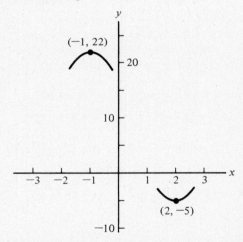

STEP 3. Inflection points are found by first setting $f''(x)$ equal to 0 and solving the resulting equation for x

$$0 = 12x - 6$$

Solution:

$$x = \tfrac{1}{2} \qquad y = \tfrac{17}{2}$$

It is not necessary to test this point to determine if it is an inflection point. In step 2, we learned that the concavity to the left, that is, at $x = -1$, is different from that to the right, that is, at $x = 2$; therefore, we can conclude that $(\tfrac{1}{2}, \tfrac{17}{2})$ is an inflection point.

Incorporating the inflection point into our graph, we can draw the segment of the curve between the two critical points as shown on the accompanying graph. The y intercept $(0, 15)$ is also included because it is very easy to locate.

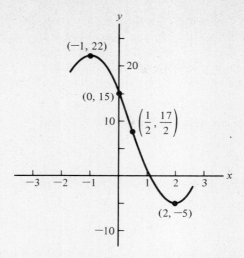

STEP 4. The graph can be completed once we determine the shape of the curve to the left of $(-1, 22)$ and to the right of $(2, -5)$. Because we have found the coordinates of all points where the first and second derivative change sign, the behavior of the curve is uniform over each of the intervals, $x < -1$ and $x > 2$. The shape of the curve over each interval can be determined by finding the sign of $f'(x)$ and $f''(x)$ at arbitrary points on each as shown in the accompanying table.

INTERVAL	x	$f'(x)$	$f''(x)$	CURVE	GRAPH
$x < -1$	-2	$+24$	-30	Rising—concave down	
$x > 2$	3	$+24$	$+30$	Rising—concave up	

Based on this analysis, we can make a rough sketch of the curve as shown in the following figure.

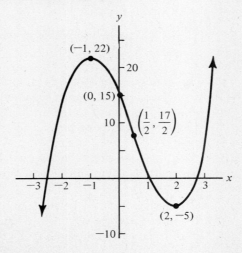

Ordinarily, we do not have to go through an analysis as detailed as that in the previous example. For a polynomial function, locating the critical and inflection points and then determining the behavior of the function in the vicinity of each is usually sufficient to enable us to make a rough sketch of the function. Because the y intercept can be found easily, it is advisable to include it also.

Example 2 | Make a rough sketch of the function

$$y = f(x) = 3x^4 - 4x^3 - 3$$

Solution | STEP 1. First, we want to locate the critical points, if any

$$f'(x) = 12x^3 - 12x^2$$

$$0 = 12x^2(x - 1)$$

Solutions:

$$x_1 = 0 \qquad y_1 = -3$$

$$x_2 = 1 \qquad y_2 = -4$$

NOTE: There are no points on the curve where $f'(x)$ does not exist.

STEP 2. The next step is to determine the behavior of the curve in the vicinity of the two critical points $(0, -3)$ and $(1, -4)$. Using the second derivative test, with

$$f''(x) = 36x^2 - 24x$$

we get

a. $f''(1) = +12$; concave up, so $(1, -4)$ is a relative minimum.
b. $f''(0) = 0$; test fails, so we return to the first derivative test
   i. Let $x_L = -1$

$$f'(x_L) = f'(-1) = -24 \qquad \text{Falling}$$

   ii. x_R can be any value in the interval $0 < x < 1$, say $x_R = \frac{1}{2}$

$$f'(x_R) = f'(\tfrac{1}{2}) = -\tfrac{3}{2} \qquad \text{Falling}$$

We conclude that $(0, -3)$ is neither a maximum nor a minimum.

Based on the results obtained to date, we can sketch the curve in the neighborhood of the critical points as shown in the figure at the top of the following page.

STEP 3. The inflection points, if any, can be found by setting $f''(x)$ equal to 0 and solving the resulting equation for 'x'

$$0 = 36x^2 - 24x$$

$$0 = 12x(3x - 2)$$

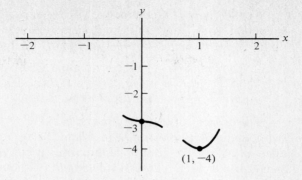

Solutions:

$$x_1 = 0 \qquad y_1 = -3$$
$$x_2 = \tfrac{2}{3} \qquad y_2 = -\tfrac{97}{27}$$

Because the shape of the curve in the vicinity of the point $(0, -3)$ is known from step 2, no further analysis at this point is required; the point is not only a critical point but is also an inflection point. Turning our attention to the point $(\tfrac{2}{3}, -\tfrac{97}{27})$, we want to find out if it is an inflection point. We select for x_L an arbitrary value in the interval $0 < x < \tfrac{2}{3}$, say $x_L = \tfrac{1}{2}$, obtaining

$$f''(x_L) = f''(\tfrac{1}{2}) = -3 \qquad \text{Concave down}$$

Letting $x_R = +1$, we know from step 2 that

$$f''(x_R) = f''(1) = +12 \qquad \text{Concave up}$$

We can conclude that $(\tfrac{2}{3}, -\tfrac{97}{27})$ is an inflection point.

STEP 4. Including this information on the graph, we can then connect the inner tails with a smooth segment and then extend the outer tails upward indefinitely generating a curve similar to that shown in the next figure.

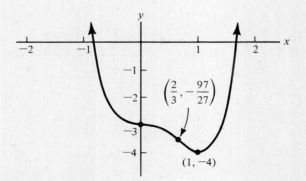

Not all types of functions are as easy to sketch as the polynomial functions encountered in Examples 1 and 2. For example, when sketching a rational function it is necessary to consider additional features such as the following:

1. The behavior of the curve in the vicinity of a discontinuity.
2. The behavior of the curve as x is assigned values that increase indefinitely ($x \rightarrow \infty$) or decrease indefinitely ($x \rightarrow -\infty$).

The process of assigning to the variable x values that *increase indefinitely* (without limit) is illustrated in Figure 4.18. Mathematically, this process is denoted as $x \rightarrow \infty$, read "x approaches infinity." When the variable x is assigned values that *decrease indefinitely* (without limit or bound), as shown in Figure 4.19, the process is denoted as $x \rightarrow -\infty$, read "x approaches minus infinity."

Figure 4.18

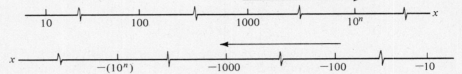

Figure 4.19

The usefulness of this type of information is illustrated in the following example where the rational function has neither critical points nor inflection points.

Example 3 | Make a rough sketch of the function

$$y = f(x) = \frac{2x}{x - 1}$$

Solution | STEP 1. First, let us note that the function is not defined at $x = 1$, and that the curve passes through $(0, 0)$.

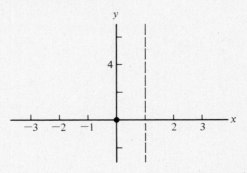

STEP 2. We look for the critical points by setting $f'(x)$ equal to 0. The quotient rule is employed to find $f'(x)$

$$f'(x) = \frac{2(x - 1) - (2x)(1)}{(x - 1)^2} = \frac{-2}{(x - 1)^2}$$

$$0 = \frac{-2}{(x-1)^2}$$

This equation has no solutions, so we conclude that the function has no critical points. However, we have not come up empty-handed; if we note that $-2/(x-1)^2$ is negative everywhere on the curve, we can conclude that the curve is falling everywhere.

STEP 3.   Next, we look for inflection points by setting $f''(x)$ equal to 0 and solving for x; $f''(x)$ itself can be found more quickly by writing $f'(x)$ as

$$f'(x) = -2(x-1)^{-2}$$

Using the general power rule, we get for $f''(x)$

$$f''(x) = 4(x-1)^{-3} = \frac{4}{(x-1)^3}$$

Setting $f''(x)$ equal to 0 gives

$$0 = \frac{4}{(x-1)^3}$$

Again, we have an equation that has no solutions, indicating that the curve has no inflection points.

Our analysis indicates that any changes in the sign of either $f'(x)$ or $f''(x)$ can occur only at the discontinuity $x = 1$. Now we can determine the shape of the curve over the intervals $x < 1$ and $x > 1$ by finding the sign of $f'(x)$ and $f''(x)$ at arbitrary points on each interval as shown in the next table. This information enables us to make a rough sketch of the curve in the vicinity of $(0, 0)$.

INTERVAL	x	$f'(x)$	$f''(x)$	CURVE	GRAPH
$x < 1$	0	-2	-4	Falling—concave down	
$x > 1$	2	-2	$+4$	Falling—concave up	

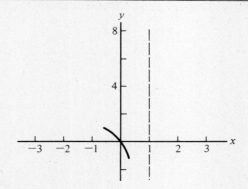

STEP 4. The behavior of the function in the vicinity of the discontinuity $x = 1$ can be found by allowing x to approach 1 from both the left and the right as described in Section 2.1. The accompanying table illustrates the process. The first two columns show that as x approaches 1 from the left, the function decreases indefinitely, written

$$\frac{2x}{x - 1} \to -\infty$$

As x approaches 1 from the right, the third and fourth columns show that the function increases indefinitely.

$$\frac{2x}{x - 1} \to \infty$$

FROM THE LEFT		FROM THE RIGHT	
x	$y = \dfrac{2x}{x - 1}$	x	$y = \dfrac{2x}{x - 1}$
0.50	−2.000	1.50	6.000
0.90	−18.000	1.10	22.000
0.99	−198.000	1.01	202.000
.	.	.	.
.	.	.	.
.	.	.	.

A graphical representation of the analysis to this point is shown in the following figure.

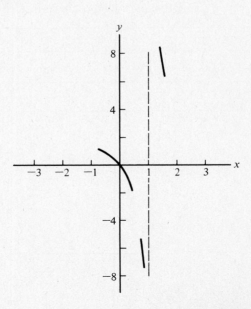

STEP 5. To see what happens to the function as $x \to \pm\infty$, we select values of x that increase indefinitely and decrease indefinitely as shown in the accompanying table.

| | $x \to -\infty$ | | | $x \to \infty$ | |
x	$y = \dfrac{2x}{x-1}$		x	$y = \dfrac{2x}{x-1}$	
-10	$\frac{0.20}{0.11} = 1.818$		10	$\frac{20}{9} = 2.222$	
-100	$\frac{200}{101} = 1.980$		100	$\frac{200}{99} = 2.020$	
-1000	$\frac{2000}{1001} = 1.998$		1000	$\frac{2000}{999} = 2.002$	

The second and fourth columns indicate that the variable y is approaching 2 as $x \to \pm\infty$. Mathematically, this result is written as follows

$$f(x) = \frac{2x}{x-1} \to 2 \qquad \text{as } x \to \pm\infty$$

or

$$\lim_{x \to \infty} \frac{2x}{x-1} = 2 \qquad \text{and} \qquad \lim_{x \to -\infty} \frac{2x}{x-1} = 2$$

The line $y = 2$ is called a *horizontal asymptote*. Incorporating the results of step 5 into our graph gives

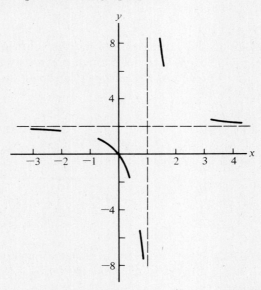

The last step is the easiest of all, that is, connecting the adjacent tails by means of smooth segments. A curve such as that shown in the following figure results.

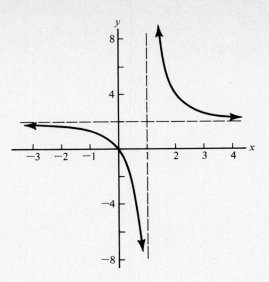

The next example is a continuation of a problem for which the critical points and the behavior of the function in the vicinity of each has been determined in an earlier example.

Example 4 Make a rough sketch of the curve

$$y = f(x) = \frac{x^2}{x - 1}$$

Solution STEP 1. The critical points and the behavior of the function in the vicinity of each were determined previously in Example 5 of Section 4.1. The result of that analysis is shown on the following figure.

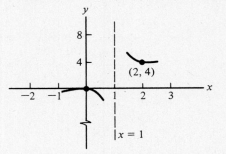

STEP 2. Next, we want to find the inflection points. Recalling that the first derivative $f'(x)$ has the form

$$f'(x) = \frac{x^2 - 2x}{(x - 1)^2}$$

the second derivative can be obtained by using the quotient rule, giving

$$f''(x) = \frac{(2x - 2)(x - 1)^2 - (x^2 - 2x)2(x - 1)}{(x - 1)^4}$$

Working through the algebra gives for the final version of $f''(x)$

$$f''(x) = \frac{2}{(x - 1)^3}$$

Setting $f''(x)$ equal to 0 yields the following equation

$$0 = \frac{2}{(x - 1)^3}$$

This equation has no solutions, indicating that there are no inflection points on the curve.

STEP 3. The behavior of the curve in the vicinity of the line $x = 1$ can be determined by allowing x to approach the line $x = 1$ from both the left and the right. The accompanying table illustrates what happens to the function. The results shown in the table indicate that

$$f(x) = \frac{x^2}{x - 1} \to -\infty \qquad \text{As } x \text{ approaches 1 from the left}$$

$$f(x) = \frac{x^2}{x - 1} \to \infty \qquad \text{As } x \text{ approaches 1 from the right}$$

FROM THE LEFT		FROM THE RIGHT	
x	$y = \dfrac{x^2}{x - 1}$	x	$y = \dfrac{x^2}{x - 1}$
0.50	−.500	1.50	4.500
0.90	−8.100	1.10	12.100
0.99	−98.010	1.01	102.010
.	.	.	.
.	.	.	.
.	.	.	.

Including these features in our graph gives the following

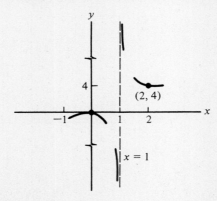

STEP 4. The behavior of the function as $x \to \pm\infty$ can be determined by letting x increase indefinitely and decrease indefinitely as shown in the accompanying table. The results shown in the table indicate that

$$f(x) = \frac{x^2}{x-1} \to -\infty \qquad \text{as } x \to -\infty$$

$$f(x) = \frac{x^2}{x-1} \to \infty \qquad \text{as } x \to \infty$$

We can include these features in our graph, resulting in the following figure.

$x \to -\infty$		$x \to \infty$	
x	$y = \dfrac{x^2}{x-1}$	x	$y = \dfrac{x^2}{x-1}$
-10	-9.0900	10	11.1111
-100	-99.0099	100	101.010

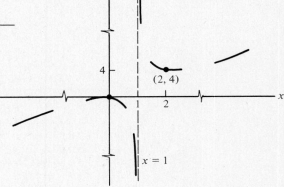

STEP 5. Because the function is differentiable everywhere on the intervals $x < 1$ and $x > 1$, we can connect adjacent segments on each branch by smooth curves resulting in the following graph.

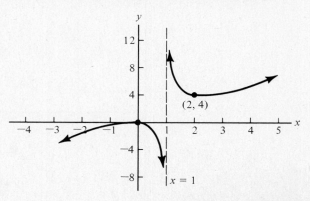

PROCEDURE FOR SKETCHING A GRAPH

The procedures used to sketch the graph of a function can be broken down into the following steps:

1. Find all the critical points.
2. Determine the behavior of the function in the vicinity of all critical points and sketch a graph of the curve in the vicinity of each.
3. Locate and plot all the inflection points.
4. Determine the behavior of the curve in the vicinity of each value of x for which the function is discontinuous; sketch the curve in the vicinity of each.
5. Determine the behavior of the function as $x \to \pm\infty$ and incorporate this feature into the graph.
6. Finally, connect all adjacent segments, adding any additional points, such as intercepts, that will improve the accuracy of the graph.

EXERCISE 4.4

Using the techniques described in this section, make a rough sketch of each of the functions given in Problems 1–17. Each has been encountered earlier, in either Exercise 4.1 or 4.3; therefore, some of the analysis should have already been carried out.

1. $y = f(x) = 2x^2 - 4x - 3$ (Exercise 4.1, problem 1)

2. $y = f(x) = x^2 + 5x - 1$ (Exercise 4.1, problem 3)

3. $y = f(x) = x^3 + 3x^2 - 1$ (Exercise 4.1, problem 5)

4. $y = f(x) = x^4 + 4x - 2$ (Exercise 4.1, problem 6)

5. $y = f(x) = 2x^3 - 6x^2 - 18x + 14$ (Exercise 4.1, problem 7)

6. $y = f(x) = x^4 - 2x^2 + 1$ (Exercise 4.1, problem 9)

7. $y = f(x) = (x - 1)^2(x - 4)$ (Exercise 4.1, problem 11)

8. $y = f(x) = \dfrac{x^2}{x - 2}$ (Exercise 4.1, problem 13)

9. $y = f(x) = x + \dfrac{1}{x}$ (Exercise 4.1, problem 14)

10. $y = f(x) = 3x^5 - 5x^3 + 1$ (Exercise 4.1, problem 15)

11. $y = f(x) = 4x - x^2 - 1$ (Exercise 4.3, problem 2)

12. $y = f(x) = \dfrac{x^3}{3} - \dfrac{x^2}{2} - 2x + 1$ (Exercise 4.3, problem 5)

13. $y = f(x) = \dfrac{x^4}{4} - \dfrac{x^2}{2} + 2$ (Exercise 4.3, problem 7)

14. $y = f(x) = x + \dfrac{4}{x}$ (Exercise 4.3, problem 8)

15. $y = f(x) = 2x^4 - 4x^2 + 3$ (Exercise 4.3, problem 9)

16. $y = f(x) = \dfrac{x^2}{x + 2}$ (Exercise 4.3, problem 10)

17. $y = f(x) = x^3 - 9x^2 + 15x + 6$ (Exercise 4.3, problem 11)

Using the techniques described in this section, make a rough sketch of each of the functions shown in problems 18–23.

18. $y = f(x) = 2 + \dfrac{x^2}{4} - \dfrac{x^3}{3}$

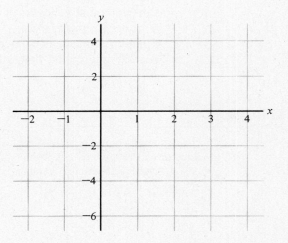

19. $y = f(x) = \dfrac{x^4}{4} + x^3 + 1$

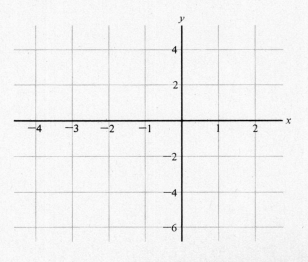

20. $y = f(x) = \dfrac{3x^2}{3x^2 + 1}$

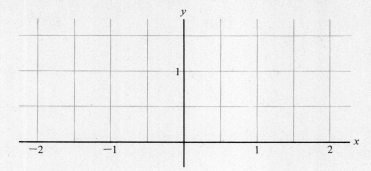

21. $y = f(x) = 2\sqrt{x} - x$

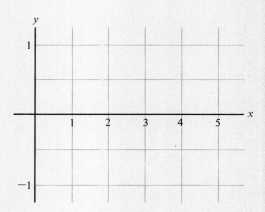

22. $y = f(x) = \dfrac{x^5}{5} - \dfrac{x^3}{3} - 10$

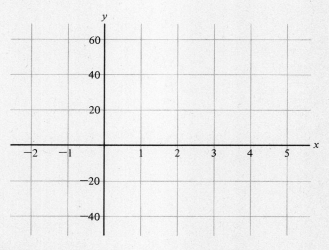

23. $y = f(x) = 4x + \dfrac{1}{x}$

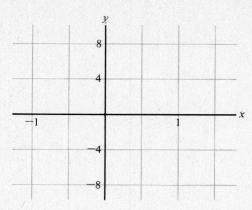

5

Exponential and Logarithmic Functions

CHAPTER 5 ❧ INTRODUCTION

In addition to the functions already studied in the text, there are many other equally important functions used in treating problems, not only in business and economics, but also in the fields of science and engineering. Among them are the two types that will be investigated in this chapter: *exponential* and *logarithmic* functions. The first two sections are devoted to a study of the graphs and the properties of these functions. In Section 5.3 the irrational number *e*, which represents the base of the Naperian or natural logarithmic system, is introduced.

❧ 5.1 Exponential Functions

A simple *exponential* function is one that has the form

$$y = f(x) = b^x \tag{5.1.1}$$

where *b* is a positive constant. The following are examples of this type of function:

$$y = f(x) = 4^x$$
$$y = f(x) = (\tfrac{1}{2})^x$$
$$y = f(x) = \left(\sqrt{3}\right)^x$$

To see how exponential functions arise in practice, let us digress for a moment and consider how banks and other financial institutions compound interest on money deposited with them. In addition to paying depositors interest for the use of the money initially deposited, called the principal, banks also pay interest on any interest that has accumulated, thus the name "compound" interest. Suppose that $500 is deposited in a bank that compounds interest once a year at the rate of 6 percent per year. Assuming that no withdrawals or additional deposits occur, the amount of money in the account will grow year by year as shown in Table 5.1, where the arithmetic has purposely not been completed. The amount in the account at the end of *n* years is designated *A* and is given by the equation

$$A = 500(1 + 0.06)^n = 500(1.06)^n \tag{5.1.2}$$

Table 5.1

YEAR	BALANCE AT BEGINNING	ANNUAL INTEREST	BALANCE AT END
1	500	500(0.06)	500(1 + 0.06)
2	500(1 + 0.06)	500(1 + 0.06)(0.06)	500(1 + 0.06)^2
3	500(1 + 0.06)^2	500(1 + 0.06)^2(0.06)	500(1 + 0.06)^3
.	.	.	.
.	.	.	.
.	.	.	.
n	500(1 + 0.06)$^{n-1}$	500(1 + 0.06)$^{n-1}$(0.06)	500(1 + 0.06)^n

It should be pointed out that unlike the power functions encountered previously, the *base* in Equation 5.1.2 is *constant* while the *exponent* is subject to *variation*. Equation 5.1.2 is used to determine the amount in the account at the end of any given number of years. For example, suppose that the amount in the account at the end of three years is desired; n is set equal to 3, yielding

$$A = 500(1.06)^3 = 500(1.06)(1.06)(1.06)$$

$$A = \$595.51$$

(5.1.3)

The prospect of evaluating the quantity $(1.06)^n$ by hand when the exponent n is large appeals to very few people. Fortunately, there are a number of methods by which the repeated multiplication can be avoided.

1. Tables containing values of the compound interest factor $(1 + i)^n$ for various combinations of n and i have been tabulated. Table A at the end of the book is an example. Suppose that $n = 20$ and $i = 0.06$; the value of the quantity $(1.06)^{20}$ is located in the row corresponding to $n = 20$ and the column corresponding to $i = 0.06$.

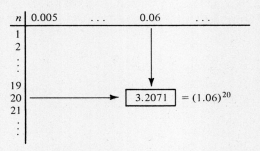

2. Logarithms, which will be introduced in the next section, can be used to evaluate $(1.06)^{20}$.
3. Best of all, a calculator with an exponential key y^x will give you the value of the factor $(1.06)^{20}$ quickly and easily. After entering 1.06 as y, 20 is entered as x. Pressing the y^x key will cause the value of $(1.06)^{20}$ to appear in the display.

The generalized form of Equation 5.1.2 is

$$A = P(1 + i)^n \tag{5.1.4}$$

where P is the principal, i is the interest rate for each interest-compounding period, and n is the number of periods. A more complete discussion and analysis of this equation will be undertaken later in this section; we will return now to the problem of examining the properties of exponential functions.

Before becoming deeply involved in the process of graphing exponential functions, it might be well to recall some of the basic properties of exponents that will play a major role in handling not only exponential functions but also logarithmic functions, which will be considered in the next section.

1. $b^m \cdot b^n = b^{m+n}$ 2. $\dfrac{b^m}{b^n} = b^{m-n}$

3. $(b^m)^n = b^{mn}$ 4. $(a \cdot b)^m = a^m \cdot b^m$

5. $\left(\dfrac{a}{b}\right)^m = \dfrac{a^m}{b^m}$ $b \neq 0$ 6. $b^0 = 1$ $b \neq 0$

7. $b^{-n} = \dfrac{1}{b^n}$ $b \neq 0$ 8. $b^{1/n} = \sqrt[n]{b}$ Provided b is not negative when n is an even integer

A complete discussion of these properties, together with examples and exercises, is contained in Appendix B.

Example 1 | Graph the exponential function

$$y = f(x) = 2^x$$

Solution | Selected integral values of x and the associated values of y are used to plot a sufficient number of points so that a smooth curve can be drawn as shown in Figure 5.1.

x	$y = 2^x$
-3	$2^{-3} = \frac{1}{8}$
-2	$2^{-2} = \frac{1}{4}$
-1	$2^{-1} = \frac{1}{2}$
0	$2^0 \;= 1$
1	$2^1 \;= 2$
2	$2^2 \;= 4$
3	$2^3 \;= 8$

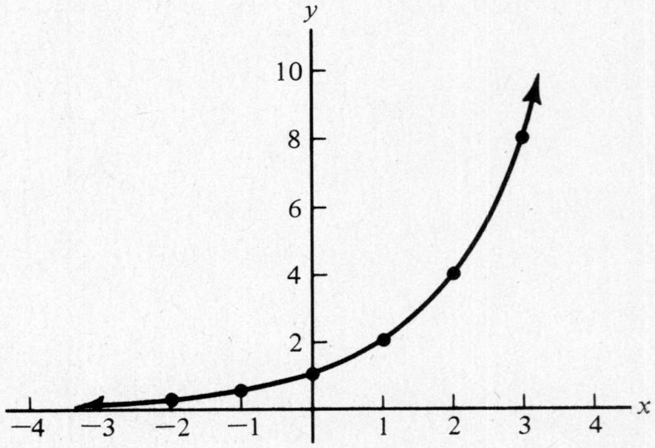

Figure 5.1

Example 2 | Graph the exponential function

$$y = f(x) = (\tfrac{1}{2})^x = 2^{-x}$$

Solution | The procedure used in Example 1 is also followed here; the graphical result is shown in Figure 5.2.

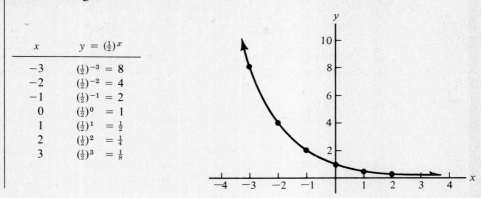

x	$y = (\frac{1}{2})^x$
-3	$(\frac{1}{2})^{-3} = 8$
-2	$(\frac{1}{2})^{-2} = 4$
-1	$(\frac{1}{2})^{-1} = 2$
0	$(\frac{1}{2})^{0} = 1$
1	$(\frac{1}{2})^{1} = \frac{1}{2}$
2	$(\frac{1}{2})^{2} = \frac{1}{4}$
3	$(\frac{1}{2})^{3} = \frac{1}{8}$

Figure 5.2

Figures 5.1 and 5.2 illustrate some of the more important features that exponential functions of the form $y = f(x) = b^x$ possess:

I. The point $(0, 1)$ is common to *all* exponential functions.

II. The domain of the function is the set of all real numbers; the range the set of all positive numbers.

III. If $b > 1$, the curve is rising or increasing for all values of x, and

$$\lim_{x \to -\infty} b^x = 0 \qquad \text{and} \qquad \lim_{x \to \infty} b^x = \infty$$

IV. If $0 < b < 1$, the curve is falling or decreasing for all values of x, and

$$\lim_{x \to -\infty} b^x = \infty \qquad \text{and} \qquad \lim_{x \to \infty} b^x = 0$$

These features are summarized graphically in Figure 5.3

Equation 5.1.4 can be written in another form that reflects the variety of ways financial institutions compound interest. Because most financial institutions compound interest more frequently than once per year, the variable i in

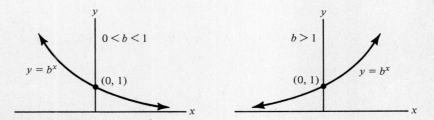

Figure 5.3

Equation 5.1.4 rarely equals the quoted or *nominal* annual rate of interest r. If interest is paid m times per year, then i and r are related as

$$i = \frac{r}{m}$$

For example, if a bank pays a 6 percent ($r = 0.06$) annual rate of interest compounded semiannually, then interest is paid on each account every six months at the rate

$$i = \frac{0.06}{2} = 0.03 \qquad \text{or 3 percent}$$

The exponent n in Equation 5.1.4 represents the total number of interest periods and, in general, does not equal the number of years that the money is left on deposit. If the length of time in years is designated t, the relationship between t and n is given by

$$n = mt$$

For example, if money is left for $t = 5$ years in a bank that compounds interest quarterly ($m = 4$), then the number of interest periods equals

$$n = (4)(5) = 20$$

Thus, we see that the compound interest formula can also be written in the form

$$A = P \left(1 + \frac{r}{m}\right)^{mt} \tag{5.1.5}$$

Example 3 Two thousand dollars is deposited in an account paying 8 percent per annum (the nominal rate). How much money is in the account at the end of three years if interest is compounded (a) annually, (b) semiannually, (c) quarterly?

Solution a. In this case, $i = r = 0.08$, $m = 1$, $n = t = 3$, so A can be written as

$$A = 2000(1.08)^3$$

Using Table A at the end of the book, we get

$$A = 2000(1.260)$$
$$= \$2520$$

b. Because interest is compounded twice each year

$$m = 2 \qquad i = \frac{0.08}{2} = 0.04 \qquad n = mt = (2)(3) = 6$$

and A becomes

$$A = 2000(1.04)^6$$
$$= \$2530$$

c. Interest is compounded four times each year

$$m = 4 \qquad i = \frac{0.08}{4} = 0.02 \qquad n = mt = (4)(3) = 12$$

And A becomes

$$A = 2000(1.02)^{12}$$
$$= \$2536$$

Equation 5.1.4 can also be written in another form in which P is expressed in terms of A, i, and n

$$P = \frac{A}{(1 + i)^n}$$

$$\boxed{P = A(1 + i)^{-n}} \qquad\qquad \textbf{(5.1.6)}$$

When written in this form, P is referred to as the *present value* of the future amount A. This terminology strikingly reflects the time value of money inherent in the concept of interest. Equation 5.1.6 yields the monetary value today of an amount A due n periods from now when the rate of return per period is i. The concept of present value plays an important role in deciding among two or more investment alternatives whose monetary rewards will not be realized until some time in the future.

Example 4 | An eccentric aunt has set up a trust fund the proceeds of which will pay you $1500 three years from now. However, you find yourself in desperate need of cash immediately. A "friend" offers to purchase the rights to the proceeds of the trust fund; if he expects a rate of return of 10 percent annually on his money, how much money should he be willing to pay you today for the rights?

Solution | Equation 5.1.6 is used to solve this problem, where the present value of $1500 due three years from now is to be found at a rate of return of 10 percent. Substituting 1500 for A, 0.10 for i, and 3 for n in Equation 5.1.6, we get

$$P = 1500(1.10)^{-3}$$
$$= \frac{1500}{(1.10)^3}$$

From Table A, we find that

$$(1.10)^3 = 1.331$$

so that P is found to be

$$P = \$1126.97$$

Variations of Equation 5.1.4 can be used to describe the growth of systems other than monetary. Growth of populations such as people, animals, and bacteria can often be described by an equation of the form

$$P = P_0(1 + r)^t \tag{5.1.7}$$

where t is the time, P_0 is the population when $t = 0$, and r is the rate at which the population is growing.

Example 5   Today world population is approximately 4 billion people. Assuming an annual rate of growth of 2 percent, what will be the population of the world 10 years from now?

Solution   Using Equation 5.1.7, we get for P 10 years from now

$$P = 4(1.02)^{10} = (4)(1.2189) = 4.88 \text{ billion}$$

Accountants often use a method called *double-declining balance* to calculate the depreciation of equipment. Annual depreciation is calculated by multiplying the book value of an asset at the beginning of a year by the fraction $2/T$ where T is the useful life of the asset. To see how the book value V of a piece of equipment declines year by year, let us consider the following case.

Darling Beer Company purchases a stainless steel vat that costs $50,000 and has a useful life of 20 years with no salvage value. The annual depreciation and the book value at the end of each year are shown in Table 5.2 where again the arithmetic has purposely not been completed.

$$\frac{2}{T} = \frac{2}{20} = 0.1$$

The value V of the vat at the end of n years is given by the equation

$$V = 50,000(1 - 0.1)^n$$
$$= 50,000(0.9)^n$$

Table 5.2

YEAR	VALUE AT BEGINNING	ANNUAL DEPRECIATION	VALUE AT END
1	$50,000$	$50,000(0.1)$	$50,000(1 - 0.1)$
2	$50,000(1 - 0.1)$	$50,000(1 - 0.1)(0.1)$	$50,000(1 - 0.1)^2$
3	$50,000(1 - 0.1)^2$	$50,000(1 - 0.1)^2(0.1)$	$50,000(1 - 0.1)^3$
.	.	.	.
.	.	.	.
.	.	.	.
n	$50,000(1 - 0.1)^{n-1}$	$50,000(1 - 0.1)^{n-1}(0.1)$	$50,000(1 - 0.1)^n$
.	.	.	.
.	.	.	.
.	.	.	.

The total depreciation D over n years equals

$$D = 50{,}000 - V = 50{,}000 - 50{,}000(0.9)^n$$
$$= 50{,}000[1 - (0.9)^n]$$

These results can be generalized for a piece of equipment whose initial cost is C, whose lifetime is T, and whose salvage value is 0; we get the following equations.

$$V = C \left(1 - \frac{2}{T}\right)^n \tag{5.1.8}$$

$$D = C \left[1 - \left(1 - \frac{2}{T}\right)^n\right] \tag{5.1.9}$$

EXERCISE 5.1

1. A person deposits \$3000 in a savings account compounding interest once a year at 5 percent annually. How much money is in the account at the end of three years?

2. Eight hundred dollars is deposited in a savings account compounding interest once a year at 7 percent annually. How much money is in the account at the end of eight years?

3. On the following coordinate system, make a rough sketch of the functions

$$y = f(x) = 3^x \quad \text{and} \quad y = f(x) = \left(\tfrac{1}{3}\right)^x = 3^{-x}$$

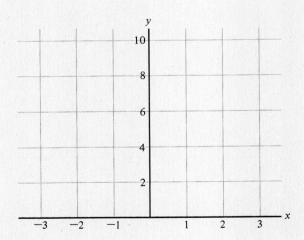

4. What does the graph of the function $y = f(x) = 1^x$ look like?

5. On the following coordinate system, make a rough sketch of the functions

$$y = f(x) = (\tfrac{3}{2})^x$$

$$y = f(x) = (\tfrac{2}{3})^x$$

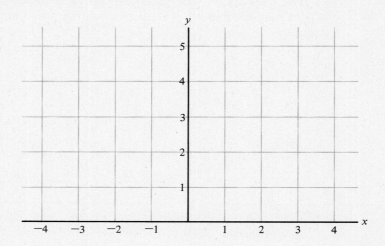

6. On the following coordinate system, graph the function

$$y = f(x) = 2^{x/2}$$

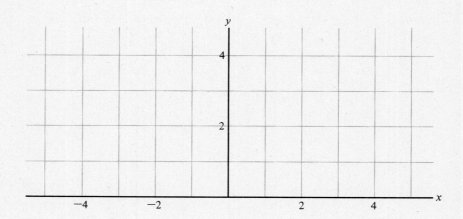

7. Using the properties of exponents, rewrite each of the following functions in the form $y = f(x) = a^x$:

a. $y = f(x) = 4^{x/2}$ **b.** $y = f(x) = 3^{2x}$

c. $y = f(x) = 3^{-x}$ **d.** $y = f(x) = 9^{-x/2}$

e. $y = f(x) = 2^{-3x}$

8. An office copying machine with a useful life of eight years is purchased for $160,000. If depreciation is calculated by the double-declining balance method, find the value of the machine at the end of three years. On the following coordinate system, plot the value of the machine as a function of time and compare with the value of the machine if straight-line depreciation had been used (assumed salvage value is zero).

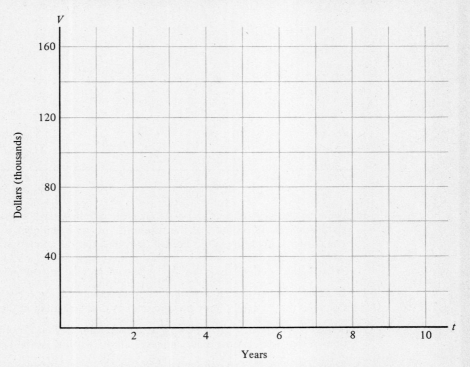

9. On the accompanying coordinate systems, plot graphs of the functions

a. $y = f_1(x) = x^2$ and $y = f_2(x) = 2^x$

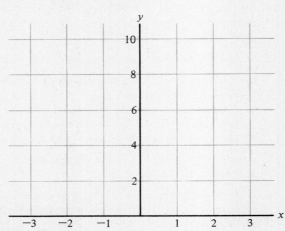

b. $y = f_1(x) = \sqrt{x} = x^{1/2}$ and $y = f_2(x) = (\tfrac{1}{2})^x$

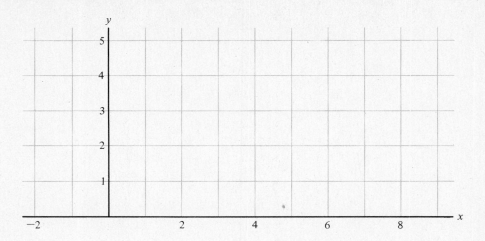

10. A manufacturer of gauges and dial indicators purchases five gear cutters at a cost of $12,000 each. Assuming that the useful life of the equipment is 10 years, find the equation that gives the book value of the five cutters as a function of time if the company uses the double-declining balance method to calculate annual depreciation. What is the book value of the five cutters at the end of two years?

11. If $2000 is deposited in a 90-day notice account paying 6 percent interest compounded semiannually, how much money is in the account at the end of six years?

12. If $2000 is deposited in a 90-day notice account paying 6 percent interest compounded quarterly, how much money is in the account at the end of six years?

13. The number of bacteria in a culture increases at a rate of 10 percent every hour. If the culture now contains 10^6 bacteria, how many bacteria will be present at the end of 5 hours? At the end of 10 hours?

14. The annual tuition at Ivy College is $3000 for the current academic year. If tuition increases at the rate of 6 percent per year, what will be the annual tuition 10 years from now?

15. A man has $10,000 to invest and is considering two possibilities

   a. A four-year certificate of deposit for which the annual rate of interest is 6 percent compounded quarterly.
   b. A mutual fund, specializing in corporate bonds, paying 8 percent interest compounded semiannually; the fund charges an initial, one-time, investment fee equal to 4 percent of the initial investment.

   Assuming interest rates do not change, how much would each investment be worth at the end of four years?

16. A stock analyst predicts that the per-share price of IBM will equal $150 five years from now. At a 10 percent rate of return, what is the present value of each share if the analyst is correct?

17. What is the present value of a $1000 promissory note payable three years from now at an annual rate of interest of 8 percent compounded quarterly?

18. Use Equation 5.1.8 to obtain an expression for the annual depreciation during year n.

5.2 Logarithmic Functions

If a graph of the equation

$$x = 2^y \qquad\qquad (5.2.1)$$

were desired, it could be obtained from the graph of the equation $y = 2^x$ (Figure 5.1) by interchanging the x and y coordinates to produce the results shown in Figure 5.4. To retain the format that has been used until now, namely expressing the variable y in terms of x, we introduce a new function called the *logarithmic* function to express the relationship between the x and y coordinates for each point on the solid curve shown in Figure 5.4; this function has the form

$$y = f(x) = \log_2 x \qquad\qquad (5.2.2)$$

x	y
$\frac{1}{8}$	-3
$\frac{1}{4}$	-2
$\frac{1}{2}$	-1
1	0
2	1
4	2
8	3

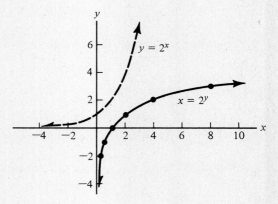

Figure 5.4

Because any set of ordered pairs (x, y) that satisfies Equation 5.2.1 also satisfies Equation 5.2.2, the two equations are equivalent, that is

$$y = \log_2 x \qquad \text{is equivalent to} \quad x = 2^y$$

Generalizing this to any base $b > 0$ gives the definition:

$$y = \log_b x \qquad \text{is equivalent to} \quad x = b^y$$

At this point, it might be beneficial to show some simple equations in both their exponential and logarithmic forms. Gaining proficiency in transforming from the logarithmic to the exponential form can be useful in solving some simple logarithmic equations.

EXPONENTIAL FORM	LOGARITHMIC FORM
$4^3 = 64$	$\log_4 64 = 3$
$5^{-2} = \frac{1}{25}$	$\log_5 \frac{1}{25} = -2$
$(\frac{1}{3})^4 = \frac{1}{81}$	$\log_{1/3} \frac{1}{81} = 4$
$(\frac{1}{7})^{-2} = 49$	$\log_{1/7} 49 = -2$
$9^0 = 1$	$\log_9 1 = 0$
$6^1 = 6$	$\log_6 6 = 1$

Example 1 | Find the solution to the equation

$$\log_6 x = 4$$

Solution | Writing the equation as an equivalent exponential equation gives

$$x = 6^4 = 1296$$

Example 2 | Solve the equation

$$\log_x 125 = 3$$

Solution | Writing the equation in its exponential form

$$x^3 = 125$$

and noting that $125 = 5^3$, we have

$$x^3 = 5^3$$

from which the solution $x = 5$ is obtained.

Example 3 | Find the solution to the equation

$$\log_{1/2} 32 = x$$

Solution | Rewriting the equation as

$$(\tfrac{1}{2})^x = 32$$

the solution can be obtained by noting that

$$32 = 2^5 = \frac{1}{2^{-5}} = (\tfrac{1}{2})^{-5}$$

The equation can now be written as

$$(\tfrac{1}{2})^x = (\tfrac{1}{2})^{-5}$$

yielding the solution $x = -5$.

The graph of the function $y = \log_2 x$ was obtained from the graph of the function $y = 2^x$ by interchanging the x and y coordinates of each point on the curve $y = 2^x$. This technique can be applied to all exponential and logarithmic functions, enabling us to graph the function

$$y = f(x) = \log_b x \qquad\qquad\qquad (5.2.3)$$

using the graph of the function $y = b^x$ (Figure 5.3) as a reference. The results are shown in Figure 5.5. The domain of the function is the set of all positive real numbers, the range the set of all real numbers.

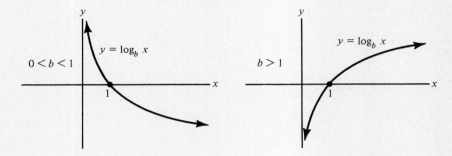

Figure 5.5

PROPERTIES OF LOGARITHMS

Logarithmic functions possess a number of important properties that are very useful in carrying out calculations and in solving simple exponential and logarithmic equations. These properties together with their exponential counterparts are illustrated next.

$$\text{I. } \log_b (M \cdot N) = \log_b M + \log_b N \qquad M, N > 0 \qquad\qquad (5.2.4)$$

The logarithm of a product equals the sum of the logarithms. This property is based on the following law of exponents:

$$b^m b^n = b^{m+n}$$

Example 4

$$\log_4 (16 \cdot 64) = \log_4 16 + \log_4 64$$
$$= \boxed{2 + 3} = 5$$

The exponential version of the same relationship is

$$16 \cdot 64 = 4^2 4^3 = 4^{\boxed{2+3}} = 4^5$$

$$\text{II. } \log_b \left(\frac{M}{N}\right) = \log_b M - \log_b N \qquad M, N > 0 \qquad\qquad (5.2.5)$$

The logarithm of a ratio equals the logarithm of the numerator minus the logarithm of the denominator. This property is based on the following law of exponents:

$$\frac{b^m}{b^n} = b^{m-n}$$

Example 5

$$\log_2 \tfrac{128}{16} = \log_2 128 - \log_2 16$$
$$= \boxed{7 - 4} = 3$$

The exponential version is

$$\tfrac{128}{16} = \frac{2^7}{2^4} = 2^{\boxed{7-4}} = 2^3$$

$$\boxed{\text{III. } \log_b(M^N) = N \cdot \log_b M \qquad M > 0}$$

(5.2.6)

The logarithm of a power expression equals the exponent multiplied by the logarithm of the base. This property is based on the following law of exponents:

$$(b^m)^n = b^{mn}$$

Example 6

$$\log_2 16^3 = 3 \cdot \log_2 16$$
$$= \boxed{3 \cdot 4} = 12$$

The exponential version is

$$16^3 = (2^4)^3 = 2^{\boxed{4 \cdot 3}} = 2^{12}$$

Two other important properties that follow directly from the definition of logarithms are

$$\boxed{\text{IV. } \log_b 1 = 0}$$

(5.2.7)

$$\boxed{\text{V. } \log_b b = 1}$$

(5.2.8)

Example 7   If $\log_{10} 2 = 0.3010$ and $\log_{10} 3 = 0.4771$, use properties I–V to evaluate each of the following: (a) $\log_{10} 6$, (b) $\log_{10} 1.5$, (c) $\log_{10} 81$.

Solution   a. $\log_{10} 6 = \log_{10} 2 \cdot 3 = \log_{10} 2 + \log_{10} 3$
$$= 0.3010 + 0.4771 = 0.7781$$

b. $\log_{10} 1.5 = \log_{10} \frac{3}{2} = \log_{10} 3 - \log_{10} 2$
$= 0.4771 - 0.3010 = 0.1761$

c. $\log_{10} 81 = \log_{10} 3^4 = 4 \log_{10} 3$
$= 4(0.4771) = 1.9084$

These properties can also be used to solve simple exponential and logarithmic equations.

Example 8   If $\log_{10} 2 = 0.3010$ and $\log_{10} 3 = 0.4771$, use properties I–V to solve the equation $2^x = 3$.

Solution   Taking the logarithm to base 10 of both sides gives

$$\log_{10} 2^x = \log_{10} 3$$

Using property III, the left-hand side can be written as

$$\log_{10} 2^x = x \log_{10} 2$$

so that the equation now reads

$$x \log_{10} 2 = \log_{10} 3$$

Solution:

$$x = \frac{\log_{10} 3}{\log_{10} 2} = \frac{0.4771}{0.3010} = 1.585$$

Example 9   Use properties I–III to solve the following equation for x

$$\log_3 x + \log_3 (x - 8) = 2$$

Solution   Using property I, we can write the left-hand side as a single term as follows:

$$\log_3 x + \log_3 (x - 8) = \log_3 x(x - 8) = \log_3 (x^2 - 8x)$$

so the equation reads

$$\log_3 (x^2 - 8x) = 2$$

The exponential version of this equation is

$$x^2 - 8x = 3^2 = 9$$

Writing this as a standard quadratic equation gives

$$x^2 - 8x - 9 = 0$$
$$(x - 9)(x + 1) = 0$$

Solution:

$$x = 9$$

Why is the solution $x = -1$ not acceptable?

EXERCISE 5.2

1. On the following coordinate system, plot each of the logarithmic functions shown.

   a. $y = f(x) \log_3 x$ **b.** $y = f(x) = \log_{1/3} x$

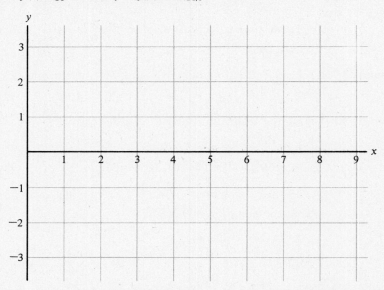

Write the exponential equations in Problems 2–9 as equivalent logarithmic equations.

2. $7^2 = 49$ 3. $(\frac{2}{3})^4 = \frac{16}{81}$ 4. $6^{-2} = \frac{1}{36}$ 5. $(\frac{5}{4})^{-3} = \frac{64}{125}$

6. $5^x = 8$ 7. $x^3 = 20$ 8. $6^{x-1} = 11$ 9. $x^{-4} = \frac{5}{3}$

Solve the equations in Problems 10–25 for x.

10. $\log_2 16 = x$ 11. $\log_{1/2} x = 3$ 12. $\log_x 49 = 2$

13. $\log_x 625 = 4$ 14. $\log_x \frac{16}{25} = 2$ 15. $\log_x \frac{27}{8} = 3$

16. $\log_x \frac{1}{4} = -1$ 17. $\log_4 2 = x$ 18. $\log_4 \frac{1}{2} = x$

19. $\log_{2/3} \frac{4}{9} = x$ 20. $\log_{4/5} x = -3$ 21. $\log_5 \sqrt{5} = x$

22. $\log_8 4 = x$ 23. $\log_2 (x + 1) = 3$ 24. $\log_3 (x^2 + 2) = 3$

If $\log_b 4 = 1.2042$ and $\log_b 3 = 0.9542$, evaluate the quantities in Problems 25–33 using properties I–III of logarithms.

25. $\log_b 12$ 26. $\log_b 0.75$ 27. $\log_b 16$

28. $\log_b 48$ 29. $\log_b 9$ 30. $\log_b 2$

31. $\log_b 36$ 32. $\log_b \frac{16}{9}$ 33. $\log_b \frac{1}{27}$

Using properties I–V of logarithms, solve Problems 34–38 for x.

34. $\log_3 x + \log_3 x = 4$ 35. $\log_2 (x + 1) + \log_2 6 = 1$

36. $\log_6 x = 2 \cdot \log_6 3 + \log_6 2$ **37.** $\log_7 x = 2 \cdot \log_7 5 - 3 \cdot \log_7 2$

38. $\log_5 x^2 = 2 \cdot \log_5 12 - \log_5 4$

5.3 *e* and Natural Logarithms

Our work with exponential and logarithmic functions has been restricted to those whose bases are rational numbers. In differential and integral calculus, the most important exponential and logarithmic functions are those whose base equals *e*, an irrational number whose value to five decimal places is 2.71828. Logarithms in the base *e* are called *natural logarithms*.

To see why the number *e* is important, let us return to the compound interest formula, Equation 5.1.5, and try to see what form it assumes as interest is compounded more and more frequently. In order to concentrate on the essential items in the process to be described, the quantities *P* and *t* in Equation 5.1.5 will be set equal to \$1 and one year respectively, giving

$$A = \left(1 + \frac{r}{m}\right)^m \tag{5.3.1}$$

This equation tells us the amount to which \$1 will grow in one year at the nominal rate *r* compounded *m* times per year. We want to see what happens as *m* increases without bound, that is, as $m \to \infty$. Before we do this, note that there are two competing processes occurring in Equation 5.3.1 as $m \to \infty$

1. The base $\left(1 + \dfrac{r}{m}\right) \to 1$.

2. The exponent *m*, which indicates the number of times the base is multiplied by itself, increases without limit.

The right-hand side of Equation 5.3.1 is not in its optimum form to indicate clearly what happens as $m \to \infty$. Multiplying the exponent by (r/r), Equation 5.3.1 becomes

$$A = \left(1 + \frac{r}{m}\right)^{m \cdot r/r} = \left[\left(1 + \frac{r}{m}\right)^{m/r}\right]^r$$

The quantity we now want to concentrate our attention on is

$$\left(1 + \frac{r}{m}\right)^{m/r} \tag{5.3.2}$$

Because *r* in the expression 5.3.2 remains constant, the expression m/r also increases without limit as $m \to \infty$. In the interest of simplicity, a new variable *w*, defined as

$$w = \frac{m}{r}$$

is introduced, so that the expression 5.3.2 becomes

$$\left(1 + \frac{1}{w}\right)^w$$

Table 5.3

w	$\left(1 + \dfrac{1}{w}\right)^w$
1	2.00000
2	2.25000
3	2.37037
4	2.44141
5	2.48832
10	2.59374
100	2.70481
1000	2.71692
10,000	2.71815
.	.
.	.
.	.

To see what happens as $w \to \infty$, let us examine Table 5.3 where the expression $(1 + 1/w)^w$ is evaluated for selected values of w.

The values in the right-hand column do appear to be approaching some limiting value. This limit is defined as e, that is,

$$e = \lim_{w \to \infty} \left(1 + \frac{1}{w}\right)^w = \lim_{m \to \infty} \left(1 + \frac{r}{m}\right)^{m/r} \tag{5.3.3}$$

Thus we find that the \$1 deposit has grown, with continuous compounding, in one year to become

$$A = e^r$$

For arbitrary values of P and t, the compound interest formula, Equation 5.1.5, assumes the form

$$A = Pe^{rt} \tag{5.3.4}$$

when compounding occurs continuously.

CALCULATOR EXERCISE

Using a calculator complete the accompanying table. The values of w were purposely selected to be powers of 2 so that the calculations can be carried out by either

1. Using the x^2 key repeatedly if the calculator has one, or
2. By multiplying the number in the display by itself repeatedly; for example, when $w = 8$

$$\left(1 + \frac{1}{8}\right)^8 = 1.125^8$$

$$= [(1.125^2)^2]^2$$

$$= (1.26563^2)^2 \qquad \text{and so on}$$

w	$\left(1 + \dfrac{1}{w}\right)^w$
1	
2	
4	
8	
16	
32	
64	
128	

In order to carry out calculations with exponential expressions whose base is e, tables containing values of the quantity e^x have been developed. Table B at the back of the book represents a typical example where values of both e^x and $e^{-x} = 1/e^x$ are shown. Figure 5.6 shows the graph of the two functions $y = f_1(x) = e^x$ and $y = f_2(x) = e^{-x}$.

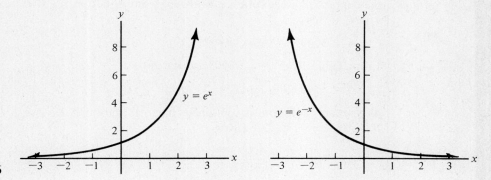

Figure 5.6

To see how Table B is used in making calculations, let us consider a few simple cases.

Example 1 | What is the dollar value at the end of three years of $500 deposited in a savings account for which interest is compounded continuously at a rate of 6 percent per year?

Solution | Noting that $rt = (0.06)(3) = 0.18$, Equation 5.3.4 becomes

$$A = 500e^{0.18}$$

From Table B, we note that

$$e^{0.18} = 1.1972$$

so that A becomes

$$A = \$598.60$$

Example 2 | If the price of suburban land is increasing continuously at a rate of 10 percent per year, how much money should you pay for an acre of land today if the expected value of the land two and a half years from now is $30,000?

Solution | Equation 5.3.4 is used with $A = 30,000$, $i = 0.10$, $t = 2.5$, $P = ?$

$$30,000 = Pe^{0.25}$$

or

$$P = 30,000e^{-0.25} = 30,000(0.7788) = \$23,364$$

Equation 5.3.4 can also be used as a good approximation for those cases where interest is compounded on a daily basis.

Example 3 | Three thousand dollars is deposited in an account paying 7 percent interest per year compounded daily. How much money is in the account at the end of two years?

Solution | The approximate solution is obtained from Equation 5.3.4 with $r = 0.07$, $t = 2$, and $P = 3000$.

$$A = 3000e^{(0.07)(2)} = 3000e^{0.14}$$

From Table B, $e^{0.14} = 1.1503$, so we get

$$A = \$3450.90$$

The exact solution is obtained from Equation 5.1.5 with $P = 3000$, $i = 0.07/365$, $n = 365(2)$, giving

$$A = 3000 \left(1 + \frac{0.07}{365}\right)^{730} = \$3450.78$$

The *present value P* of a future amount A due t years from now when the nominal rate r is compounded continuously can be obtained from Equation 5.3.4.

$$\boxed{P = \frac{A}{e^{rt}} = Ae^{-rt}}$$

(5.3.5)

Example 4 Find the present value of a work of art if its value five years from now is expected to be $40,000. Assume a 10 percent rate of return and that interest is compounded continuously.

Solution The present value can be found by setting A equal to $40,000, t equal to 5, and $r = 0.10$, giving

$$P = 40000e^{-(0.10)(5)} = 40000e^{-0.50} = 40000(0.6065) = \$24,260$$

Our attention has been focused exclusively on the exponential function to the base e. However, equally important is the associated logarithmic function

$$y = f(x) = \log_e x$$

called the *natural logarithmic* function. Because this function appears frequently in a variety of applications, its designation has been shortened to

$$y = f(x) = \ln x$$

that is

$$\boxed{y = f(x) = \ln x = \log_e x} \tag{5.3.6}$$

Table C at the back of the book contains values of the quantity $\ln x$ for selected values of x. Using these values, it is possible to sketch the graph of the natural logarithmic function $y = f(x) = \ln x$ shown in Figure 5.7.

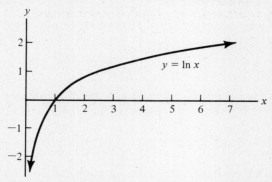

Figure 5.7

Properties I–III of logarithms described in Section 5.2 also apply to natural logarithms.

I. $\ln (u \cdot v) = \ln (u) + \ln (v)$

II. $\ln \left(\dfrac{u}{v}\right) = \ln (u) - \ln (v)$

III. $\ln (u^v) = v \cdot \ln (u)$

These properties can be used to calculate the natural logarithm of numbers not contained in Table C, as shown in the next example.

Example 5 | Find ln (350).

Solution | First the number is written in scientific notation

$$350 = (3.5)(10^2)$$

Next, using property I we can write

$$\ln (3.5 \cdot 10^2) = \ln 3.5 + \ln (10^2)$$

Property III can be used to rewrite the second term giving

$$\ln (350) = \ln 3.5 + 2 \cdot \ln (10)$$

Finally, Table C is used to find ln (3.5) and ln (10), giving

$$\ln 350 = 1.2528 + 2(2.3026) = 5.8580$$

These properties are also useful in solving simple exponential and logarithmic equations, as shown in the next two examples.

Example 6 | Solve the equation $3^x = 11$.

Solution | Taking the natural logarithm of both sides gives

$$\ln (3^x) = \ln (11)$$

Using property III, the equation can be rewritten as

$$x(\ln 3) = \ln (11)$$

Solving for x gives

$$x = \frac{\ln 11}{\ln 3} = \frac{2.3979}{1.0986} = 2.18$$

Example 7 | Solve the equation $\ln (x + 4) - \ln x = 2$.

Solution | The left-hand side of the equation can be rewritten with the aid of property II to give

$$\ln (x + 4) - \ln x = \ln \left(\frac{x + 4}{x}\right)$$

The equation now reads

$$\ln \left(\frac{x + 4}{x}\right) = 2$$

Transforming the equation to its equivalent exponential form gives

$$\frac{x + 4}{x} = e^2$$

This equation can now be solved for x

$$x + 4 = e^2 x$$

$$4 = x(e^2 - 1)$$

$$x = \frac{4}{e^2 - 1}$$

EXERCISE 5.3

Using Table B at the back of the book as a reference, plot the functions in Problems 1–4.

1. $y = f(x) = \dfrac{e^x}{2}$

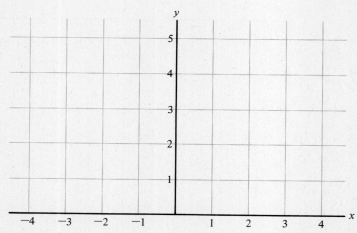

2. $y = f(x) = e^{x-1}$

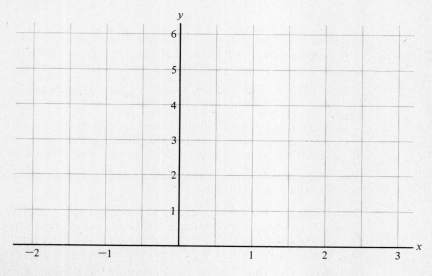

3. $y = f(x) = e^{-x/2}$

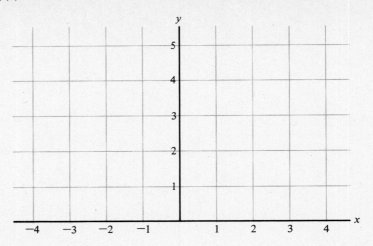

4. $y = f(x) = e^{2-x}$

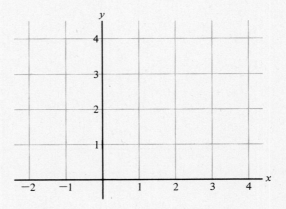

5. The world population, which today is approximately 4 billion people, is growing continuously at an annual rate of 2 percent. Assuming that this rate does not change, how many people will inhabit the earth 10 years from now?

6. If $5000 is deposited in a savings account paying $5\frac{1}{4}$ percent interest compounded daily, how much money will be in the account at the end of four years?

7. Sales for a small cable TV company have been growing continuously at an annual rate of 20 percent. Assuming that this rate of growth remains constant, find the dollar volume of sales five years from now if present annual sales are $50 million.

8. One hundred dollars is deposited in a savings account that pays an annual rate of 5 percent compounded daily.

    a. How much is in the account at the end of one year?
    b. How much money would be in the account if interest were compounded once a year?

Using Table C at the back of the book as a reference, plot the functions in Problems 9–11.

9. $y = f(x) = \dfrac{\ln x}{2}$

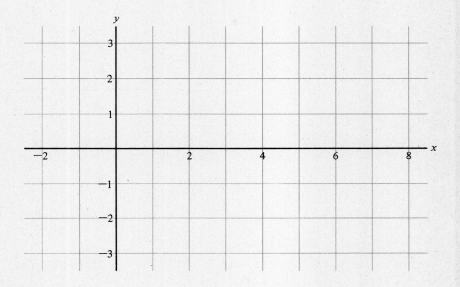

10. $y = f(x) = 3 \cdot \ln (1 + x)$

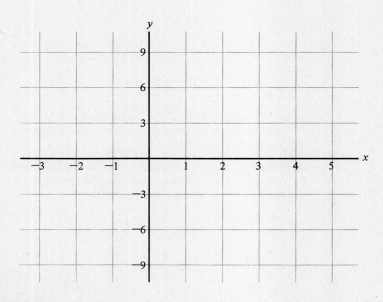

11. $y = f(x) = \ln(1 - x)$

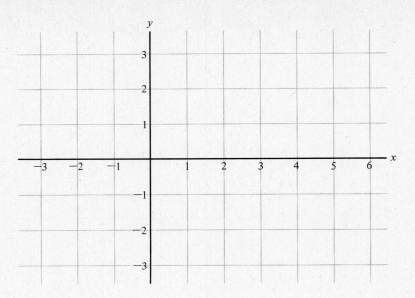

Find each of the following:

12. $\ln 470$ **13.** $\ln 0.89$ **14.** $\ln 1000$ **15.** $\ln 6500$

16. $\ln 0.0046$

Solve each of the following equations for x:

17. $5^x = 12$ **18.** $3^{2-x} = 1$ **19.** $6^{x-3} = 24$ **20.** $2^{x^2} = 9$

21. $4^{x-1} = 2.5$ **22.** $5^x = 210$

23. If depreciation occurs continuously when using the double-declining balance method, the equation giving the value V of an asset as a function of time t assumes the form

$$V(t) = Ce^{-2t/T}$$

where C is the cost and T the useful life of the asset.

   A restaurant purchases a dishwasher for $2000; if the expected life of the dishwasher is 10 years, find

a. The value of the dishwasher at the end of three years.
b. The value four and one-half years after purchase.
c. The value nine months after purchase.

6

Derivatives of Exponential and Logarithmic Functions

CHAPTER 6 ❧ INTRODUCTION

The material presented in Chapter 5 was intended to familiarize you with the properties and characteristics of exponential and logarithmic functions. It is now time to find the derivatives of these types of functions. However, it is important that you keep in mind the interpretation and meaning of the first and second derivative as the slope of the tangent line and the concavity of the curve, respectively. In addition, some applications of these functions will be introduced in Section 6.3 to indicate why the exponential and the logarithmic functions are important.

❧ 6.1 First Derivative of Logarithmic Functions

The first derivative of the natural logarithmic function

$$y = f(x) = \ln x \qquad (6.1.1)$$

has a very simple form

$$\boxed{\frac{dy}{dx} = f'(x) = \frac{1}{x}} \qquad (6.1.2)$$

that is, the slope of the line tangent to the curve $y = \ln x$ at the point $(a, \ln a)$ equals $1/a$, *the reciprocal of the x coordinate* of the point; this result is shown graphically in Figure 6.1. The first derivative of the general logarithmic function

$$y = f(x) = \log_b x \qquad (6.1.3)$$

is slightly more complex

$$\boxed{\frac{dy}{dx} = f'(x) = \frac{\log_b e}{x}} \qquad (6.1.4)$$

These results can be derived from the definition of the first derivative, Equation 3.2.1, as shown below. On first reading, it might be advisable to skim lightly over the derivation, returning to it later for a more thorough analysis.

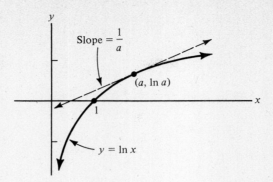

Figure 6.1

Into the definition of the first derivative

$$\frac{dy}{dx} = f'(x) = \lim_{h \to 0} \frac{f(x + h) - f(x)}{h}$$

we substitute Equation 6.1.1, getting

$$\frac{dy}{dx} = f'(x) = \lim_{h \to 0} \frac{\ln (x + h) - \ln x}{h}$$

Rewriting the right-hand side will enable us to find the limit more easily. First, using property II of logarithms permits us to write

$$\lim_{h \to 0} \frac{\ln (x + h) - \ln x}{h} = \lim_{h \to 0} \frac{\ln [(x + h)/x]}{h} = \lim_{h \to 0} \frac{\ln (1 + h/x)}{h}$$

The next step, multiplying the denominator by x/x, can be justified at this stage by assuring you that it is needed to generate the final form of $f'(x)$. The remaining steps are carried out as follows:

$$\lim_{h \to 0} \frac{\ln \left(1 + \dfrac{h}{x}\right)}{x \left(\dfrac{h}{x}\right)} = \lim_{h \to 0} \frac{1}{x} \left\{ \frac{x}{h} \ln \left(1 + \frac{h}{x}\right) \right\} = \lim_{h \to 0} \frac{1}{x} \ln \left(1 + \frac{h}{x}\right)^{x/h}$$

where the last step was carried out using property III of logarithms (Section 5.2). Next we write

$$\lim_{h \to 0} \frac{1}{x} \ln \left(1 + \frac{h}{x}\right)^{x/h} = \frac{1}{x} \lim_{h \to 0} \ln \left(1 + \frac{h}{x}\right)^{x/h}$$

where the last step follows from the fact that the variable x, although arbitrary, remains constant as h approaches 0. To proceed further, a new variable w, defined as

$$w = \frac{x}{h}$$

is introduced. Noting that $w \to \infty$ as h assumes smaller and smaller positive values, we can write

$$\frac{1}{x}\left[\lim_{h\to 0} \ln\left(1 + \frac{h}{x}\right)^{x/h}\right] = \frac{1}{x}\left[\lim_{w\to\infty} \ln\left(1 + \frac{1}{w}\right)^{w}\right]$$

Recalling from Section 5.3 that

$$\lim_{w\to\infty} \ln\left(1 + \frac{1}{w}\right)^{w} = \ln e$$

and that

$$\ln e = 1$$

we obtain the desired result, that is

$$\frac{dy}{dx} = f'(x) = \frac{1}{x}(\ln e) = \frac{1}{x}$$

The same procedure is followed for the logarithmic function, Equation 6.1.3. Because the $\log_b e$ factor is not equal to 1, unless $b = e$, it appears explicitly in the final result. Equation 6.1.4 can be written in a more convenient form as follows:

Let

$$M = \log_b e$$

Next, write this equation as an equivalent exponential equation

$$b^M = e$$

Finally, taking the natural logarithm of both sides gives

$$\ln (b^M) = \ln e, \text{ or } M \ln b = 1$$

from which we get

$$M = \log_b e = \frac{1}{\ln b}$$

Using this result, we can express the first derivative of the function $y = f(x) = \log_b x$ as follows:

If $y = f(x) = \log_b x$, then

$$\frac{dy}{dx} = f'(x) = \frac{1}{x \ln b} \tag{6.1.5}$$

Example 1 | Find the slope and the equation of the line tangent to the curve

$$y = f(x) = \ln x$$

at the point whose x coordinate equals 2.

Solution | The slope of the line is given by $f'(x)$, which according to Equation 6.1.2 equals $\frac{1}{2}$ at the point $(2, \ln 2)$. The equation of the tangent line can be found from the point-slope formula

$$y - y_1 = m(x - x_1)$$

where $m = \frac{1}{2}$, $x_1 = 2$, and $y_1 = \ln 2$, so we get

$$y - \ln 2 = \tfrac{1}{2}(x - 2)$$

or

$$2y = x + 2 \ln 2 - 2$$

Table C at the end of the book can be used to find $\ln 2$, which equals 0.6931, so that the equation of the tangent line becomes

$$2y = x - 0.6138$$

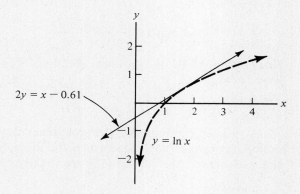

It should be noted that all the techniques, such as the sum and difference, product, quotient, and power rules that have been developed earlier, are applicable in differentiating functions that contain logarithmic terms and/or factors.

Example 2 | Find the first and second derivatives of the function

$$y = f(x) = x^2 \ln x$$

Solution | Using the product rule (Formula 3.4.1) with

$$u(x) = x^2 \qquad v(x) = \ln x$$

$$u'(x) = 2x \qquad v'(x) = \frac{1}{x}$$

we can combine the expressions in the usual way to give

$$\frac{dy}{dx} = f'(x) = x^2 \left(\frac{1}{x}\right) + 2x \ln x$$

$$= x + 2x \ln x$$

The second derivative is found by using the sum rule together with the product rule. The result is

$$\frac{d^2y}{dx^2} = f''(x) = 1 + 2x\left(\frac{1}{x}\right) + 2\ln x = 2\ln x + 3$$

Example 3 Find the first derivative of the function

$$y = f(x) = (\ln x)^3$$

Solution Here we have a situation where the general power rule (Formula 3.5.2) is called for, with

$$u(x) = \ln x \qquad u'(x) = \frac{1}{x} \qquad n = 3$$

Thus we get

$$\frac{dy}{dx} = f'(x) = 3(\ln x)^2\left(\frac{1}{x}\right) = \frac{3(\ln x)^2}{x}$$

NOTE: Property III of logarithms (Section 5.2) cannot be used to find the derivative because

$$(\ln x)^3 \neq 3\ln x$$

Also keep in mind that the techniques developed in Chapter 4 on curve sketching are also applicable when the function contains logarithmic expressions. However, the calculations are slightly more involved and the equation

$$f'(x) = 0$$

may assume a more formidable appearance.

Example 4 Find the critical points of the function

$$y = f(x) = x\ln x$$

In addition, make a rough sketch of the function.

Solution Before finding the critical points, it is worthwhile to note that the domain of the function is limited to positive values of x. Awareness of this feature will make you less prone to accept spurious solutions of the equation

$$f'(x) = 0$$

The critical points of a function are found by setting $f'(x)$ equal to 0 and then solving the resulting equation for x. In this case, the product rule must be used to find $f'(x)$, yielding

$$f'(x) = \ln x + 1$$

$$0 = \ln x + 1$$

or

$$-1 = \ln x$$

The value of x that satisfies this equation can be found by using Table C, or preferably, by writing the equation in its equivalent exponential form to yield the solution

$$x = e^{-1} = \frac{1}{e}$$

$$y = \left(\frac{1}{e}\right) \ln \left(\frac{1}{e}\right) = \frac{-1}{e}$$

To determine how the function behaves in the neighborhood of $(1/e, -1/e)$, the second derivative test is applied. The expression for $f''(x)$ is

$$f''(x) = \frac{1}{x}$$

$$f'' \left(\frac{1}{e}\right) = e > 0 \qquad \text{Concave up}$$

So we can conclude that $(1/e, -1/e)$ is a relative minimum. A graph of the function in the vicinity of $(1/e, -1/e)$ is shown in Figure 6.2.

Figure 6.2

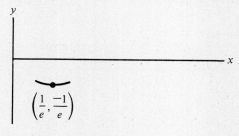

In order to complete the sketch, it is necessary to

1. Locate all inflection points, if any exist.
2. Determine how the function behaves as $x \to \infty$ and as $x \to 0$.

First, if any inflection points exist, they can be found by solving the equation

$$f''(x) = 0$$

For this case, we get

$$\frac{1}{x} = 0$$

Multiplying both sides by x yields the result

$$1 = 0$$

which indicates that no solutions exist; so we can conclude that there are no inflection points.

Next, we want to see what happens to the function as $x \to \infty$. Since

$$\lim_{x \to \infty} x = \infty \qquad \text{and} \qquad \lim_{x \to \infty} \ln x = \infty$$

the limit of the product, that is, $x \cdot \ln x$, also increases without limit as x increases without limit

$$\lim_{x \to \infty} (x \cdot \ln x) = \infty$$

The more interesting case is discovering what happens to the function as $x \to 0$. The result is not clear because

$$\lim_{x \to 0} x = 0 \qquad \lim_{x \to 0} \ln x = -\infty$$

A clue as to the behavior of the function as $x \to 0$ is provided by Figure 6.2 where the curve to the left of $(1/e, -1/e)$ appears to be directed toward the origin. To see what happens as $x \to 0$, let us look at Table 6.1. From this table, it appears that $x \to 0$ more rapidly than $\ln x \to -\infty$, so that

$$\lim_{x \to 0} (x \cdot \ln x) = 0$$

On the basis of this analysis, we can sketch the remainder of the curve as shown in Figure 6.3.

Table 6.1

x	$y = x(\ln x)$
0.30	$(0.30)(-1.20) = -0.36$
0.20	$(0.20)(-1.61) = -0.32$
0.10	$(0.10)(-2.30) = -0.23$
0.01	$(0.01)(-4.61) = -0.05$
.	.
.	.
.	.

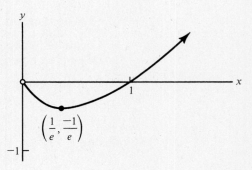

Figure 6.3

There are many occasions when functions of the type

$$y = f(x) = \ln u(x)$$

(6.1.6)

are encountered, for example

$$y = f(x) = \ln (2x^3 + 5) \qquad \text{where } u(x) = 2x^3 + 5$$

$$y = f(x) = \ln (1 - x^2 - 3x^4) \qquad \text{where } u(x) = 1 - x^2 - 3x^4$$

$$y = f(x) = \ln (7 + \sqrt{x}) \qquad \text{where } u(x) = 7 + \sqrt{x}$$

The first derivative of functions having this form is

$$\frac{dy}{dx} = f'(x) = \frac{1}{u(x)} \cdot u'(x) = \frac{u'(x)}{u(x)} \qquad \textbf{(6.1.7)}$$

This result can be obtained by applying the chain rule, Equation 3.5.3, to Equation 6.1.6. Noting that $dy/du = 1/u$, we get

$$\frac{dy}{dx} = \frac{dy}{du}\frac{du}{dx} = \frac{1}{u(x)} \cdot u'(x)$$

Example 5 | Find the first derivative of the function

$$y = f(x) = \ln (8x^2 + 5x - 3)$$

Solution | Here

$$u(x) = 8x^2 + 5x - 3$$
$$u'(x) = 16x + 5$$

so

$$\frac{dy}{dx} = \frac{16x + 5}{8x^2 + 5x - 3}$$

Example 6 | Find the equation of the line tangent to the curve whose equation is

$$y = f(x) = \ln (2e - x)$$

at the point $(e, 1)$.

Solution | The slope of the tangent line is found by evaluating $f'(x)$ at $x = e$. Using Equation 6.1.7 with

$$u(x) = 2e - x$$

$$u'(x) = -1$$

REMINDER: The quantity e represents a number and is therefore treated as a constant when finding the first derivative.

The first derivative $f'(x)$ then becomes

$$\frac{dy}{dx} = f'(x) = \frac{-1}{2e - x}$$

while, at $x = e$, the slope $f'(e)$ of the tangent line equals

$$f'(e) = \frac{-1}{e}$$

Next the point-slope formula $y - y_1 = m(x - x_1)$ with

$$m = \frac{-1}{e} \qquad x_1 = e \qquad y_1 = 1$$

is used to find the equation of the tangent line. The result is

$$y - 1 = \frac{-1}{e}(x - e)$$

or simplifying gives

$$ey + x = 2e$$

Often the first derivative of a logarithmic function can be found by: (a) using Equation 6.1.7 directly, or (b) first rewriting the expression using properties I–V of logarithms (Section 5.2) and then differentiating.

Example 7 | Find the first derivative of the function

$$y = f(x) = \ln\left(\frac{x}{x - 1}\right) \qquad x > 1$$

Solution | I. Using Equation 6.1.7 with

$$u(x) = \frac{x}{x - 1} \qquad u'(x) = \frac{-1}{(x - 1)^2}$$

we can write

$$\frac{dy}{dx} = f'(x) = \frac{1}{x/(x - 1)} \cdot \frac{-1}{(x - 1)^2} = \frac{-1}{x(x - 1)}$$

II. First, rewrite the function using property II

$$y = f(x) = \ln x - \ln(x - 1)$$

Next, differentiate term by term obtaining

$$\frac{dy}{dx} = f'(x) = \frac{1}{x} - \frac{1}{x - 1} = \frac{-1}{x(x - 1)}$$

which agrees with the result obtained in part I.

COMMENT: The restriction $x > 1$ is needed for part II because $\ln(x - 1)$ is not defined when $x \leq 1$. If this restriction were removed, the domain of the function would be the set $\{-\infty < x < 0\} \cup \{x > 1\}$ shown on the next figure, and the method used in solution II for finding $f'(x)$ would not be correct.

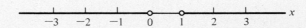

EXERCISE 6.1

Find the first derivative of the functions in Problems 1–17.

1. $y = f(x) = x^2 \ln x$ **2.** $y = f(x) = \dfrac{\ln x}{x}$

3. $y = f(x) = 3x^2 + \ln x$ **4.** $y = f(x) = \dfrac{x}{\ln x}$

5. $y = f(x) = \ln x + \dfrac{1}{\ln x}$ **6.** $y = f(x) = \ln (x^4)$

7. $y = f(x) = (\ln x)^4$ **8.** $y = f(x) = x \ln (x^2 + 3)$

9. $y = f(x) = \dfrac{\ln (1 - x)}{1 - x}$ **10.** $y = f(x) = \sqrt{x} \ln (x + 1)$

11. $y = f(x) = \ln (-x)$ **12.** $y = f(x) = \log_{10} x$

13. $y = f(x) = \log_5 (2x + 3)$ **14.** $y = f(x) = \log_7 (x^2 + 1)$

15. $y = f(x) = \ln \left(\dfrac{x + 1}{x + 2}\right)$ $x > -1$ **16.** $y = f(x) = \ln (x^2\sqrt{x - 2})$ $x > 2$

17. $y = f(x) = \ln \left(\dfrac{x^2}{x^2 + 3}\right)$

In Problems 18–23, find the slope and the equation of the line tangent to each of the curves at the indicated points.

18. $y = \ln x$ at $(e, 1)$ **19.** $y = \ln (x^2 - 3)$ at $(2, 0)$

20. $y = x \ln x$ at $(1, 0)$ **21.** $y = \dfrac{\ln x}{x}$ at $\left(e, \dfrac{1}{e}\right)$

22. $y = \ln (2x + 3)$ at $(1, \ln 5)$ **23.** $y = x^2 \ln (x + 2)$ at $(-1, 0)$

In Problems 24–27, find those points on each curve where the slope of the line tangent to the curve has the indicated value.

24. $y = f(x) = \ln x$ $m_T = \frac{1}{2}$

25. $y = f(x) = 2 \ln (3x + 1)$ $m_T = \frac{3}{2}$

26. $y = f(x) = x \ln x$ $m_T = 1$

27. $y = f(x) = \dfrac{\ln x}{x}$ $m_T = 0$

Make a rough sketch of the functions in Problems 28–31 on the accompanying coordinate systems.

28. $y = f(x) = x^2 - 2 \ln x$

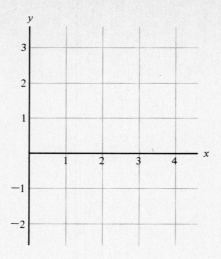

29. $y = f(x) = x^2 \ln x$

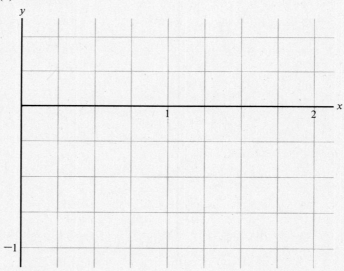

30. $y = f(x) = (\ln x)^2$

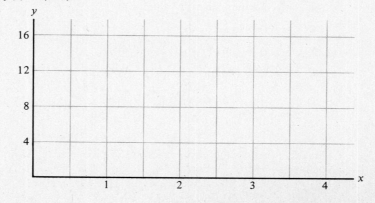

31. $y = f(x) = \ln x - x$

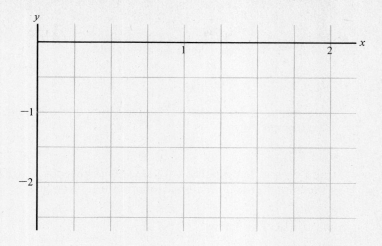

6.2 First Derivative of Exponential Functions

The first derivative of the natural exponential function

$$y = f(x) = e^x \tag{6.2.1}$$

has a very simple form

$$\frac{dy}{dx} = f'(x) = e^x \tag{6.2.2}$$

that is, the slope of the line tangent to the curve $y = e^x$ at the point (a, e^a) equals *the y coordinate, e^a,* of the point; this result is shown graphically in Figure 6.4.

The first derivative of an arbitrary exponential function

$$y = f(x) = b^x \tag{6.2.3}$$

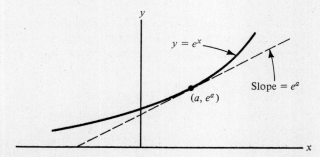

Figure 6.4

has the form

$$\frac{dy}{dx} = f'(x) = b^x(\ln b)$$ **(6.2.4)**

Equation 6.2.2 can be obtained by writing Equation 6.2.1 in its equivalent logarithmic form

$$\ln y = x$$

and then differentiating both sides with respect to x. Differentiating implicitly (Section 3.8) and recalling Equation 6.1.7, we get

$$\frac{1}{y} \left(\frac{dy}{dx} \right) = 1$$

solving for dy/dx gives

$$\frac{dy}{dx} = y = e^x$$

This same technique can also be applied to obtain the result shown in Equation 6.2.4.

CALCULATOR EXERCISE

The first derivative of the function $y = f(x) = e^x$ can also be found by using the definition of the first derivative

$$f'(x) = \lim_{h \to 0} \left[\frac{f(x + h) - f(x)}{h} \right]$$

For the function $y = e^x$, we get

$$f'(x) = \lim_{h \to 0} \left(\frac{e^{x+h} - e^x}{h} \right) = \lim_{h \to 0} \left(\frac{e^x e^h - e^x}{h} \right)$$

$$= \lim_{h \to 0} \left[\frac{e^x(e^h - 1)}{h} \right] = e^x \lim_{h \to 0} \left(\frac{e^h - 1}{h} \right)$$

The limit can be determined by employing a calculator with a $\sqrt{x}$ key in completing the table on the following page. $e = \mathbf{2.71828}$. NOTE: The values of h were selected so that repeated use of the $\sqrt{x}$ key would permit you to calculate e^h, for example

$$e^{1/4} = (e^{1/2})^{1/2} = \sqrt{\sqrt{e}}$$

The entries in the table should indicate that $(e^h - 1)/h \to 0$ as $h \to 0$; so again we get $f'(x) = e^x$.

h	$\dfrac{e^h - 1}{h}$	h	$\dfrac{e^h - 1}{h}$
-1		1	
$-\frac{1}{2}$		$\frac{1}{2}$	
$-\frac{1}{4}$		$\frac{1}{4}$	
$-\frac{1}{8}$		$\frac{1}{8}$	
$-\frac{1}{16}$		$\frac{1}{16}$	
$-\frac{1}{32}$		$\frac{1}{32}$	

Example 1   Find the slope and the equation of the line tangent to the curve

$$y = f(x) = e^x$$

at the point whose x coordinate is 1.

Solution   The slope of the tangent line at any point is given by $f'(x)$, which according to Equation 6.2.2 equals e at the point $(1, e)$. Using the point-slope formula $y - y_1 = m(x - x_1)$ with

$$m = e \qquad x_1 = 1 \qquad y_1 = e$$

we get

$$y - e = e(x - 1)$$

or

$$y = ex$$

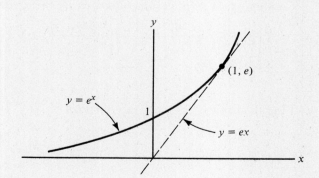

As in the case of the logarithmic function, one or more of the differentiation techniques learned previously may be needed to find the first derivative of a function containing exponential expressions.

Example 2   Find the first and second derivatives of the function

$$y = f(x) = x^2 e^x$$

Solution | Using the product rule with

$$u(x) = x^2 \qquad v(x) = e^x$$

$$u'(x) = 2x \qquad v'(x) = e^x$$

Combining in the customary way gives

$$\frac{dy}{dx} = f'(x) = x^2 e^x + 2xe^x$$

The second derivative is found by first using the sum rule and then the product rule in differentiating each term, producing in this case the result

$$\frac{d^2y}{dx^2} = f''(x) = (x^2 + 4x + 2)e^x$$

Example 3 | Find the first derivative of the function

$$y = f(x) = \frac{e^x - 1}{e^x + 1}$$

Solution | The quotient rule is used with

$$u(x) = e^x - 1 \qquad v(x) = e^x + 1$$

$$u'(x) = e^x \qquad v'(x) = e^x$$

Combining the expressions indicated by the arrows, we get

$$\frac{dy}{dx} = f'(x) = \frac{e^x(e^x + 1) - e^x(e^x - 1)}{(e^x + 1)^2}$$

$$= \frac{2e^x}{(e^x + 1)^2}$$

Example 4 | Find the critical points of the function

$$y = f(x) = xe^x$$

and make a rough graph of the function.

Solution | Because there are no values of x for which the factors e^x and x are not continuous, the function xe^x is continuous for all values of x. To locate the critical points, we first find $f'(x)$ using the product rule. The result is

$$\frac{dy}{dx} = f'(x) = xe^x + e^x = (1 + x)e^x = (x + 1)e^x$$

Next setting $f'(x)$ equal to 0 gives

$$0 = (1 + x)e^x$$

Because $e^x > 0$ for all values of x, the only solution is

$$x = -1 \qquad y = -(e^{-1}) = \frac{-1}{e}$$

The second derivative test is applied to determine the behavior of the function in the vicinity of $(-1, -1/e)$; $f''(x)$ is found to be

$$f''(x) = e^x(x + 2)$$

$$f''(-1) = e^{-1} = \frac{1}{e} > 0 \qquad \text{Concave up}$$

Thus, the point $(-1, -1/e)$ is a relative minimum. To complete the graph, we want to

1. Locate all inflection points, if any exist.
2. Determine the behavior of the function as $x \to \pm \infty$.

First, the inflection points are found by setting $f''(x)$ equal to 0 and solving the equation

$$0 = e^x(x + 2)$$

Solution:

$$x = -2 \qquad y = \frac{-2}{e^2}$$

To determine whether $(-2, -2/e^2)$ is an inflection point, two values of x, x_L and x_R, located to the left and right respectively of $x = -2$, are selected. Then $f''(x_L)$ and $f''(x_R)$ are evaluated to determine whether the concavity on one side differs from that on the other:

$$x_L = -3 \qquad f''(-3) = (-1)e^{-3} = \frac{-1}{e^3} < 0 \qquad \text{Concave down}$$

$$x_R = -1 \qquad f''(-1) = e^{-1} = \frac{1}{e} > 0 \qquad \text{Concave up}$$

The test indicates that $(-2, -2/e^2)$ is an inflection point. The results obtained to this stage are shown graphically in Figure 6.5.

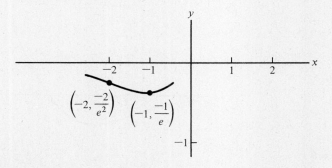

Figure 6.5

Next, we want to see what happens to the function as $x \to \pm \infty$. Since

$$\lim_{x \to \infty} x = \infty \qquad \text{and} \qquad \lim_{x \to \infty} e^x = \infty$$

the limit of the product xe^x also increases without limit as $x \to \infty$, that is,

$$\lim_{x \to \infty} xe^x = \infty$$

Again, an interesting case arises as $x \to -\infty$ because

$$\lim_{x \to -\infty} x = -\infty \qquad \text{while} \qquad \lim_{x \to -\infty} e^x = 0$$

Figure 6.5 indicates that the function appears to be approaching a finite value as $x \to -\infty$. To see more precisely what is occurring, we can use a table such as Table 6.2. From this table, we can conclude that

$$\lim_{x \to -\infty} (xe^x) = 0$$

Table 6.2

x	$y = xe^x$
-5	$-5e^{-5} = -\dfrac{5}{e^5} = -0.03369$
-10	$-10e^{-10} = -\dfrac{10}{e^{10}} = -0.00045$
-100	$-100e^{-100} = -\dfrac{100}{e^{100}} = 0.00000^*$
.	.
.	.
.	.

* Calculated to five decimal places.

The remainder of the curve can be sketched easily once we note that the curve passes through the origin (see Figure 6.6).

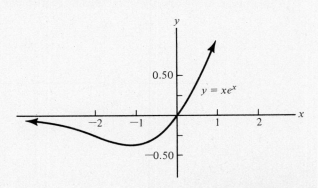

Figure 6.6

Very often, functions of the type

$$y = f(x) = e^{u(x)} \qquad \text{(6.2.5)}$$

must be differentiated, for example

$$y = f(x) = e^{(x^2-3x+4)} \quad \text{where} \quad u(x) = x^2 - 3x + 4$$

$$y = f(x) = e^{(1-\ln x)} \quad \text{where} \quad u(x) = 1 - \ln x$$

$$y = f(x) = e^{x/(x+1)} \quad \text{where} \quad u(x) = \frac{x}{x+1}$$

The first derivative of such functions is given by the formula

$$\boxed{\frac{dy}{dx} = f'(x) = u'(x) \cdot e^{u(x)}} \qquad \text{(6.2.6)}$$

This result can be obtained by applying the chain rule, Equation 3.5.3, to Equation 6.2.5. Noting that $dy/du = e^u$, we get

$$\frac{dy}{dx} = \frac{dy}{du}\frac{du}{dx} = e^{u(x)}u'(x)$$

Example 5 Find the first derivative of the function

$$y = f(x) = e^{x^2-3x+4}$$

Solution Formula (6.2.6) can be used with

$$u(x) = x^2 - 3x + 4$$

$$u'(x) = 2x - 3$$

so that

$$\frac{dy}{dx} = f'(x) = (2x - 3)e^{x^2-3x+4}$$

Example 6 Find those points on the curve

$$y = f(x) = e^{2x}$$

where the slope of the line tangent to the curve equals 4.

Solution The first derivative $f'(x)$ is obtained from Equation 6.2.6 by noting that

$$u(x) = 2x$$

$$u'(x) = 2$$

then

$$f'(x) = 2e^{2x}$$

Setting $f'(x)$ equal to 4 gives

$$4 = 2e^{2x}$$

$$2 = e^{2x}$$

Converting to the logarithmic version of this equation gives

$$\ln 2 = \ln (e^{2x}) = 2x(\ln e)$$

Solving for x gives

$$x = \frac{\ln 2}{2} = 0.35 \qquad y = e^{\ln 2} = 2$$

EXERCISE 6.2

Find the first derivative of the functions in Problems 1–14.*

1. $y = f(x) = x^2 e^x$

2. $y = f(x) = e^{-2x}$

3. $y = f(x) = e^x + e^{-x}$

4. $y = f(x) = \dfrac{e^x}{x + 1}$

5. $y = f(x) = 3e^{x^2+3x}$

6. $y = f(x) = e^{x+\ln x}$

7. $y = f(x) = \sqrt{x + e^x}$

8. $y = f(x) = (x + 1)(e^{-x} + 2)$

9. $y = f(x) = 10^x$

10. $y = f(x) = 3^{4x-5}$

11. $y = f(x) = \ln (e^x + 1)$

12. $y = f(x) = \dfrac{e^x}{e^x + 1}$

13. $y = f(x) = 6^{2-x}$

14. $y = f(x) = e^{-x} + \ln (2x + 9)$

In Problems 15–19, find the slope and the equation of the line tangent to each of the curves at the indicated points.

15. $y = f(x) = e^x$ at $(1, e)$

16. $y = f(x) = 3e^{x^2+2x+1}$ at $(-1, 3)$

17. $y = f(x) = x^2 e^{-x}$ at $(-1, e)$

18. $y = f(x) = \dfrac{e^x}{x}$ at $\left(-1, \dfrac{-1}{e}\right)$

19. $y = f(x) = x^2 + e^{-x}$ at $(0, 1)$

* In differentiating exponential functions for the first time many students, through sheer habit, incorrectly use the simple power rule (Formula 3.3.4) to find the first derivative; we want to emphasize that $d/dx\ e^x \neq x\ e^{x-1}$.

In Problems 20–23, find those points on each curve where the slope of the line tangent to the curve has the indicated value.

20. $y = f(x) = e^x$ $m_T = 1$ **21.** $y = f(x) = e^{-x}$ $m_T = -2$

22. $y = f(x) = x^2 e^x$ $m_T = 0$ **23.** $y = f(x) = e^{-x} + 3x$ $m_T = 2$

In Problems 24–27, make a rough sketch of each function on the accompanying coordinate systems.

24. $y = f(x) = xe^{-x}$

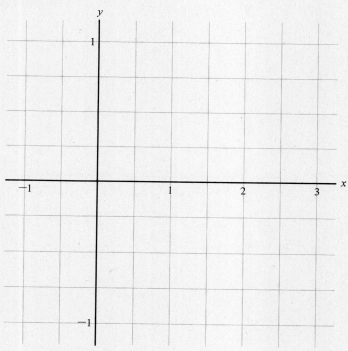

25. $y = f(x) = x^2 e^x$

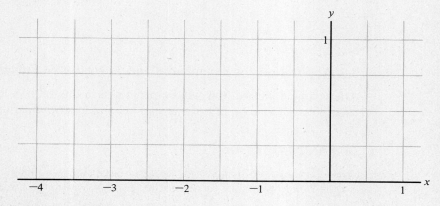

26. $y = f(x) = \dfrac{e^x}{x}$

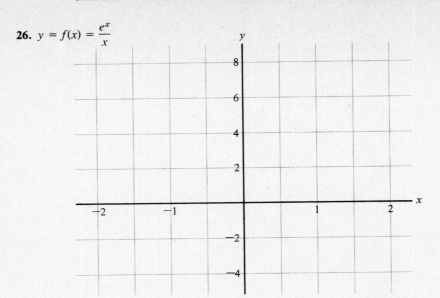

27. $y = f(x) = 2x - e^x$

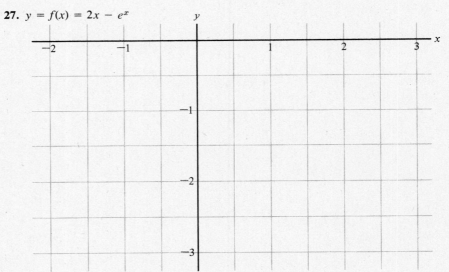

6.3 Applications of Exponential and Logarithmic Functions

A few simple illustrations will be presented in this section to demonstrate the usefulness and versatility of exponential and logarithmic functions.

DEMAND EQUATIONS

When the concept of the linear demand equation was introduced (Section 1.4), it was found necessary to restrict its domain to a narrow interval because the linear demand equation possesses an unattractive and unrealistic feature, namely, the number of items demanded does not increase dramatically as the price drops to very low levels. When the demand equation is given a simple exponential form, this undesirable property is removed.

Example 1 | Further analysis by the MacDougall Hamburger Company (Section 1.4) indicates that the relationship between the unit price p and the weekly sales q for its new Gluttonburger is given by the equation

$$p(q) = 2.5e^{-0.02q} \tag{6.3.1}$$

which is shown graphically in Figure 6.7. This model indicates that demand for the product will increase dramatically once the price drops to the \$1 level.

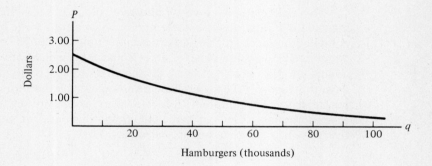

Figure 6.7

It is also possible to construct the revenue function $R(q)$ and to then determine the level of demand q at which revenue is maximized

$$R(q) = q \cdot p = 2.5qe^{-0.02q}$$

Using the techniques developed in Chapter 4 for locating maxima, we first find $R'(q)$, getting

$$R'(q) = 2.5e^{-0.02q} - 0.05qe^{-0.02q}$$

Next setting $R'(q)$ equal to 0 gives

$$0 = (2.5 - 0.05q)e^{-0.02q}$$

Because $e^{-0.02q} > 0$ for all values of q, the only solution is obtained by solving

$$0 = 2.5 - 0.05q$$

solution:

$$q = 50 \text{ thousand burgers}$$

$$R(50) = (2.5)(50)e^{-(0.02)(50)}$$

$$= 45.98 \text{ thousand dollars}$$

Proving that the profit is maximized when $q = 50$ is left as an exercise.

RELATIVE RATE OF CHANGE

When the first derivative of the quantity

$$\ln [u(x)]$$

was found in Section 6.2, the result was

$$\frac{d}{dx} [\ln u(x)] = \frac{u'(x)}{u(x)}$$

The quantity $[u'(x)/u(x)]$ is called the *relative rate of change* of $u(x)$ with respect to x. The *percentage rate of change* is defined as

$$\frac{u'(x)}{u(x)} \cdot 100$$

This concept plays a very important role in decision making as the following situation illustrates. A stock market analyst for a large brokerage house has compiled a list of stocks he is recommending as "buys" for the near future. Among the stocks are those of firms A and B for which the expected increase in annual earnings per share is \$1.00 and \$0.25, respectively. It might be inferred from this that A is the faster growing company and therefore its stock has better potential for price appreciation. However, suppose that further investigation reveals the data shown in Table 6.3. Using this data, we get the following relative rates of change in annual earnings

Company A　　$\dfrac{1.00}{5.00} = 0.20$　　or 20%

Company B　　$\dfrac{0.25}{0.75} = 0.33$　　or 33%

From this perspective, the earnings of company B are growing at a faster annual rate and its stock would appear to be the more attractive candidate for price appreciation.

Table 6.3

COMPANY	CURRENT ANNUAL EARNINGS (Dollars per share)	ESTIMATED INCREASE IN ANNUAL EARNINGS
A	5.00	1.00
B	0.75	0.25

Example 2 It is estimated that the number of employees of the Malady Medical Company, a manufacturer of disposable syringes, thermometers, and so on, will grow linearly as described by the following equation

$$N(t) = 300 + 50t$$

where t represents the time in years. What is the relative rate of growth at (a) the end of one year, (b) the end of five years?

Solution | The annual rate of change in the number of employees is uniform, that is

$$N'(t) = 50 \text{ employees/year}$$

However, the relative rate of change, $N'(t)/N(t)$, is far from constant

$$\frac{N'(t)}{N(t)} = \frac{50}{300 + 50t}$$

a. At the end of one year

$$\frac{N'(1)}{N(1)} = \frac{50}{350} = 0.14 \qquad \text{or } 14\%$$

b. At the end of five years

$$\frac{N'(5)}{N(5)} = \frac{50}{550} = 0.09, \qquad \text{or } 9\%$$

Thus, although the annual increase is uniform, the relative rate of change is decreasing with time as shown in Figure 6.8.

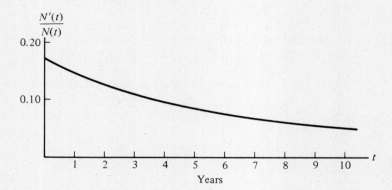

Figure 6.8

ADVERTISING

When a firm fails to promote its products, sales generally decline due in part to the advertising efforts of competing companies. Often the sales decline can be described by an exponential function of the form

$$S(t) = S_0 e^{-\lambda t}$$

where $S(t)$ = the level of sales at any time t
S_0 = the level of sales when $t = 0$
λ = a positive constant, called the "sales decay constant"

The value of λ will vary according to the nature of the product. For those products in a very competitive market, λ will be large while λ will be small for products in a weak competitive environment. The relative rate of decline has a very simple form

$$\frac{S'(t)}{S(t)} = \frac{-\lambda S_0 e^{-\lambda t}}{S_0 e^{-\lambda t}} = -\lambda$$

so that λ describes the constant relative rate of decline of sales under this model.

Example 3 | The manufacturer of Sparkle, an all-purpose detergent, decides to cancel all promotional activities. A local consultant predicts that annual sales will decline steadily according to the equation

$$S(t) = 25e^{-0.10t}$$

where S is given in millions of dollars.

    a. What will be the sales level at the end of five years?
    b. What is the rate at which sales are declining at the end of the first year?
    c. What is the relative rate of change in sales?

Solution | a. The level of sales is found by evaluating S when $t = 5$, giving

$$S(5) = 25e^{-0.5} = 15.2 \text{ million dollars}$$

b. The rate at which sales are declining at the end of year 1 is given by $S'(1)$

$$S'(t) = -2.5e^{-0.10t}$$

The rate of decline when $t = 1$ is found next

$$S'(1) = -2.5e^{-0.10} = -2.3 \text{ million dollars/year}$$

c. The relative rate of change in sales equals $[S'(t)]/[S(t)]$.

$$\frac{S'(t)}{S(t)} = \frac{-2.5e^{-0.10t}}{25e^{-0.10t}} = -.10 = -10 \text{ percent per year}$$

EXERCISE 6.3

1. Show that the point (50, 45.98) in Example 1 is a maximum by applying the second derivative test, that is, find $R''(q)$ and show that $R''(50) < 0$.

2. Solve the Gluttonburger problem (Example 1) by writing Equation 6.3.1 in its equivalent logarithmic form and expressing the revenue function R in terms of p alone. Find the unit price that maximizes the revenue.

3. Suppose that the value V, in dollars, of a 1940 Batman comic book is given by the equation

$$V(t) = 125 + 10t^{3/2}$$

where t is the time in years from today.

    a. What will be the value of a Batman comic book four years from now?
    b. What will be the relative rate of growth of the value of the comic book four years from now?

4. The number of homes in a planned 3000-acre vacation community is expected to grow according to the equation

$$N(t) = 2000(1 - e^{-0.1t})$$

a. Plot this function on the following coordinate system.
b. What is the relative rate of growth at any time?
c. What is the relative rate of growth when $t = 1$?

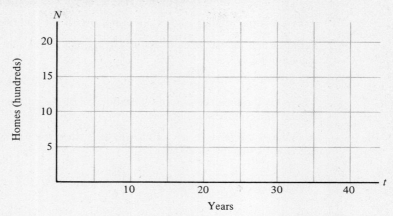

5. The Jiffy Button Company canceled all its promotional activities. As a result of this decision, its sales are expected to decline according to the equation

$$S(t) = 15e^{-0.01t}$$

where $S(t)$ represents the company's sales in millions.

a. How fast are sales declining when $t = 2$?
b. What is relative rate of decline when $t = 2$?

6. A sporting goods company has developed a new golf ball that carries further than that of any competing model. An intense advertising campaign is instituted to promote sales of the ball. As a result, sales increase according to the equation

$$S(t) = \frac{10}{1 + 4e^{-t}}$$

where t is the time in years and $S(t)$ is the sales in millions of dollars.

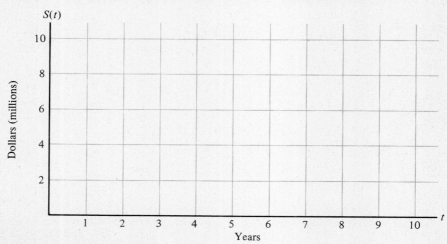

a. Plot the function on the accompanying coordinate system. HINT: There are no critical points. However, there is an inflection point; finding it together with determining $\lim_{t \to \infty} \dfrac{10}{1 + 4e^{-t}}$ will ease construction of the curve.

b. What is the relative rate of growth of $S(t)$ at the end of two years?

7. Suppose that annual sales of posters of a popular TV star can be described by the equation

$$S(t) = 3t^2 e^{-t}$$

where t is the time (years) from introduction of the poster to the public and S is the annual sales (millions).

a. What is the relative rate of growth of S when $t = 1$?

b. At what time do sales peak?

7

Additional Applications of the First Derivative

The use of the first derivative in determining the rate of change of a function and in finding maxima and minima has been detailed. In this chapter two additional applications using the first derivative will be studied. Section 7.1 will be devoted to defining the differential and to showing how it can be used to approximate the value of a function in the vicinity of a point where the value of the function is known. Section 7.2 will indicate how one or more roots of an equation can be approximated by employing a technique known as Newton's method.

❧ 7.1 The Differential

When the average rate of change of a function $y = f(x)$ was introduced in Section 3.1, the change in the independent variable x was denoted by the symbol h. For our immediate purposes and for future work in integration, it is more convenient to designate an arbitrary change in the independent variable x by the quantity dx, called the *differential* of the variable x, that is

$$dx = x_2 - x_1$$

An associated quantity dy, known as the *differential* of the variable y, is defined as

$$dy = f'(x) \cdot dx \qquad (7.1.1)$$

A geometrical representation of the quantities dx and dy is shown in Figure 7.1 where the line segment PR coincides with the line tangent to the curve at $[x, f(x)]$. In the figure, the base PQ of the triangle PQR represents the magnitude of dx, while the height QR represents the magnitude of dy. Basically, the quantity dy represents the change in the y coordinate on the tangent line when x is changed by an amount dx. The following examples illustrate how dy is found for a given function and how it is evaluated for specific values of x and dx.

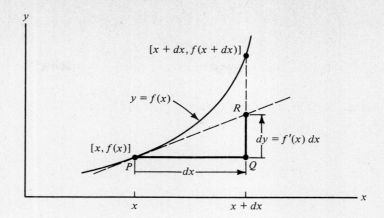

Figure 7.1

Example 1   a. Find the differential dy for the function

$$y = f(x) = x^2$$

b. Evaluate the differential when $x = 1$ and $dx = 0.75$.

Solution   a. The first derivative $f'(x)$ is given by

$$f'(x) = 2x$$

so that the differential dy can be written as

$$dy = 2x \cdot dx$$

b. When $x = 1$, and $dx = 0.75$, dy becomes

$$dy = 2(1)(0.75) = 1.50$$

The differentials dx, dy, together with the function are displayed in Figure 7.2.

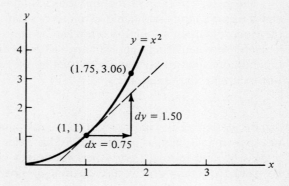

Figure 7.2

Example 2 a. Find the differential dy for the function

$$y = f(x) = \sqrt{x} = x^{1/2}$$

b. Evaluate the differential dy when $x = 4$ and $dx = -3$.

Solution a. The first derivative $f'(x)$ has the form

$$f'(x) = \tfrac{1}{2}x^{-1/2} = \frac{1}{2\sqrt{x}}$$

Applying Equation 7.1.1 gives

$$dy = \frac{1}{2\sqrt{x}}\,dx = \frac{dx}{2\sqrt{x}}$$

for arbitrary values of x and dx.

b. When $x = 4$ and $dx = -3$, dy becomes

$$dy = \frac{1}{2\sqrt{4}}(-3) = -0.75$$

The differentials dx, dy, and the curve $y = f(x) = \sqrt{x}$ are shown in Figure 7.3.

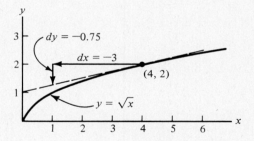

Figure 7.3

The graphs in each of the previous examples indicate that the tangent line and the curve almost coincide when dx is small. For this reason, the change in the y coordinate of a function often can be approximated by dy. This technique is useful in calculating approximate values of the y coordinates when a given function does not easily yield exact values of y except for selected values of x. For example, the function

$$y = f(x) = \sqrt[3]{x}$$

can be plotted easily and quickly if the independent variable x is assigned values that are classified as perfect cubes, for example, 0, ± 1, ± 8, ± 27, ± 64, $\pm \tfrac{1}{8}$, $\pm \tfrac{1}{27}$, $\pm \tfrac{27}{8}$, and so forth. For other values of x, approximation techniques must be used to obtain $\sqrt[3]{x}$ to any desired number of decimal places. A simple approximation technique that produces satisfactory results uses the tangent line to the curve at the closest point where x is a perfect cube, in place of the curve itself to determine the value of y corresponding to the given value of x.

To put these remarks in perspective, suppose that $\sqrt[3]{8.5}$ is desired. The behavior of the function $y = f(x) = \sqrt[3]{x}$ together with the tangent line in the vicinity of $(8, 2)$ is shown in Figure 7.4 (the curvature of the curve has been exaggerated purposely). Because the tangent line and the curve almost coincide when x is assigned values close to 8, the change in the function can be approximated by the differential dy for a given change dx in the x coordinate. In this case

$$dx = 8.5 - 8.0 = 0.5$$

$$f'(x) = \frac{1}{3(\sqrt[3]{x})^2} \qquad f'(8) = \frac{1}{3(\sqrt[3]{8})^2} = \frac{1}{12}$$

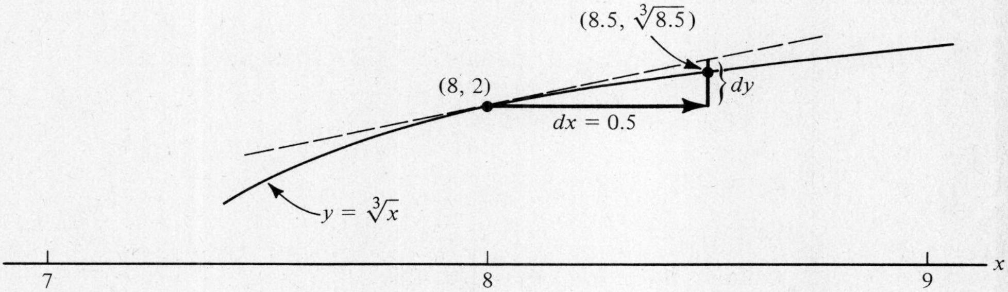

Figure 7.4

The differential dy then becomes

$$dy = f'(8) \cdot dx = \tfrac{1}{12}(0.5) = 0.042$$

so that we can now write

$$\sqrt[3]{8.5} \approx \sqrt[3]{8} + dy = 2 + 0.042 = 2.042$$

The exact value of $\sqrt[3]{8.5}$, correct to three decimal places, is 2.041. Thus the procedure provides a good estimate in this case. The generalized version of this procedure takes the form

$$f(x + dx) \approx f(x) + dy = f(x) + f'(x) \cdot dx \qquad \textbf{(7.1.2)}$$

and this result is shown graphically in Figure 7.5.

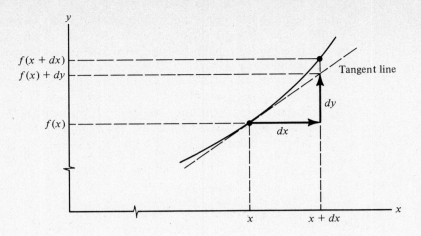

Figure 7.5

Example 3 | Using differentials, find ln (1.10).

Solution | Noting that ln 1 = 0, Equation 7.1.2 can be applied with

$$f(x) = \ln x \qquad f'(x) = \frac{1}{x}$$

$$x = 1 \qquad dx = 0.10$$

We then get

$$\ln (1.10) \approx \ln (1.0) + \tfrac{1}{1}(0.10)$$
$$\approx 0.10$$

which is close to the value given in Table C at the back of the book for ln (1.10), that is, 0.0953.

The differential can also be used to give an approximate value for a polynomial function when the independent variable x is assigned a value close to an integer.

Example 4 | Given the function

$$y = f(x) = 2x^4 - 5x^2 + 9x$$

use differentials to approximate $f(3.04)$.

Solution | Because 3.04 differs slightly from 3.00, Equation 7.1.2 is used with

$$x = 3.00 \qquad dx = 0.04 \qquad f'(x) = 8x^3 - 10x + 9$$

$$f(3.04) \approx f(3.0) + f'(3.0) \cdot (0.04)$$
$$\approx 144 + 195(0.04) = 151.80$$

The exact value of $f(3.04)$, correct to two decimal places, equals 151.97, so the error obtained by using the approximation is slight. The wide availability of calculators has considerably reduced the need for this approach when x assumes nonintegral values.

For most applications, the values assigned to the independent variable are not known precisely. For example, measurements of industrial output, income levels, revenue from taxes, and so forth, are subject to error even when the observations are carried out carefully. When the error is not large, the differential can be used to determine the error in the dependent variable when the approximate error or uncertainty in the independent variable is given. The error in the independent variable is represented by dx while the corresponding error in the dependent variable is represented by dy.

More important than the errors themselves are quantities called *relative errors*. The relative error in the independent variable x is defined as dx/x, while the relative error in the dependent variable y is defined as dy/y. For example, when a company forecasts annual sales of 6 million units, it is not expected that exactly 6 million units will be sold. If the uncertainty associated with the forecast is 300,000 or 0.3 million units, then the number of units expected to be sold will be somewhere between 5.7 million and 6.3 million. The relative error in the number of units sold then becomes

$$\frac{dx}{x} = \pm \frac{0.30}{6.00} = \pm 0.05 \qquad \text{or} \qquad \pm 5\%$$

The next step is to determine what effect the relative error in sales will have on, say, the revenue or profit functions for the firm. Suppose that the profit P is related to the number of units sold via the equation

$$P(x) = 14x - x^2 - 8$$

where $P(x)$ is expressed in millions of dollars. The company's profit, based on the forecast of 6 million units, is given by

$$P(6) = 84 - 36 - 8 = \$40 \text{ million}$$

The relative error in the profit function dP/P can be found by first finding dP

$$dP = P'(x) \cdot dx = (14 - 2x) \cdot dx$$

so that

$$\frac{dP}{P} = \frac{(14 - 2x) \cdot dx}{14x - x^2 - 8} = \frac{(14x - 2x^2)}{14x - x^2 - 8} \cdot \frac{dx}{x}$$

The last step was carried out because in most problems, the relative error dx/x, or percentage relative error $100(dx/x)$, is specified. The relative error dP/P then becomes

$$\frac{dP}{P} = \frac{84 - 72}{40} (\pm 0.05) = \frac{12}{40} (\pm 0.05) = \pm 0.015 \quad \text{or} \quad \pm 1.5\%$$

This result indicates that profits will be somewhere within a range of approximately 1.5 percent above to 1.5 percent below the predicted $40 million if the sales level turns out to be somewhere between 5 percent above to 5 percent below the sales forecast of 6 million units.

Example 5 | The relationship between p, the price per bushel of wheat, and x, the number of bushels harvested annually, is given by the equation

$$p(x) = 60e^{-1.5x}$$

where p is expressed in dollars and x in billions of bushels. Agricultural economists are forecasting a two billion bushel harvest. The percentage error in their forecasts is usually about 10 percent, that is

$$\frac{dx}{x} = \pm 0.10$$

What is the relative error or uncertainty in the per-bushel price of wheat?

Solution | The differential dp is found first

$$dp = 60(-1.5)e^{-1.5x}\, dx$$

Next, the quantity dp/p is formed giving

$$\frac{dp}{p} = \frac{60(-1.5)e^{-1.5x}(dx)}{60e^{-1.5x}} = -1.5 \cdot dx = -1.5x\frac{dx}{x}$$

Substituting 2 for x and $\pm(0.10)$ for dx/x, we get for dp/p

$$\frac{dp}{p} = (-1.5)(2)(\pm 0.10) = \mp 0.30$$

From this, we can conclude that if the wheat harvest is 10 percent above predicted levels, the per-bushel price would drop about 30 percent and vice versa.

It was stated in Section 3.6 that the additional revenue obtained when production or sales increases from x units to $x + 1$ units can be approximated by $R'(x)$. The rationale for the statement stems from the fact that

$$R(x + dx) - R(x) \approx dR$$

when dx is small. Using the definition of the differential dR, it is possible to write

$$R(x + dx) - R(x) \approx R'(x) \cdot dx$$

When $dx = 1$ and x is large, the approximation takes the form

$$R(x + 1) - R(x) \approx R'(x)$$

where the left-hand side represents the additional revenue when sales increase by one unit and the right-hand side is the marginal revenue.

EXERCISE 7.1

For the functions given in Problems 1–10, (a) find the differential dy, and (b) evaluate dy for the given values of x and dx.

1. $y = f(x) = 6x^3$ $x = -1, dx = 0.10$

2. $y = f(x) = \dfrac{5}{x - 1}$ $x = 2, dx = -0.25$

3. $y = f(x) = x^2 - 5x - 2$ $x = 3, dx = 1$

4. $y = f(x) = \sqrt{x^2 + 1}$ $x = 0, dx = 0.01$

5. $y = f(x) = \dfrac{x}{3 - 2x}$ $x = 2, dx = -0.50$

6. $y = f(x) = e^{x^2}$ $x = -1, dx = 0.20$

7. $y = f(x) = \ln (x + 1)$ $x = 3, dx = -\frac{1}{8}$

8. $y = f(x) = xe^{-x}$ $x = -1, dx = 2$

9. $y = f(x) = \dfrac{\ln x}{x}$ $x = 1, dx = 0.30$

10. $y = f(x) = x^2(2 + x)$ $x = 4, dx = -2$

Using differentials, find approximate values for the quantities in Problems 11–16.

11. $\sqrt{26}$ **12.** $\sqrt{8.9}$

13. $\sqrt[3]{28}$ **14.** $\sqrt[3]{63}$

15. $\sqrt[4]{15.7}$ **16.** $\sqrt[3]{124}$

17. A 100 ft by 100 ft plot of land is being enclosed by a chain link fence. If the relative error in measuring the length of each side is 1 percent, what is the relative error in calculating the area enclosed by the fence?

18. Annual automobile production next year is expected to be 9 million units. The Rain and Snow Tire Company, which produces original equipment tires for the major automobile companies, has found that its annual profit P is related to automobile production x by the equation

$$P(x) = 9.5x - \frac{x^2}{2} - 3.5$$

where x is expressed in millions of units and P in millions of dollars. If the forecast has a relative error of ± 15 percent, what is the corresponding relative error in company profits?

19. Extend the analysis carried out in Example 5 to determine the relative error in the total farm revenue from wheat sales.

For the functions in Problems 20–24

    a. Find the differential dy and

    b. Evaluate dy for the given values of x and dx

    c. Sketch the triangle whose base equals dx, whose height equals dy, and whose hypotenuse coincides with the tangent line to the curve at the given value of x.

20. $y = f(x) = \dfrac{x^2}{2} + x - 1$ $x = 1,\ dx = 2$

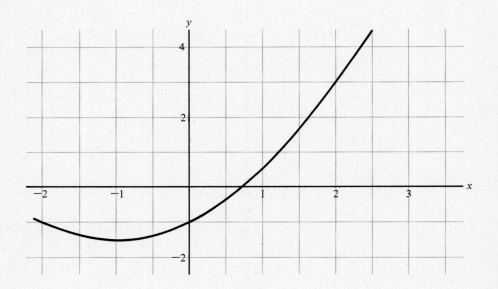

21. $y = f(x) = e^x$ $x = 2,\ dx = -1$

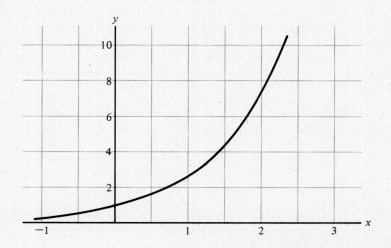

22. $y = f(x) = \ln (x + 1)$ $x = 1, dx = -2$

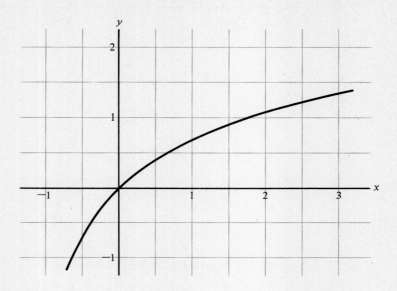

23. $y = f(x) = \sqrt{1 - x}$ $x = 0, dx = -3$

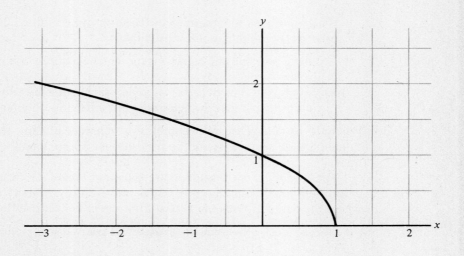

24. $y = f(x) = \dfrac{2}{x}$ $x = -1,\ dx = \frac{3}{2}$

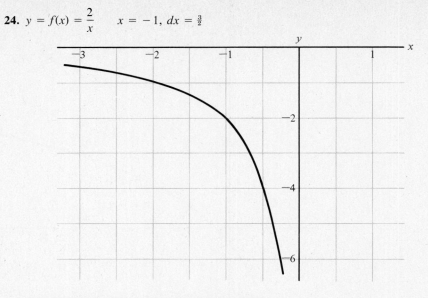

7.2 Solving Equations by Newton's Method*

In many situations it is necessary to find those values of x that satisfy an equation of the form

$$f(x) = 0 \qquad\qquad\qquad (7.2.1)$$

for example

$$x^2 + 3x - 4 = 0$$

$$6x^5 - 7x^4 + 2 = 0$$

$$xe^x + 7x + \ln x = 0$$

When solutions to equations of the form in Equation 7.2.1 were sought in previous work, the equations were either

1. Linear or quadratic equations for which solutions could be obtained easily, or
2. Equations for which $f(x)$ was factorable, thus enabling the multiplication principle (see Appendix B) to be used in obtaining the solutions.

When the equations are more complex, a technique utilizing the first derivative, called *Newton's method*, can be used to find approximate values of the solutions.

The essentials of the method are shown in Figure 7.6, where a rough graph of a differentiable function $y = f(x)$ is plotted in the vicinity of the point $(t, 0)$ where the curve crosses the x axis. First, an initial estimate of t is made by se-

* A calculator is recommended for the examples and exercises in this section.

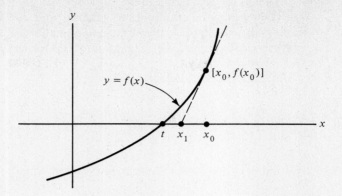

Figure 7.6

lecting a value of x close to t, usually one of the two closest integers on either side. This estimate is designated x_0. Next, the point $[x_0, f(x_0)]$ is located on the curve and the line tangent to the curve at that point drawn. The next estimate, called x_1, is provided by the point where the tangent line intersects the x axis. The equation used to find x_1 can be derived by noting that

1. The slope of the tangent line is given by $f'(x_0)$, and
2. $[x_0, f(x_0)]$ and $[x_1, 0]$ are two points on the tangent line.

Using Equation 1.3.1, it is possible to write the following equation

$$f'(x_0) = \frac{f(x_0) - 0}{x_0 - x_1} = \frac{f(x_0)}{x_0 - x_1}$$

We now solve this equation for x_1 as follows:

$$(x_0 - x_1)f'(x_0) = f(x_0) \qquad \text{or} \qquad x_0 - x_1 = \frac{f(x_0)}{f'(x_0)}$$

from which we get

$$x_1 = x_0 - \frac{f(x_0)}{f'(x_0)}$$

The process can be repeated as shown in Figure 7.7 to yield still another estimate, called x_2. The equation that defines x_2 has the form

$$x_2 = x_1 - \frac{f(x_1)}{f'(x_1)}$$

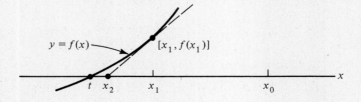

Figure 7.7

The process can be continued to generate still another estimate, x_3, defined in terms of $x_2, f(x_2)$, and $f'(x_2)$. The generalization of this technique enables us to express x_{n+1} in terms of $x_n, f(x_n)$, and $f'(x_n)$ for any integer $n \geq 0$. The result is

$$x_{n+1} = x_n - \frac{f(x_n)}{f'(x_n)}$$

(7.2.2)

A number of examples illustrating Newton's method will be presented next.

Example 1 | Find the approximate solution of the equation

$$x^3 - 5 = 0$$

Solution | For this case, $f(x) = x^3 - 5$ and $f'(x) = 3x^2$. A rough sketch of the function shows that the solution is located between $x = 1$ and $x = 2$. Because the solution is closer to 2 than to 1, x_0 will be set equal to 2. The estimate x_1 is found using Equation 7.2.2

$$x_1 = x_0 - \frac{f(x_0)}{f'(x_0)} = 2 - \frac{f(2)}{f'(2)} = 2 - \frac{3}{12} = 1.7500$$

where calculations are carried out to four decimal places.

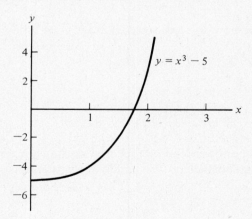

Repeating the process to obtain x_2, we get

$$x_2 = x_1 - \frac{f(x_1)}{f'(x_1)} = 1.75 - \frac{0.3594}{9.1875} = 1.7109$$

while for x_3, the method yields

$$x_3 = x_2 - \frac{f(x_2)}{f'(x_2)} = 1.7109 - \frac{0.0081}{8.7815} = 1.7100$$

Continuing, x_4 becomes

$$x_4 = x_3 - \frac{f(x_3)}{f'(x_3)} = 1.7100 - \frac{0.0002}{8.7723} = 1.7100$$

Because x_3 and x_4 are identical, expressed to four decimal places, there is no need to continue the process any further. The approximate value of t is

$$t = 1.7100$$

To carry out the process efficiently, it is advisable to set up a table such as the accompanying to obtain the successive approximations to t. It should be noted that the process just described is equivalent to finding the cube root of 5.

n	x_n	$f(x_n) = x^3 - 5$	$f'(x) = 3x^2$	x_{n+1}
0	2.0000	3.0000	12.0000	1.7500
1	1.7500	0.3594	9.1875	1.7109
2	1.7109	0.0081	8.7815	1.7100
3	1.7100	0.0002	8.7723	1.7100

Example 2   Among the roots of the equation $x^4 + 5x - 3 = 0$, find the one that lies between $x = 0$ and $x = 1$.

Solution   Noting that $f(x) = x^4 + 5x - 3$ and $f'(x) = 4x^3 + 5$, we can carry out the process as shown in the accompanying table where $x_0 = 0$.

n	x_n	$f(x_n) = (x_n)^4 + 5(x_n) - 3$	$f'(x_n) = 4(x_n)^3 + 5$	x_{n+1}
0	0.0000	-3.0000	$+5.0000$	0.6000
1	0.6000	$+0.1296$	$+5.8640$	0.5779
2	0.5779	$+0.0010$	$+5.7720$	0.5777
3	0.5777	-0.0001	$+5.7712$	0.5777

Therefore, to four decimal places, the root is 0.5777.

This technique can also be used to determine the interest rate per period i using the compound interest formula, Equation 5.1.4, when P, A, and n are given.

Example 3   What annual rate of interest will cause a $500 deposit to grow to $750 in three years if interest is compounded semiannually?

Solution   The equation to be solved has the form

$$750 = 500(1 + i)^6$$

or

$$(1 + i)^6 - 1.5 = 0$$

The value of i that satisfies this equation lies between $i = 0$ and $i = 1$. For this case

$$f(i) = (1 + i)^6 - 1.5 \qquad f'(i) = 6(1 + i)^5$$

Using Equation 7.2.2 with $i_0 = 0$, the process is carried out as illustrated in the following table.

n	i_n	$f(i_n) = (1 + i_n)^6 - 1.5$	$f'(i_n) = 6(1 + i_n)^5$	i_{n+1}
0	0.0000	−0.5000	6.0000	0.0833
1	0.0833	0.1162	8.9515	0.0703
2	0.0703	0.0033	8.4271	0.0699
3	0.0699	−0.0001	8.4114	0.0699

We then get an annual interest rate of $(0.0699)(2) = 0.1398 = 13.98\%$.

This technique can also be used to locate the x coordinates of critical points when the expression for the first derivative cannot be factored.

Example 4   Find the critical points for the function

$$y = g(x) = x^4 - 3x^2 + 5x - 2$$

Solution   The critical points are found by solving the equation

$$g'(x) = 0$$

for x. Differentiating $g(x)$ term by term gives

$$4x^3 - 6x + 5 = 0$$

as the equation to be solved. Newton's method, with $f(x) = 4x^3 - 6x + 5$, can be applied to find the solution(s), if any exist. A rough graph of $f(x)$ indicates that one solution exists in the interval between $x = -2$ and $x = -1$. NOTE: The graph does not represent $g(x)$ but $g'(x) = f(x)$.

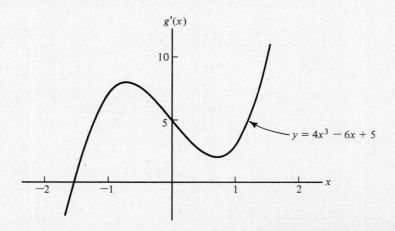

The following table summarizes the results obtained using Newton's method. There is one critical point for the function $y = g(x)$, located at $(-1.523, -11.193)$.

n	x_n	$f(x_n)$	$f'(x_n)$	x_{n+1}
0	-2	-1.5	42	-1.643
1	-1.643	-2.883	26.393	-1.534
2	-1.534	-0.235	22.238	-1.523
3	-1.523	0.007	21.834	-1.523

Newton's method can be used to determine a quantity known as the *internal rate of return* on a capital or long-term investment, that is, one whose revenues and expenses are expected to continue for many years. Projects such as plant expansion, purchase of equipment, or development of a new product are examples of capital investments. If the initial cost of the project is designated C and the net cash flows for years 1 through m are designated R_1, R_2, R_3, . . . , R_m, the net present value NPV of the investment is defined as

$$NPV = \frac{R_1}{(1 + r)} + \frac{R_2}{(1 + r)^2} + \frac{R_3}{(1 + r)^3} + \cdots + \frac{R_m}{(1 + r)^m} - C$$

where r is the rate of return the firm seeks to earn on the investment. The internal rate of return is defined as the rate of return r for which NPV equals 0, or the value of r that is a solution to the equation

$$0 = \frac{R_1}{(1 + r)} + \frac{R_2}{(1 + r)^2} + \frac{R_3}{(1 + r)^3} + \cdots + \frac{R_m}{(1 + r)^m} - C \qquad \textbf{(7.2.3)}$$

If the value of r that satisfies Equation 7.2.3 is greater than the rate of return the firm requires on its investments, then the project is considered a profitable one.

To find the value of r that satisfies Equation 7.2.3, we can use Newton's method with $f(r)$ and $f'(r)$ defined as

$$f(r) = \frac{R_1}{(1 + r)} + \frac{R_2}{(1 + r)^2} + \cdots + \frac{R_m}{(1 + r)^m} - C \qquad \textbf{(7.2.4)}$$

$$f'(r) = -\frac{R_1}{(1 + r)^2} - \frac{2R_2}{(1 + r)^3} - \cdots - \frac{mR_m}{(1 + r)^{m+1}} \qquad \textbf{(7.2.5)}$$

For most projects, the internal rate of return r falls between 0 and 1.

Example 5   A firm purchases a minicomputer whose initial cost is \$25,000. The net cash flows provided by this acquisition over the next three years are expected to be \$10,000 annually. Find the internal rate of return on the investment.

Solution   The internal rate of return is determined by finding the value of r that satisfies the equation

$$0 = \frac{10{,}000}{(1 + r)} + \frac{10{,}000}{(1 + r)^2} + \frac{10{,}000}{(1 + r)^3} - 25{,}000$$

Newton's method can be used to solve this equation. Defining $f(r)$ and $f'(r)$ as

$$f(r) = \frac{10,000}{(1 + r)} + \frac{10,000}{(1 + r)^2} + \frac{10,000}{(1 + r)^3} - 25,000$$

$$f'(r) = -\frac{10,000}{(1 + r)^2} - \frac{20,000}{(1 + r)^3} - \frac{30,000}{(1 + r)^4}$$

Equation 7.2.2 can be used to find r. Beginning with $r_0 = 0$, the successive steps in the approximation are shown in the following table. The internal rate of return is found to be 9.70 percent. The management of the firm must now decide if this rate of return is sufficient to warrant approval of the project.

n	r_n	$f(r_n)$	$f'(r_n)$	r_{n+1}
0	0.0000	5,000.00	−60,000	0.0833
1	0.0833	618.00	−46,035	0.0967
2	0.0967	13.70	−44,201	0.0970
3	0.0970	0.45	−44,175	0.0970

Although Newton's method generates values $x_0, x_1, x_2, x_3, \ldots, x_n, \ldots$ that move successively closer to the desired solution $x = t$ for most of the functions encountered, there are occasions when each successive value may not be closer to the solution than its predecessor. Situations of this sort may arise if

1. The first derivative $f'(x)$ equals 0 at one or more points on the curve between $x = x_0$ and $x = t$. An illustration of this case is provided in the graph shown in Figure 7.8, where x_2 falls between x_0 and x_1.
2. The second derivative $f''(x)$ changes sign at one or more points on the curve close to the root $x = t$. This case is illustrated graphically in Figure 7.9, where x_2 is located further than x_0 or x_1 from the solution $x = t$.

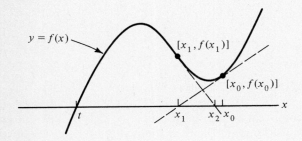

Figure 7.8

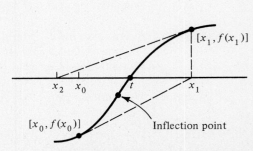

Figure 7.9

When these situations arise, it is advisable to draw a rough graph of the function in the vicinity of $x = t$ to determine the cause of the seemingly erratic behavior of the approximations. Often, selecting a new value for x_0 closer to the solution $x = t$ will remove the difficulty.

EXERCISE 7.2

1. a. Work out Example 1 using $x_0 = 1$.
   b. Work out Example 2 using $x_0 = 1$.
   c. Work out Example 3 using $i_0 = 1$.

2. a. Work out Example 4 using $x_0 = -1$.
   b. Determine the nature of the critical point found and make a rough sketch of the curve on the accompanying coordinate system.

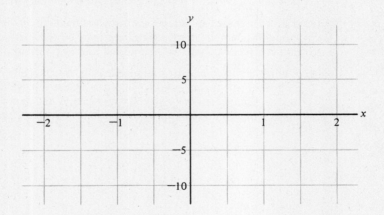

Using Newton's method, find the root located between $x = a$ and $x = b$ for Problems 3–7. Express your answer correct to three decimal places.

3. $y = f(x) = x^2 + 2x - 10 \qquad a = 2, b = 3$

4. $y = f(x) = x^3 + x^2 - 5x + 1 \qquad a = -3, b = -2$

5. $y = f(x) = x^4 + x - 3 \qquad a = 1, b = 2$

6. $y = f(x) = \dfrac{2 - x^2}{x} \qquad a = 1, b = 2$

7. $y = f(x) = x^5 - 40 \qquad a = 2, b = 3$

Using Newton's method, locate the critical points for the functions in Problems 8–10.

8. $y = g(x) = x^3 - 4x^2 + 5x - 1$

9. $y = g(x) = x^4 + 6x + 2$

10. $y = g(x) = 9x^2 - x^4 + 2$

Use Newton's method in solving Problems 11–16.

11. Five thousand dollars invested in the Doggie Bag Company grows to $9200 in five years. What is the average annual rate of return on this investment?

12. The neighborhood loan shark loans money at rather exorbitant interest rates. A $100 loan is to be repaid in seven days by paying him $110. Assuming that interest is compounded daily, determine the annual interest rate he is charging. (Calculate i to four decimal places.)

13. A two-year certificate of deposit (CD), purchased for $1000, is worth $1655 two years later. Find the annual rate of interest paid on the CD.

14. A retail store plagued by shoplifters and petty thieves installs a closed circuit video security system at a cost of $25,000. The net annual savings are expected to be $8000 the first year and $15,000 per year thereafter. If the company is planning to move to new quarters three years from now, at which time a new system will have to be purchased, find the internal rate of return on the present purchase. Assume that salvage value is 0 at the end of the three-year period.

15. A firm purchases an old, poorly insulated building that it plans to use for five years. Proper insulation will reduce heating costs by $500 per year. If installation of insulation costs $1000, what is the internal rate of return on the project?

16. A company is considering purchasing the patent rights to an electronic game from its inventor. If the inventor is willing to accept $50,000 for the patent rights and net income is expected to be $15,000 over each of the next five years at which time the rights expire, find the internal rate of return on the investment.

8

Integration

Until now, our attention has been directed toward finding the derivatives of given functions in order to determine rates of change or to find critical values of an independent variable for the purpose of either optimizing some quantity or sketching a curve. An equally important type of problem arises when the first derivative $f'(x)$ is given and the function $f(x)$ is sought. Problems of this type are classified in the general category of *integration*.

An example of a problem requiring integration is determining the amount of money the United States will spend on imported oil during the next five years if the equations for the number of barrels imported per day and the price per barrel are known functions of the time. Integration techniques will also yield the present value of all monies spent on imported oil during this period.

The antiderivative or indefinite integral of a function will be introduced in Section 8.1 and applied in Section 8.2 to determine the area under a curve. Section 8.3 will be devoted to defining the definite integral and presenting the fundamental theorem of calculus. Using the definite integral to find the area between two curves will be studied in Section 8.4; additional applications of the definite integral will be given in Section 8.5.

ᔈ 8.1 Antiderivative, Indefinite Integral

Previously, one of our main concerns centered on finding the first derivative of a given function. Beginning now, our attention will also be focused on the problem in reverse:

Given the function $y = f(x)$, find a second function $y = F(x)$ whose first derivative equals $f(x)$, that is, $F'(x) = f(x)$. $F(x)$ is called an *antiderivative* of $f(x)$.

This type of problem falls under the general heading of antidifferentiation or, as it is more commonly called, *integration*. For example, suppose we are given the function

$$f(x) = x^2$$

and we seek a second function $F(x)$ whose first derivative $F'(x)$ equals x^2. By working the simple power rule, Equation 3.3.2, in reverse, that is, adding 1 to the exponent, we might expect $F(x)$ to have the form

$$F(x) = x^3$$

However, when we differentiate $F(x)$, we obtain

$$F'(x) = 3x^2 \neq x^2$$

This discrepancy is not difficult to correct: dividing our initial attempt, that is, x^3, by 3 yields a correct version of $F(x)$

$$F(x) = \frac{x^3}{3} \qquad F'(x) = \frac{3x^2}{3} = x^2$$

There are many other forms of $F(x)$ that will give $f'(x)$; for example, the first derivative of each of the following functions equals x^2:

$$F(x) = \frac{x^3}{3} + 11$$

$$F(x) = \frac{x^3}{3} - \frac{7}{4}$$

$$F(x) = \frac{x^3}{3} - \sqrt{5}$$

These examples indicate that adding or subtracting a constant term does not affect $F'(x)$, that is,

$$F'(x) = x^2$$

For this reason, the general antiderivative of $f(x) = x^2$ is written as

$$F(x) = \frac{x^3}{3} + C$$

where C is an arbitrary constant.

The antiderivative of almost all simple power functions can be found in the same way as that for $f(x) = x^2$; for example, suppose

$$f(x) = \frac{1}{x^5} = x^{-5}$$

Adding 1 to the exponent gives as a first attempt at finding $F(x)$

$$F(x) = x^{-4} + C = \frac{1}{x^4} + C$$

Again differentiating this function gives

$$F'(x) = -4x^{-5} = \frac{-4}{x^5} \neq \frac{1}{x^5} = f(x)$$

This discrepancy can be corrected by dividing our initial estimate x^{-4} by -4 to give

$$F(x) = -\frac{1}{4} x^{-4} + C = -\frac{1}{4x^4} + C$$

Checking this result by differentiating gives

$$F'(x) = \left(-\frac{1}{4}\right)(-4)x^{-4-1} = x^{-5} = \frac{1}{x^5} = f(x)$$

This approach also works when the exponents are fractions, for example

$$f(x) = \sqrt[3]{x} = x^{1/3}$$

Adding 1 to the exponent and dividing by the new exponent gives

$$F(x) = \frac{x^{1+1/3}}{1 + \frac{1}{3}} + C = \frac{3x^{4/3}}{4} + C = \frac{3\sqrt[3]{x^4}}{4} + C$$

as the general antiderivative of $f(x) = \sqrt[3]{x}$. Differentiating $F(x)$ gives

$$F'(x) = \frac{3}{4} \cdot \frac{4}{3} x^{(4/3)-1} = x^{1/3} = f(x)$$

The procedure just described can be generalized as follows:

If $f(x) = x^n$, $n \neq -1$, then the general antiderivative of $f(x)$ is

$$F(x) = \frac{x^{n+1}}{n + 1} + C \qquad\qquad\qquad\qquad (8.1.1)$$

where C is an arbitrary constant, called the constant of integration.

The result in Equation 8.1.1 can be checked very quickly. Differentiating $F(x)$ gives

$$F'(x) = \frac{n + 1}{n + 1} x^{n+1-1} + 0 = x^n = f(x)$$

The operation of finding the general antiderivative of a function is also denoted by

$$\int f(x) \, dx \qquad\qquad\qquad\qquad (8.1.2)$$

called the *indefinite integral* of $f(x)$. This notation may appear bizarre at first, but for the moment it might be advisable to look on the quantity 8.1.2 as merely another way of indicating that the general antiderivative of $f(x)$ is sought, that is

---

$$\int f(x) \, dx = F(x) \qquad \text{where } F'(x) = f(x) \qquad\qquad (8.1.3)$$

---

where $f(x)$ is called the *integrand* in this format. The notation itself will take on more meaning once the definite integral and the fundamental theorem of calculus have been presented. Examples of Equation 8.1.3 are

$$\int x^5 \, dx = \frac{x^{5+1}}{5 + 1} + C = \frac{x^6}{6} + C$$

$$\int \frac{1}{x^2}\, dx = \int x^{-2}\, dx = \frac{x^{-2+1}}{-2+1} + C = \frac{-1}{x} + C$$

$$\int \sqrt[4]{x}\, dx = \int x^{1/4}\, dx = \frac{x^{(1/4)+1}}{(1/4)+1} + C = \frac{4x^{5/4}}{5} + C$$

$$\boxed{\int x^n\, dx = \frac{x^{n+1}}{n+1} + C \qquad n \neq -1} \qquad\qquad \textbf{(8.1.4)}$$

In addition, other simple functions whose antiderivatives you should know are e^{ax} and $1/x$:

$$\boxed{\int e^{ax}\, dx = \frac{e^{ax}}{a} + C} \qquad\qquad \textbf{(8.1.5)}$$

For example $\int e^{3x}\, dx = (e^{3x}/3) + C$. Equation 8.1.5 can be checked directly

$$\frac{d}{dx}\left(\frac{e^{ax}}{a} + C\right) = \frac{1}{a}(a)e^{ax} + 0 = e^{ax}$$

The antiderivative of $(1/x)$ has a form that surprises many people when it is first encountered

$$\boxed{\int \frac{1}{x}\, dx = \ln|x| + C} \qquad\qquad \textbf{(8.1.6)}$$

The absolute value of x appears because the algebraic form of the first derivative is identical for both $\ln x$ and $\ln(-x)$, that is

$$\frac{d}{dx}(\ln x) = \frac{1}{x} \qquad \frac{d}{dx}[\ln(-x)] = \frac{1}{-x}(-1) = \frac{1}{x}$$

It should be pointed out that only positive values may be substituted for x in $\ln x$ and only negative values in $\ln(-x)$.

Just as there are techniques which enable us to obtain the first derivative of a function quickly and efficiently, so too are there similar methods for integrating. Two of the simpler ones are

$$\boxed{\text{I.} \int C \cdot f(x)\, dx = C \int f(x)\, dx} \qquad\qquad \textbf{(8.1.7)}$$

where C is a constant. In words, this property states that *the indefinite inte-gral of a constant times a function equals the constant times the indefinite integral of the function.* This property is the reverse of the property stated in Equation 3.3.3

$$\frac{d}{dx}[CF(x)] = C\left[\frac{d}{dx}F(x)\right]$$

Example 1 $\qquad \displaystyle\int 3x^4\,dx = 3\int x^4\,dx = \frac{3x^5}{5} + C$

Example 2 $\qquad \displaystyle\int \frac{2}{7x}\,dx = \frac{2}{7}\int \frac{1}{x}\,dx = \frac{2}{7}\ln|x| + C$

Example 3 $\qquad \displaystyle\int 8\sqrt{x}\,dx = 8\int \sqrt{x}\,dx = \frac{16}{3}x^{3/2} + C$

$$\text{II.} \int [f(x) \pm g(x)]\,dx = \left[\int f(x)\,dx\right] \pm \left[\int g(x)\,dx\right] \qquad \textbf{(8.1.8)}$$

In words, this property states that *the integral of a sum or difference of two functions equals the sum or difference of the integrals.* This property is the re-verse of the property stated in Equation 3.3.5

$$\frac{d}{dx}[F(x) \pm G(x)] = \frac{d}{dx}F(x) \pm \frac{d}{dx}G(x)$$

Example 4 $\qquad \displaystyle\int (7x^3 + 5x)\,dx = \int 7x^3\,dx + \int 5x\,dx = 7\int x^3\,dx + 5\int x\,dx$

$$= \frac{7x^4}{4} + C_1 + \frac{5x^2}{2} + C_2 = \frac{7x^4}{4} + \frac{5x^2}{2} + C$$

where the two arbitrary constants C_1 and C_2 have been combined into one, that is, $C = C_1 + C_2$. In addition it is important at this stage to check all your answers by differentiating; for example,

$$\frac{d}{dx}\left(\frac{7x^4}{4} + \frac{5x^2}{2} + C\right) = \frac{7(4)x^3}{4} + \frac{5(2)x}{2} + 0 = 7x^3 + 5x$$

Example 5 $\qquad \displaystyle\int \left(e^x - \frac{3}{x^2}\right)dx = \int e^x\,dx - \int \frac{3}{x^2}\,dx$

$$= e^x - 3\int \frac{1}{x^2}\,dx = e^x - 3\int x^{-2}\,dx = e^x - \frac{3x^{-1}}{-1} + C$$

$$= e^x + \frac{3}{x} + C$$

Check the answer

$$\frac{d}{dx}\left(e^x + \frac{3}{x} + C\right) = \frac{d}{dx}\,e^x + \frac{d}{dx}\,(3x^{-1}) + \frac{d}{dx}\,C$$

$$= e^x + (-3)x^{-2} + 0 = e^x - \frac{3}{x^2}$$

Example 6

$$\int \frac{x^2 - 4}{x^3}\,dx = \int\left(\frac{1}{x} - \frac{4}{x^3}\right)dx = \int \frac{1}{x}\,dx - \int \frac{4}{x^3}\,dx$$

$$= \ln|x| - \int 4x^{-3}\,dx = \ln|x| - \frac{4x^{-3+1}}{-3+1} + C$$

$$= \ln|x| + \frac{2}{x^2} + C$$

Check the answer

$$\frac{d}{dx}\left(\ln|x| + \frac{2}{x^2} + C\right) = \frac{d}{dx}\,(\ln|x|) + \frac{d}{dx}\,(2x^{-2}) + \frac{d}{dx}\,C$$

$$= \frac{1}{x} + (-4x^{-3}) + 0 = \frac{1}{x} - \frac{4}{x^3}$$

The appearance of the arbitrary constant C, when integrating, is due to the fact that the expression for the slope of the line tangent to a curve does not uniquely specify the curve. Inspection of Figure 8.1 indicates that the slopes of the tangent lines are equal for each value of x for all curves whose equations have the form

$$y = F(x) = x^2 + C$$

$$F'(x) = 2x$$

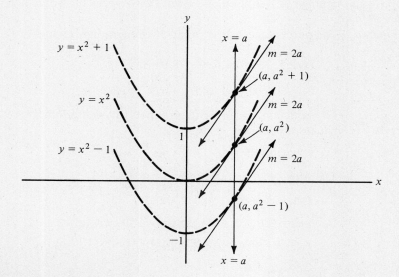

Figure 8.1

Additional information in the form of the coordinates of one point on the curve must be supplied in order to specify the curve when $F'(x)$ is given.

Example 7 | The slope of the line tangent to the curve $y = F(x)$ is described by the equation

$$F'(x) = 3x^2$$

If the curve passes through the point $(-1, 6)$, find the equation of the function $y = F(x)$.

Solution | The form of $F(x)$ is found by integrating, yielding

$$y = F(x) = \int 3x^2 \, dx = \frac{3x^3}{3} + C = x^3 + C$$

The value of C can be found by substituting -1 for x and 6 for y

$$6 = (-1)^3 + C$$

giving for C

$$C = 7$$

The equation defining $F(x)$ then becomes

$$y = F(x) = x^3 + 7$$

Example 8 | The percentage P of the U.S. population 65 years of age or older is increasing at a rate given by the equation

$$\frac{dP}{dt} = 0.01t + 0.01$$

where t is the time in years from 1960 ($t = 0$). If $P(0) = 9.2$ percent, what percentage of the population will be 65 years of age or older by 1980?

Solution | The percentage $P(t)$ is found by integrating, giving

$$P(t) = \int 0.01t \, dt + \int 0.01 \, dt$$
$$= 0.005t^2 + 0.01t + C$$

The constant C can be determined by noting that $P = 9.2$ when $t = 0$, so the equation becomes

$$9.2 = 0 + 0 + C \qquad \text{giving } C = 9.2$$

Therefore $P(t)$ has the form

$$P(t) = 0.005t^2 + 0.01t + 9.2$$

Because 1980 corresponds to $t = 20$, we get

$$P(20) = 0.005(20)^2 + 0.01(20) + 9.2$$
$$= 11.40\%$$

The rate of growth of the percentage $P(t)$ has serious implications for the continued financial viability of the social security system. In order to keep costs under control, there has been discussion about raising the minimum age for full benefits from 65 to 68.

Example 9   As workers become more familiar and experienced, the length of time required to perform a given task decreases. Suppose the rate at which a new employee assembles sabre saws for the Superior Tool Company is given by the equation

$$\frac{dT}{dx} = \frac{20}{\sqrt[3]{x}} = 20x^{-1/3}$$

where T is the time, in minutes, required to assemble x units. Find the total time to assemble 64 units.

Solution   The expression for $T(x)$ can be found by finding the antiderivative of the right-hand side giving

$$T(x) = \int 20x^{-1/3}\, dx = \frac{20x^{2/3}}{2/3} + C = 30(\sqrt[3]{x})^2 + C$$

The constant C can be found by noting that it takes no time to assemble 0 units. Therefore we have

$$T(x) = 30(\sqrt[3]{x})^2$$

The length of time required to assemble 64 units is therefore

$$T(64) = 30(\sqrt[3]{64})^2 = 480 \text{ minutes}$$

EXERCISE 8.1

Find each of the integrals in Problems 1–19; in addition, check each of your answers by differentiation.

1. $\int x^5\, dx$

2. $\int 4\, dx$

3. $\int x\, dx$

4. $\int \frac{1}{x^7}\, dx$

5. $\int (7x^2 + 4x - 9)\, dx$

6. $\int \left(3x^4 - \frac{2}{x}\right) dx$

7. $\int (10 - 5x^2)\, dx$

8. $\int \left(3e^x + \frac{1}{x^2}\right) dx$

9. $\int (8x^3 - 4e^{2x} + 2)\, dx$

10. $\int 6\sqrt[3]{x^2}\, dx$

11. $\int (5x^4 + 2\sqrt{x})\, dx$

12. $\int (e^{4x} + \sqrt[4]{x})\, dx$

13. $\int \left(\dfrac{3}{\sqrt{x}} + 4 \right) dx$ **14.** $\int (x + 1)^2 \, dx$

15. $\int \left(\dfrac{1}{x^2} - \dfrac{5}{x} \right) dx$ **16.** $\int \left(2e^{-x} + \dfrac{3}{x} \right) dx$

17. $\int \left(6x - \dfrac{4}{x} + e^{2x} \right) dx$ **18.** $\int \dfrac{x^2 + 3}{x} \, dx$

19. $\int \dfrac{2x^4 + 6x^3 + 8}{2x^2} \, dx$

20. The slope of the line tangent to the curve $y = F(x)$ is given by the equation

$$F'(x) = 2x - 5$$

If the curve passes through $(3, -1)$, find the equation of the function $y = F(x)$.

21. The slope of the line tangent to the curve $y = F(x)$ is given by the equation

$$F'(x) = \sqrt{x} + 3$$

If the curve passes through $(9, 2)$, find the equation of the function $y = F(x)$.

22. The slope of the line tangent to the curve $y = F(x)$ is given by the equation

$$F'(x) = 1 + e^x$$

If the curve passes through $(0, 4)$, find the equation of the function $y = F(x)$.

23. The slope of the line tangent to the curve $y = F(x)$ is given by

$$F'(x) = 4x^3 - \dfrac{1}{2x^3}$$

If the curve passes through $(1, 3)$, find the equation of the function $y = F(x)$.

24. The marginal cost function for the Acme Can Company is given as

$$C'(x) = 0.05x - 0.003x^2$$

where x is the level of production in thousands of units and C is in dollars. If fixed costs are \$5000, find the total cost function $C(x)$.

25. The rate at which a new employee can assemble three-speed bicycles for the Supreme Sporting Goods Company is given by the equation

$$\frac{dT}{dx} = \frac{90}{\sqrt[3]{x^2}}$$

where T is the total assembly time in minutes and x is the number of bicycles assembled. How long does it take for the employee to assemble 27 bicycles?

✍ 8.2 Area Under a Curve

One of the more important problems in calculus is determining the area of a region in the xy plane that is bounded

1. *Above* by a given function $y = f(x)$.
2. *Below* by the x axis.

3. On the *left* by the vertical line $x = a$.
4. On the *right* by the vertical line $x = b$.

The shaded area in Figure 8.2 graphically depicts the problem.

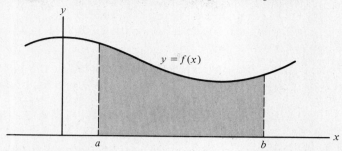

Figure 8.2

Determining the area has importance beyond the field of geometry alone. Suppose a new toy has been placed on the market in July for the Christmas season. If predicted weekly sales are described by the curve in Figure 8.3, the area beneath the curve between any two values of t, say t_1 and t_2, denotes predicted total sales during the period from t_1 to t_2.

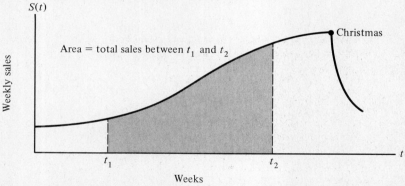

Figure 8.3

Another type of problem in which the area under a curve plays an important role is determining the length of time until known reserves of some natural resource are exhausted, assuming that annual consumption continues on its present trend. Suppose the curve in Figure 8.4 describes the worldwide

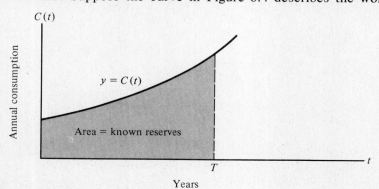

Figure 8.4

annual consumption of natural gas as a function of time. If we want to find out how long until all known reserves are exhausted, we determine the value of t, labeled T in Figure 8.4, for which the shaded area under the annual consumption curve equals total known reserves.

Having indicated why determining the area is an important operation, we will now turn our attention to the evaluation process. The development will involve two phases:

I. Showing the relationship between the curve $y = f(x)$ and the rate at which the area A changes as a function of the position of one of the vertical boundary lines, and

II. Using the result of step I to develop the equation that gives the area under the curve.

I. We will assume that (a) the given function $y = f(x)$ is continuous and nonnegative over the interval $a \le x \le b$, and (b) we can find an antiderivative $F(x)$. Suppose we denote the area beneath the curve $y = f(x)$ between $x = a$ and an arbitrary value of x, $x > a$, as $A(x)$, shown in Figure 8.5. Next, the right-hand boundary line is shifted slightly to the right to $x + h$ and the resulting area under the curve between a and $x + h$ is denoted $A(x + h)$ as shown in Figure 8.6

For the function shown in Figure 8.6 we see that the area of the strip $[A(x + h) - A(x)]$ is greater than or equal to the area of the rectangle whose base is h and whose height equals $f(x)$, that is

$$h \cdot f(x) \le A(x + h) - A(x)$$

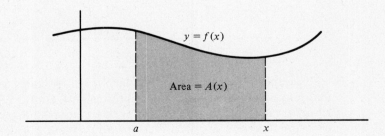

Figure 8.5

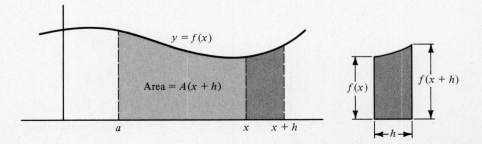

Figure 8.6

At the same time, we note that the area of the strip is less than or equal to the area of the rectangle whose base is h and whose height equals $f(x + h)$, that is

$$A(x + h) - A(x) \leq h \cdot f(x + h)$$

These two statements can be combined into one

$$h \cdot f(x) \leq A(x + h) - A(x) \leq h \cdot f(x + h)$$

Dividing through by h gives, for $h > 0$,

$$f(x) \leq \frac{A(x + h) - A(x)}{h} \leq f(x + h)$$

When $h < 0$, the sense of each inequality is reversed. Next, letting $h \to 0$ gives

$$\lim_{h \to 0} f(x) \leq \lim_{h \to 0} \frac{A(x + h) - A(x)}{h} \leq \lim_{h \to 0} f(x + h)$$

We can conclude from this that

$$f(x) \leq \frac{dA}{dx} \leq f(x)$$

or

$$A'(x) = \frac{dA}{dx} = f(x) \tag{8.2.1}$$

Equation 8.2.1 states that the rate at which the area beneath the curve $y = f(x)$ changes equals the y coordinate $f(x)$.

II. The second phase of the process is to use Equation 8.2.1 to obtain the expression that enables us to calculate the area under a given curve between two specified values of x, $x = a$ and $x = b$, as shown in Figure 8.7. If we let $F(x)$ denote an antiderivative of $f(x)$, we get from Equation 8.2.1 the following relation

$$A(x) = \int f(x) \, dx = F(x) + C$$

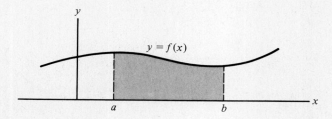

Figure 8.7

When $x = a$, the area beneath the curve equals 0, so we have

$A(a) = 0 = F(a) + C$

Solving for C gives

$C = -F(a)$

so that

$A(x) = F(x) - F(a)$

When $x = b$, the quantity $A(b)$ represents the area under the curve from $x = a$ to $x = b$ and we have

$$\boxed{\text{Area under the curve} = F(b) - F(a)}$$ (8.2.2)

To obtain the area under the curve $y = f(x)$ *between* $x = a$ *and* $x = b$, *evaluate an antiderivative* $F(x)$ *at* $x = b$ *and at* $x = a$ *and find the difference* $F(b) - F(a)$.

The result shown in Equation 8.2.2 will now be used to determine the area beneath a given curve $y = f(x)$ between $x = a$ and $x = b$ where $a < b$.

Example 1 | Find the area beneath the curve

$y = f(x) = x^2$

between $x = 1$ and $x = 2$.

Solution | The region whose area is to be determined is shown in the following figure. The general antiderivative of $f(x) = x^2$ is

$F(x) = \dfrac{x^3}{3} + C$

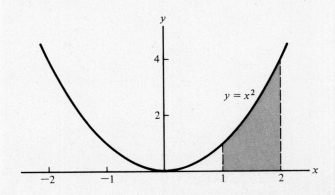

Using Equation 8.2.2, we get for the area beneath the curve

$$\text{Area} = F(2) - F(1) = (\tfrac{8}{3} + C) - (\tfrac{1}{3} + C) = \tfrac{7}{3}$$

Because the arbitrary constant C always appears in both terms in this process and therefore always cancels itself, it will be omitted in future calculations.

Example 2 Find the area beneath the curve

$$y = f(x) = \frac{3\sqrt{x}}{2} + 1$$

between $x = 4$ and $x = 9$.

Solution The region whose area is to be determined is shown in the following figure.

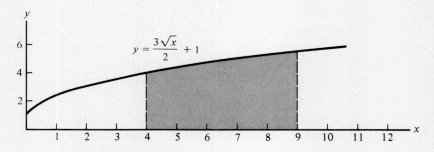

Noting that an antiderivative of $f(x)$ is

$$F(x) = x^{3/2} + x$$

we get for the area, using Equation 8.2.2

$$\text{Area} = F(9) - F(4)$$
$$= (9^{3/2} + 9) - (4^{3/2} + 4) = 24$$

Example 3 The marginal cost function $C'(x)$ of a company making stratoloungers is described by the equation

$$C'(x) = 25 + 0.20x$$

where x is daily output. The present level of production stands at 100 units per day. What additional costs are incurred if output is increased to 150 units per day?

Solution The marginal cost function is shown in the figure on the following page. The additional cost resulting from increased production is shown as the shaded area. Noting that an antiderivative of $25 + 0.2x$ is given by

$$F(x) = 25x + 0.10x^2$$

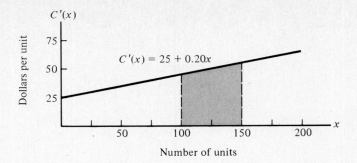

We get for the additional costs

Additional costs $= F(150) - F(100)$

$$= (3750 + 2250) - (2500 + 1000) = \$2500$$

This answer could also have been obtained through simple geometrical procedures by breaking the region up into a rectangle and triangle, as shown in the accompanying figure, and calculating the sum of the areas.

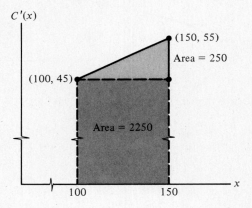

Occasionally, the area itself will be given and either a or b is to be found.

Example 4 | Find the value of b for which the area of the region beneath the curve

$$y = f(x) = \frac{1}{x^2}$$

between $x = \frac{1}{2}$ and $x = b$ equals $\frac{7}{4}$.

Solution | The curve and the region whose area has been specified are shown in the accompanying figure. Again, noting that an antiderivative of $1/x^2$ is

$$F(x) = -\frac{1}{x}$$

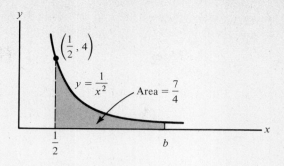

we obtain, using Equation 8.2.2

$$\text{Area} = F(b) - F(a)$$

$$\frac{7}{4} = \frac{-1}{b} - \left(\frac{-1}{\frac{1}{2}}\right)$$

from which we get

$$b = 4$$

Problems of the type illustrated in Example 4 are useful for estimating how long it would take to exhaust all known reserves of some natural resource, as illustrated in the next example.

Example 5 | Known natural gas reserves in the United States are estimated to be 225 trillion ft^3. Annual production is occurring at a rate described by the equation

$$P(t) = 20e^{-0.08t}$$

where t is measured in years from the present ($t = 0$) and P is the annual production level in trillions of cubic feet per year. How long would it take to exhaust known reserves if annual production continues according to the equation?

Solution | The curve showing the annual production P as a function of time is shown in the following figure. We want to find the value of t, called T, for which the area

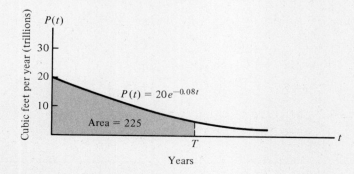

under the curve equals the known reserves, namely 225 trillion ft³. Noting that an antiderivative of $20e^{-0.08t}$ is given by

$$F(t) = \frac{20e^{-0.08t}}{-0.08} = -250e^{-0.08t}$$

we use Equation 8.2.2, with area = 225, $a = 0$, and $b = T$, getting

$$225 = F(T) - F(0)$$

$$225 = -250e^{-0.08T} + 250$$

$$-25 = -250e^{-0.08T}$$

$$0.10 = e^{-0.08T}$$

This equation can be solved for T by using the methods described in Section 5.2. Taking the natural logarithm of both sides gives

$$\ln (0.10) = \ln (e^{-0.08T}) = (-0.08T)(\ln e)$$

Noting that $\ln e = 1$ and that $\ln (0.10) = -2.3$ (Table C), we get

$$-0.08T = -2.3$$

from which we get

$$T = 28.75 \text{ years}$$

EXERCISE 8.2

For Exercises 1–12, find the area of the regions shown in the figures.

1.

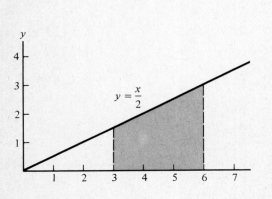

2.

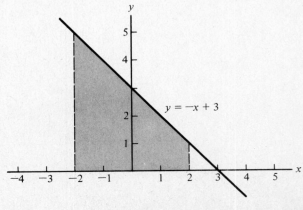

3.

4.

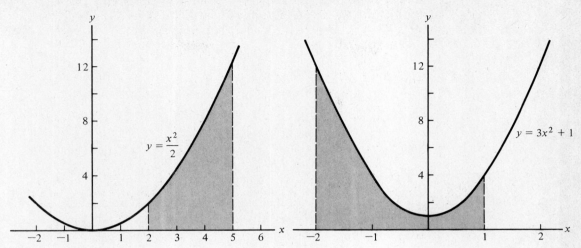

5.

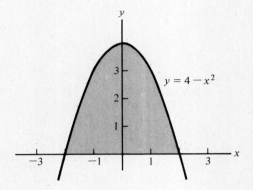

6.

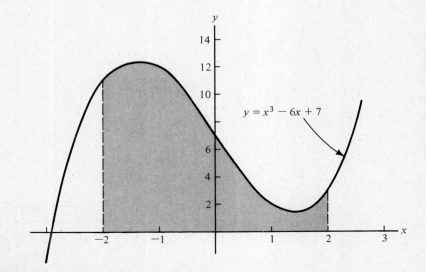

7.

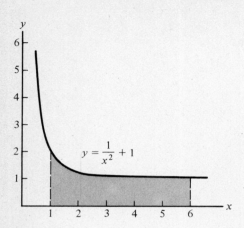

$y = \dfrac{1}{x^2} + 1$

8.

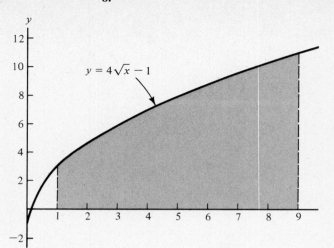

$y = 4\sqrt{x} - 1$

9.

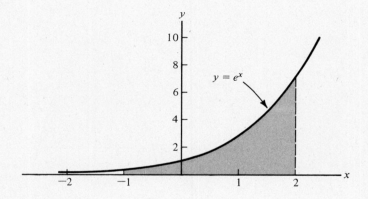

$y = e^x$

10.

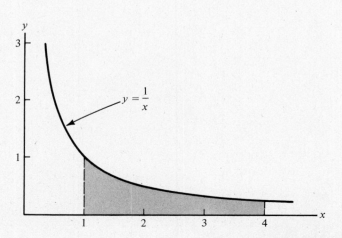

$y = \dfrac{1}{x}$

11.

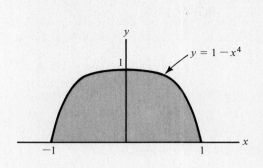

12. HINT: Use Newton's method to find one of the roots.

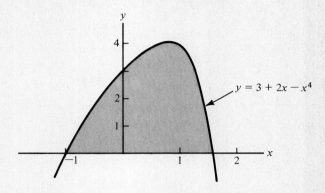

Find the areas under the curves described in Exercises 13–19. The x axis serves as the lower boundary for each region.

13. $y = f(x) = x^2 - 2x + 2$ From $x = -2$ to $x = 1$

14. $y = f(x) = 2 - 3x^3$ From $x = -3$ to $x = 0$

15. $y = f(x) = 3x - \dfrac{1}{x^4}$ From $x = 1$ to $x = 2$

16. $y = f(x) = e^x - 2x$ From $x = 0$ to $x = 1$

17. $y = f(x) = x^2 - \dfrac{1}{x}$ From $x = 1$ to $x = 3$

18. $y = f(x) = 3\sqrt{x} + 2$ From $x = 4$ to $x = 16$

19. $y = f(x) = 2$ From $x = -3$ to $x = 2$

20. The marginal profit function for a company manufacturing motorbikes is described by the equation

$$P'(x) = 300 + 6x - 0.03x^2$$

where x represents the number of units sold and $P(x)$ is the profit in dollars. How much additional profit is earned when the sales level is increased from 100 to 110 units?

21. Using continuous double-declining balance, the annual depreciation of a mobile television unit as a function of time is given by the equation

$$D(t) = 50{,}000e^{-0.20t}$$

where t is time in years (from date of purchase) and $D(t)$ is the annual depreciation in dollars per year.

(a) How much depreciation occurs between the end of the second and fifth years following purchase of the unit?

(b) How long does it take from date of purchase for the total depreciation to equal $150,000?

22. A large daily metropolitan newspaper purchased automated typesetting and printing equipment. The annual savings in labor costs is given by the equation

$$S(t) = 40 - 0.2t - 0.03t^2$$

where t represents time in years from date of purchase and S is the annual savings in thousands of dollars per year. Annual operating and maintenance costs are given by the equation

$$C(t) = 10 + 3\sqrt{t}$$

where C is expressed in thousands of dollars per year. The net annual savings $N(t)$ equals

$$S(t) - C(t)$$

Find the total net savings over the first four years of operation.

23. The marginal cost function associated with producing file cabinets for an office equipment company is described by the equation

$$C'(x) = 9 + 0.02x$$

where x is the number of units produced and $C'(x)$ is the marginal cost in dollars per unit. If total production costs have been set at $3600, how many units can be produced?

8.3 Definite Integral, Fundamental Theorem of Calculus

In the last section, it was shown that the area of a region in the plane beneath the curve $y = f(x)$ between $x = a$ and $x = b$ can be found by evaluating the difference $F(b) - F(a)$ where $F'(x) = f(x)$. This section will describe another method by which the area can be determined. As shown in Figure 8.8, the area under the curve is subdivided into a large number of rectangles, the sum of whose areas can be found directly. The area under the curve is determined by finding the limit of this sum as the number of rectangles becomes infinite. More important, this process will lead to a definition of the *definite integral*

$$\int_a^b f(x) \, dx$$

following which the *fundamental theorem of calculus* will be presented.

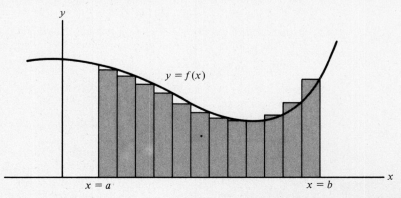

Figure 8.8

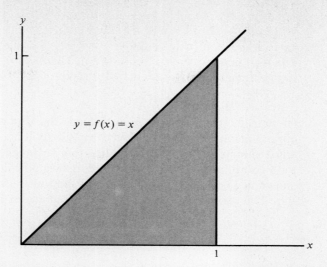

Figure 8.9

We will begin by finding the area of the triangular region beneath the curve $y = f(x) = x$ between $x = 0$ and $x = 1$ as shown in Figure 8.9. From geometry we know that the area A of any triangle is given by the formula

$$A = \tfrac{1}{2}(\text{base})(\text{height})$$

Noting that the base and height are both equal to 1, we find that A becomes

$$A = \tfrac{1}{2}$$

The process we are about to describe to obtain the area is fairly lengthy when compared to the simple geometrical analysis just employed. Our objective is to describe the process underlying the definition of the definite integral and this can be accomplished more effectively if the geometry of the region is not complex.

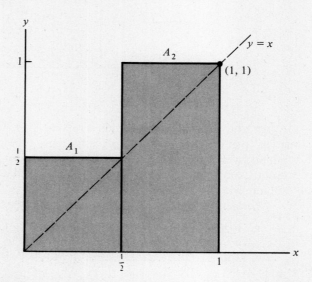

Figure 8.10

Suppose we begin by dividing the interval $[0, 1]$ into two equal subintervals $[0, \frac{1}{2}]$ and $[\frac{1}{2}, 1]$ and construct two rectangles whose heights equal the y coordinates at the right-hand endpoints, that is, $f(\frac{1}{2})$ and $f(1)$, and whose bases equal $\frac{1}{2}$. The resulting rectangles are shown in Figure 8.10. The sum of the areas of the two rectangles can be calculated directly

$$A_1 + A_2 = (\tfrac{1}{2})(\tfrac{1}{2}) + (\tfrac{1}{2})(1) = \tfrac{3}{4}$$

Obviously, this approximation exceeds the true area beneath the curve. The approximation can be improved by subdividing the interval $[0, 1]$ into four equal subintervals and constructing four rectangles whose heights are equal to the y coordinates of the curve at the right-hand endpoints of each subinterval as shown in Figure 8.11. The heights of the four rectangles are $f(\frac{1}{4})$, $f(\frac{1}{2})$, $f(\frac{3}{4})$, and $f(1)$. The sum of the areas of the four rectangles becomes

$$A_1 + A_2 + A_3 + A_4 = (\tfrac{1}{4})(\tfrac{1}{4}) + (\tfrac{1}{4})(\tfrac{1}{2}) + (\tfrac{1}{4})(\tfrac{3}{4}) + (\tfrac{1}{4})(1)$$

$$= \frac{1}{4}\left(\frac{1 + 2 + 3 + 4}{4}\right) = \frac{10}{16} = \frac{5}{8}$$

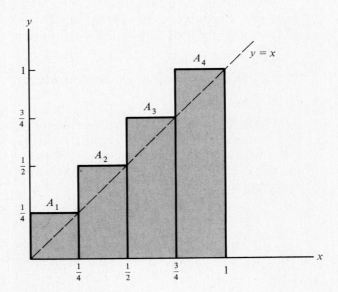

Figure 8.11

Again, the approximation yields a result that is too large, but it is an improvement over the preceding approximation. A better approximation can be obtained by subdividing the interval $[0, 1]$ into an even larger number of subintervals, say eight, as shown in Figure 8.12, and constructing rectangles on each subinterval as described previously.

The sum of the areas of the eight rectangles then becomes

$$A_1 + A_2 + \cdots + A_7 + A_8 = (\tfrac{1}{8})(\tfrac{1}{8}) + (\tfrac{1}{8})(\tfrac{1}{4}) + \cdots + (\tfrac{1}{8})(\tfrac{7}{8}) + (\tfrac{1}{8})(1)$$

$$= \frac{1}{8}\left(\frac{1 + 2 + 3 + 4 + 5 + 6 + 7 + 8}{8}\right)$$

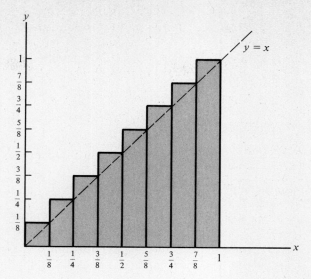

Figure 8.12

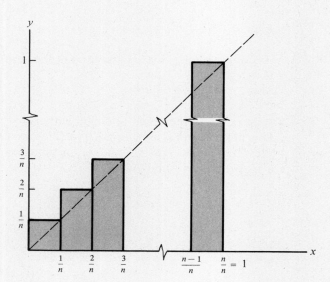

Figure 8.13

We want to call your attention to the structure of the numerator of the second factor because an expression similar to this will appear when we generalize this procedure. Carrying through the calculation gives

$$A_1 + A_2 + \cdots + A_7 + A_8 = \tfrac{1}{8}(\tfrac{36}{8}) = \tfrac{9}{16}$$

At this point, the procedure can be generalized by subdividing the interval $[0, 1]$ into n equal subintervals and constructing n rectangles whose heights equal the y coordinates corresponding to the right-hand endpoint of each subinterval as shown in Figure 8.13. The sum of the areas of the n rectangles then can be written

$$A_1 + A_2 + A_3 + \cdots + A_n = \frac{1}{n}\left(\frac{1}{n}\right) + \frac{1}{n}\left(\frac{2}{n}\right) + \frac{1}{n}\left(\frac{3}{n}\right) + \cdots$$

$$+ \frac{1}{n}\left(\frac{n-1}{n}\right) + \frac{1}{n}\left(\frac{n}{n}\right)$$

$$= \frac{1}{n^2}\left[1 + 2 + 3 + \cdots + (n-1) + n\right]$$

In Appendix D, we show that

$$1 + 2 + 3 + \cdots + (n-1) + n = \frac{n(n+1)}{2}$$

so the sum of the areas can be written

$$A_1 + A_2 + A_3 + \cdots + A_n = \frac{n+1}{2n}$$

Next, we want to determine the area under the curve by letting the number of rectangles n become infinite. Using the methods described in Section 4.4, we have

$$\text{Area} = \lim_{n \to \infty}(A_1 + A_2 + A_3 + \cdots + A_n) = \lim_{n \to \infty}\frac{n+1}{2n} = \lim_{n \to \infty}\left(\frac{1}{2} + \frac{1}{2n}\right)$$

As n becomes infinite, the second term in the parentheses goes to 0 and we get

$$\text{Area} = \lim_{n \to \infty}\left(\frac{1}{2} + \frac{1}{2n}\right) = \frac{1}{2}$$

which agrees with the result obtained earlier.

This process can also be applied to determining the area beneath the curve $y = f(x) = x^2$ between $x = 0$ and $x = 1$, shown in Figure 8.14. We will proceed, as before, by subdividing the interval $[0, 1]$ into subintervals of equal

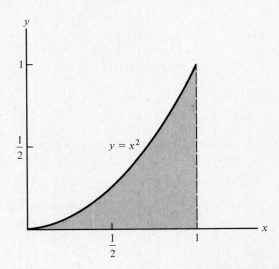

Figure 8.14

width, constructing rectangles on each subinterval, determining the sum of the areas of the rectangles, and finally obtaining the limit of this sum as the number of rectangles n becomes infinite. The process, together with the appropriate figures, is shown next.

$$A_1 + A_2 = \tfrac{1}{2}(\tfrac{1}{4}) + \tfrac{1}{2}(1) = \tfrac{5}{8}$$

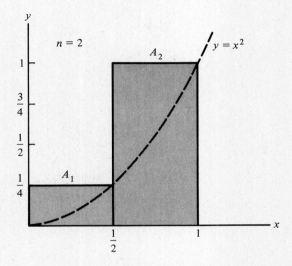

$$A_1 + A_2 + A_3 + A_4 = \tfrac{1}{4}(\tfrac{1}{16}) + \tfrac{1}{4}(\tfrac{4}{16}) + \tfrac{1}{4}(\tfrac{9}{16}) + \tfrac{1}{4}(\tfrac{16}{16})$$

$$= \frac{1}{4}\left(\frac{1 + 4 + 9 + 16}{16}\right)$$

$$= \tfrac{30}{64} = \tfrac{15}{32}$$

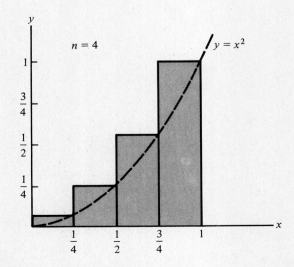

$$A_1 + A_2 + \cdots + A_8 = \tfrac{1}{8}(\tfrac{1}{64}) + \tfrac{1}{8}(\tfrac{4}{64}) + \tfrac{1}{8}(\tfrac{9}{64}) + \cdots + \tfrac{1}{8}(\tfrac{64}{64})$$

$$= \frac{1}{8}\left(\frac{1 + 4 + 9 + \cdots + 49 + 64}{64}\right) = \tfrac{204}{512} = \tfrac{51}{128}$$

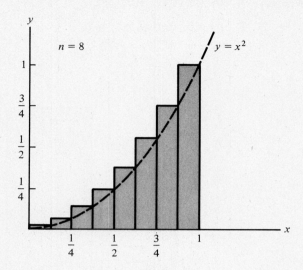

$$A_1 + A_2 + \cdots + A_n = \frac{1}{n}\left(\frac{1}{n}\right)^2 + \frac{1}{n}\left(\frac{2}{n}\right)^2 + \cdots + \frac{1}{n}\left(\frac{n}{n}\right)^2$$

$$= \frac{1}{n}\left(\frac{1^2 + 2^2 + 3^2 + \cdots + n^2}{n^2}\right)$$

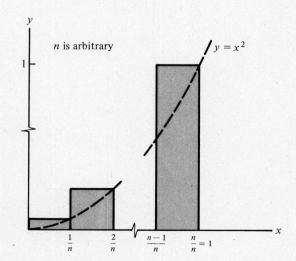

In Appendix D, we show that $(1)^2 + (2)^2 + (3)^2 + \cdots + (n)^2 = \dfrac{n(n+1)(2n+1)}{6}$ so that we have, for arbitrary n

$$A_1 + A_2 + A_3 + \cdots + A_n = \frac{1}{n}\left[\frac{n(n+1)(2n+1)}{6n^2}\right]$$

$$= \frac{(n+1)(2n+1)}{6n^2}$$

The next step is to determine the limit as $n \to \infty$. The process will be simplified somewhat if we write the sum as

$$A_1 + A_2 + A_3 + \cdots + A_n = \frac{1}{6}\left(\frac{n+1}{n}\right)\left(\frac{2n+1}{n}\right)$$

$$= \frac{1}{6}\left(1 + \frac{1}{n}\right)\left(2 + \frac{1}{n}\right)$$

The area under the curve then becomes

$$\text{Area} = \lim_{n\to\infty} \frac{1}{6}\left(1 + \frac{1}{n}\right)\left(2 + \frac{1}{n}\right)$$

Noting that $\frac{1}{n} \to 0$ as $n \to \infty$, we get for A

$$\text{Area} = \tfrac{1}{6}(1)(2) = \tfrac{1}{3}$$

When the process just described is generalized and applied to any function $y = f(x)$ without regard to the sign of $f(x)$ over the closed interval $[a, b]$, we arrive at the definition of the *definite integral*

$$\int_a^b f(x)\ dx$$

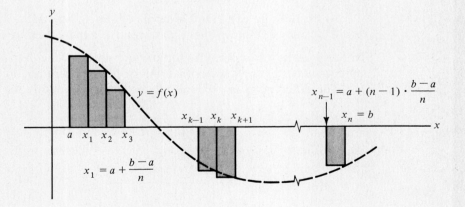

Suppose that $y = f(x)$ is a continuous function over the closed interval $a \le x \le b$. Let the interval be subdivided into n equal subintervals, each of whose widths is $(b - a)/n$. Designating the right-hand endpoint in each subinterval x_1, $x_2, \ldots, x_n = b$, the definite integral $\int_a^b f(x)\ dx$ can be defined as

$$\int_a^b f(x)\,dx = \lim_{n\to\infty} \frac{b-a}{n}\,[f(x_1) + f(x_2) + f(x_3) + \cdots + f(x_n)] \qquad \textbf{(8.3.1)}$$

As in the case of the first derivative, the effort required to obtain a result directly from Equation 8.3.1 is generally difficult and time consuming. Fortunately, the effort can be reduced considerably for those situations where the antiderivative of $f(x)$ is known by employing the fundamental theorem of calculus.

FUNDAMENTAL THEOREM OF CALCULUS

If the function $y = f(x)$ is continuous over the closed interval $[a,\,b]$, and if $F(x)$ is an antiderivative of $f(x)$, then

$$\int_a^b f(x)\,dx = F(b) - F(a) \qquad\qquad \textbf{(8.3.2)}$$

This result should not be too surprising in view of the similar result in Equation 8.2.2, where the area beneath the curve $y = f(x)$ between $x = a$ and $x = b$ was expressed as the difference $F(b) - F(a)$. Equation 8.2.2 can be regarded as a special case of the fundamental theorem of calculus when $f(x) \geq 0$ everywhere over $[a,\,b]$. Before going on, it should be noted that the fundamental theorem of calculus is useful at this stage only if

1. $F(x)$ can be expressed in terms of simple functions, and
2. It can be found without an excessive expenditure of time and effort.

In those cases where these conditions are not met, the definite integral can be approximated by techniques that will be discussed in Section 9.4.

The difference $F(b) - F(a)$ can also be expressed in the form

$$F(b) - F(a) = F(x) \Big|_a^b$$

so that the fundamental theorem of calculus takes the form

$$\int_a^b f(x)\,dx = F(x)\Big|_a^b = F(b) - F(a) \qquad\qquad \textbf{(8.3.3)}$$

Example 1 | Evaluate $\displaystyle\int_1^2 x^2\,dx$

Solution | Since $F(x) = x^3/3$, we get

$$\int_1^2 x^2\, dx = \frac{x^3}{3}\Big|_1^2 = \frac{8}{3} - \frac{1}{3} = \frac{7}{3}$$

In defining the definite integral, it is not necessary that the integrand $f(x)$ be nonnegative everywhere over the interval $a \le x \le b$. However, the definite integral cannot be interpreted as an area unless the integrand is nonnegative everywhere over $[a, b]$. In those cases where the integrand is nonpositive for all values of x between $x = a$ and $x = b$, the area of the region between the curve and the x axis is found by using the formula

$$\text{Area} = -\int_a^b f(x)\, dx \qquad \text{where } f(x) \le 0 \text{ for } a \le x \le b \qquad \textbf{(8.3.4)}$$

Example 2 | Find the area of the region between the curve $y = f(x) = x^2 - 1$ and the x axis bounded by the vertical lines $x = -1$ and $x = 1$.

Solution | A quick sketch of the curve indicates that the region whose area is to be evaluated lies totally below the x axis between $x = -1$ and $x = +1$. The area of the region then becomes

$$\text{Area} = -\int_{-1}^{1} (x^2 - 1)\, dx = -\left(\frac{x^3}{3} - x\right)\Big|_{-1}^{1}$$
$$= -(\tfrac{1}{3} - 1) + (-\tfrac{1}{3} + 1) = \tfrac{4}{3}$$

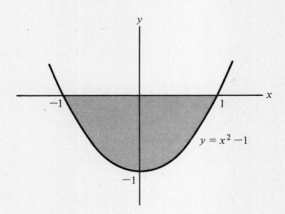

$y = x^2 - 1$

The definite integral possesses a number of properties including

I. $\displaystyle\int_a^a f(x)\, dx = 0$

II. $\displaystyle\int_a^b f(x)\, dx = -\int_b^a f(x)\, dx$

III. $\displaystyle\int_a^b cf(x)\, dx = c\int_a^b f(x)\, dx \qquad$ where c is a constant

Example 3 | Evaluate $\displaystyle\int_1^4 5x^2\,dx$

Solution | Using property III, we can write

$$\int_1^4 5x^2\,dx = 5\int_1^4 x^2\,dx$$

Because $F(x) = x^3/3$, we get

$$5\int_1^4 x^2\,dx = 5\,\frac{x^3}{3}\,\Big|_1^4 = 5\left(\frac{64}{3} - \frac{1}{3}\right) = 105$$

IV. $\displaystyle\int_a^b [f(x) \pm g(x)]\,dx = \int_a^b f(x)\,dx \pm \int_a^b g(x)\,dx$

Example 4 | Evaluate

$$\int_0^1 (e^x + x)\,dx$$

Solution | Using property IV, we can write

$$\int_0^1 (e^x + x)\,dx = \int_0^1 e^x\,dx + \int_0^1 x\,dx = e^x\,\Big|_0^1 + \frac{x^2}{2}\,\Big|_0^1$$

$$= (e^1 - e^0) + (\tfrac{1}{2} - 0) = \frac{2e - 1}{2}$$

V. $\displaystyle\int_a^b f(x)\,dx = \int_a^c f(x)\,dx + \int_c^b f(x)\,dx \qquad \text{where } a < c < b$

Property V is very useful in the following situations:

1. The integrand $f(x)$ is defined piecewise for $a \le x \le b$.

Example 5 | Evaluate

$$\int_1^4 f(x)\,dx$$

where

$$f(x) = \begin{cases} 3x^2 & 1 \le x \le 2 \\ \dfrac{48}{x^2} & 2 < x \le 4 \end{cases}$$

Solution | The easiest way to carry out the integration is to break the integral into two parts

$$\int_1^2 f(x)\,dx + \int_2^4 f(x)\,dx = \int_1^2 3x^2\,dx + \int_2^4 \frac{48}{x^2}\,dx = x^3\,\Big|_1^2 - \frac{48}{x}\,\Big|_2^4$$

$$= (8 - 1) - (12 - 24) = 19$$

The result obtained represents the total area of the two regions shown in the next figure.

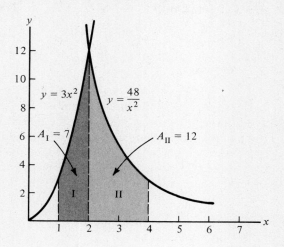

2. Part of the region whose area is to be evaluated is above the x axis while the remainder is below the x axis.

Example 6   Find the area of the region between the curve $y = f(x) = 1 - x^2$ and the x axis, bounded on the left by the vertical line $x = 0$ and on the right by the line $x = 2$.

Solution   The region whose area we wish to find lies above the x axis for $0 \le x < 1$ and below for $1 < x \le 2$, as shown in Figure 8.15. The area of the shaded region is composed of two parts

$$\int_0^1 (1 - x^2)\, dx \qquad \text{and} \qquad -\int_1^2 (1 - x^2)\, dx$$

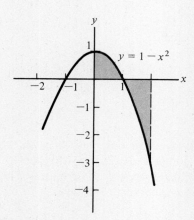

Figure 8.15

that is

$$\text{Area} = \int_0^1 (1 - x^2)\, dx - \int_1^2 (1 - x^2)\, dx$$

$$= \left(x - \frac{x^3}{3}\right)\bigg|_0^1 - \left(x - \frac{x^3}{3}\right)\bigg|_1^2$$

$$= (1 - \tfrac{1}{3}) - [(2 - \tfrac{8}{3}) - (1 - \tfrac{1}{3})]$$

$$= 2$$

Because the definite integral, unlike the area, can be either positive or negative, it is very useful in indicating the direction in which total changes occur. For example, the definite integral can be used to calculate a nation's balance of payments over a given period of time. If the definite integral is positive, the dollar value of exported goods and services exceeds that of imported goods and services, whereas the reverse is true if the definite integral is negative.

Example 7   Because of increasing dependence on foreign oil, more dollars are flowing out of the United States than are flowing in due to exports. However, as alternate sources of energy are found, the balance of payments situation is expected to improve. Suppose that the annual balance of payments over the next 10 years is expected to be governed by the equation

$$B(t) = -20.4 + 1.6t + 0.3t^2 \qquad 0 \le t \le 10$$

where $B(t)$ is the annual balance of payments in billions of dollars per year and t is the time (years) from the present.
   a. What will be the nation's balance of payments (BP) over the next three years?

Solution   The net number of dollars BP flowing into the country over the next three years if given by the equation

$$\text{BP} = \int_0^3 (-20.4 + 1.6t + 0.3t^2)\, dt$$

which equals

$$\text{BP} = (-20.4t + 0.8t^2 + 0.1t^3)\, \bigg|_0^3$$

Evaluating gives

$$\text{BP} = (-61.2 + 7.2 + 2.7) = -\$51.3 \text{ billion}$$

The negative sign indicates that over the next three years the dollar value of imported goods and services is expected to exceed the dollar value of exported goods and services by $51.3 billion. The area of the shaded region in the following figure equals $51.3 billion.

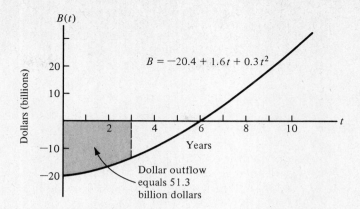

b. What is the nation's balance of payments from year 5 to year 10?

Solution   In this case the balance of payments BP is given by the equation

$$BP = \int_5^{10} (-20.4 + 1.6t + 0.3t^2)\, dt$$

$$= (-20.4t + 0.8t^2 + 0.1t^3) \Big|_5^{10}$$

$$= (-204 + 80 + 100) - (-102 + 20 + 12.5)$$

$$= \$45.5 \text{ billion}$$

Thus, over this 5-year period the dollar value of exported goods and services is expected to exceed that of imported goods and services by $45.5 billion. The next figure shows that the $45.5 billion is a net, resulting from an outflow of $2.5 billion from year 5 to 6 and a $48 billion inflow during years 6 through 10.

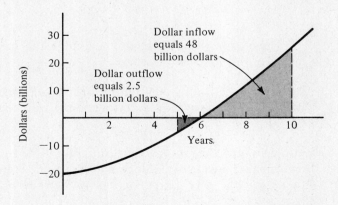

Example 8 | The marginal revenue function for a firm producing skateboards is given by

$$R'(x) = 120 - 2x$$

where x is the number of units produced, given in thousands, and $R'(x)$ is the marginal revenue expressed in dollars per unit. If production is currently at 50 thousand units, what will be the change in total revenue if production level is raised to 65 thousand units?

Solution The change in total revenue is given by

$$\int_{50}^{65} R'(x) \, dx = \int_{50}^{65} (120 - 2x) \, dx$$

which becomes

$$(120x - x^2) \Big|_{50}^{65} = [(120)(65) - 65^2] - [(120)(50) - 50^3] = \$75 \text{ thousand}$$

The change in total revenue can be viewed graphically as the net of a \$100 thousand increase as production is raised from 50 to 60 thousand units and a \$25 thousand decrease as production is raised further to 65 thousand units.

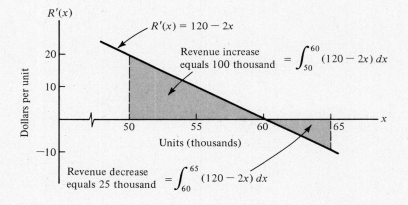

The definite integral can also be used in applications requiring the determination of the area under a curve.

Example 9 Annual purchases of goods and services by the federal government during the past few years have risen according to the equation

$$P(t) = 65e^{0.07t}$$

where t is the time in years measured from 1965 ($t = 0$), and $P(t)$ represents annual purchases in billions of dollars per year. What is the total dollar value of the goods and services purchased by the federal government between 1970 ($t = 5$) and 1975 ($t = 10$)?

Solution The total dollar value of goods and services purchased during this five-year period equals the area shown in the following figure and can be found by evaluating the following definite integral:

$$\int_5^{10} P(t) \ dt = \int_5^{10} 65e^{+0.07t} \ dt$$

$$= \frac{65e^{0.07t}}{0.07} \ \Big|_5^{10}$$

$$= 928.6(e^{0.7} - e^{0.35})$$

$$= \$552.2 \ \text{billion}$$

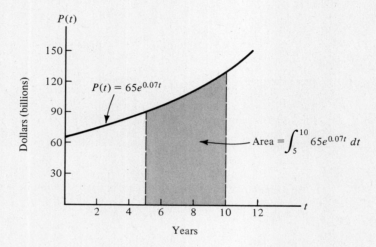

EXERCISE 8.3

1. When demonstrating the process for approximating the area under the curve $y = f(x) = x$ by subdividing the interval $[0, 1]$ into equal subintervals, the case $n = 3$ was among those not treated. On the coordinate system below, sketch the three rectangles and calculate the sum of the areas.

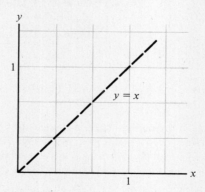

2. When finding the area under a curve, it is not necessary to restrict the analysis to those cases where the height of the rectangles corresponds to the right-hand end-

points. Other methods for accomplishing this can also be developed. Suppose that for the curve $y = f(x) = x$, the heights of the rectangles are determined at the left-hand endpoints.

a. Calculate the approximations to the area for $n = 2$, 4, and 8 as shown in the following figures.

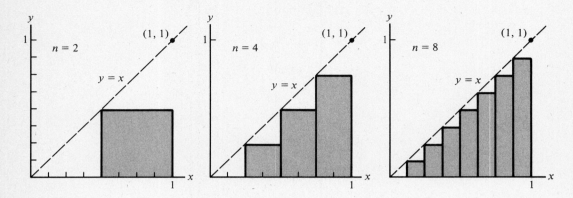

b. Generalize the result obtained in part a to the case where the number of rectangles n is arbitrary.

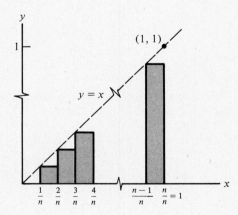

c. Find the limit of the result obtained in part b as $n \to \infty$ to obtain the area beneath the curve.

3. Extend the method used in the text to find the area beneath the curve $y = f(x) = x$ between $x = 0$ and $x = 2$. As shown in the figure on the next page, the interval $[0, 2]$ is subdivided into n equal subintervals; next n rectangles are constructed using the value of the right-hand edge of each subinterval to find the height.

a. Determine the sum of the areas $A_1 + A_2 + A_3 + \cdots + A_n$
b. Find the area under the curve by determining the limit of the result obtained in part a as $n \to \infty$.

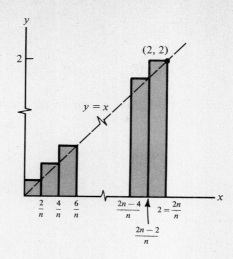

Evaluate the definite integrals in Exercises 4–19.

4. $\int_1^4 (5 + x^2) \, dx$

5. $\int_{-1}^0 (2 - x) \, dx$

6. $\int_{-2}^1 4x^2 \, dx$

7. $\int_1^3 \frac{6}{x} \, dx$

8. $\int_{-3}^3 (x^2 - 7) \, dx$

9. $\int_1^9 \left(\frac{1}{2\sqrt{x}} + x^2 - 3 \right) dx$

10. $\int_0^1 (e^{2x} + x^3 + 1) \, dx$

11. $\int_{-2}^{-1} \frac{1}{x^2} \, dx$

12. $\int_{-1}^2 (3t^2 - t + 2) \, dt$

13. $\int_1^8 2\sqrt[3]{x} \, dx$

14. $\int_1^2 \left(6e^{3x} - \frac{1}{x} \right) dx$

15. $\int_1^{16} 5\sqrt[4]{t} \, dt$

16. $\int_{1/2}^{3/4} \left(\frac{2}{x^4} + 9x \right) dx$

17. $\int_1^3 f(x) \, dx$ where $f(x) = \begin{cases} x^2 & 0 \le x \le 2 \\ 6 - x & x > 2 \end{cases}$

18. $\int_{-1}^1 f(x) \, dx$ where $f(x) = \begin{cases} 1 - x^3 & x \le 0 \\ e^x & x > 0 \end{cases}$

19. $\int_0^4 f(x) \, dx$ where $f(x) = \begin{cases} 7 - 2x & x \le 3 \\ x^2 - 2x - 2 & x > 3 \end{cases}$

Find the value of t that satisfies the equations in Exercises 20–24.

20. $\int_1^t 3x \, dx = \frac{21}{2}$

21. $\int_0^t (6x + 3) \, dx = 6$

22. $\int_1^t 3x^2\, dx = 7$ **23.** $\int_t^1 \sqrt{x}\, dx = \frac{1}{3}$

24. $\int_t^3 (8 - 3x)\, dx = \frac{33}{2}$

For Exercises 25–29, find the area of the regions shown in the graphs.

25.

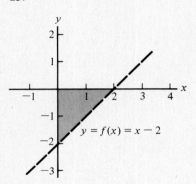

$y = f(x) = x - 2$

26.

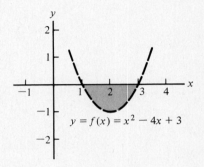

$y = f(x) = x^2 - 4x + 3$

27.

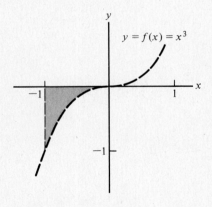

$y = f(x) = x^3$

28.

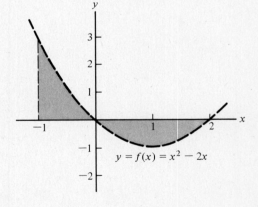

$y = f(x) = x^2 - 2x$

29.

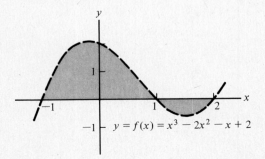

$y = f(x) = x^3 - 2x^2 - x + 2$

30. A television station has purchased a mobile unit that gets 16 miles per gallon of gasoline. The unit travels 2000 miles per month. If the price per gallon of gasoline is expected to increase according to the equation

$$p(t) = 0.95 + 0.05\sqrt{t}$$

where t is the time from the present in months, and $p(t)$ is the price per gallon in dollars, find the total amount of money that will be spent on gasoline during the next three years.

31. The marginal revenue $R'(x)$ for a company selling cash registers equipped with microprocessors is given by the equation

$$R'(x) = 300 - 2x$$

where x equals the number of units sold (thousands) and $R'(x)$ is the marginal revenue in dollars per unit. If total sales presently equal 100,000 units, what change in total revenue will occur if production is increased to 160,000 units? Plot the marginal revenue function on the following system and shade in the area corresponding to the change in total revenue.

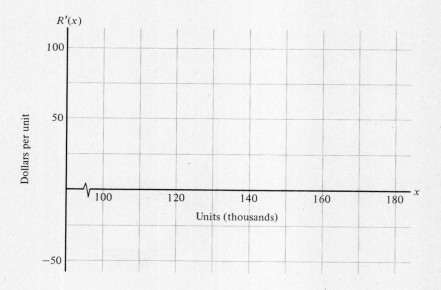

32. Annual purchases of goods and services by state and local governments are described by the equation

$$P(t) = 70e^{0.12t}$$

where t is the time in years measured from the beginning of 1965 and $P(t)$ is the annual spending in billions of dollars. How much money was spent by state and local governments between the beginning of 1971 and the end of 1974? On the following coordinate system plot $P(t)$ and shade in the area corresponding to the amount spent by state and local governments.

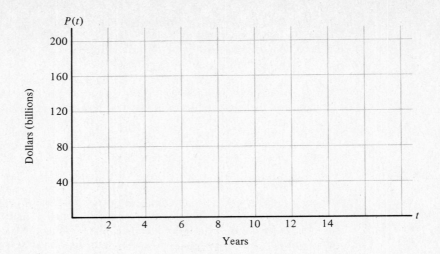

33. During the recessionary period of the mid-1970s, annual inventory accumulation $A(t)$ by U.S. manufacturers in billions of dollars per year can be described by the equation

$$A(t) = t^3 + 7t^2 - 28t + 20 \qquad 0 \leq t \leq 3$$

where $t = 0$ corresponds to the beginning of 1974. The graph of the function is shown in the following figure. What was the change in inventories between the beginning of 1974 and the end of 1975?

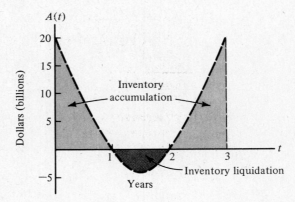

8.4 Area between Two Curves

In addition to determining the area of a region between a given curve $y = f(x)$ and the x axis, there are many situations where the area between two given curves is desired. Suppose the curve $y_1 = f(t)$ in Figure 8.16 represents the predicted annual consumption of gasoline in the United States as a function of time t expressed in years from the present. In an attempt to reduce consumption, the

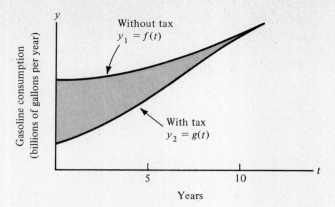

Figure 8.16

president's chief economic advisor recommends a large tax increase on gasoline. An analysis of the effect of the tax on gasoline consumption yields the
curve $y_2 = g(t)$, also shown in Figure 8.16. Assuming that no further tax increases occur, the figure indicates that the tax increase becomes less effective
with time so that 10 years from now, annual consumption will return to its un-
taxed levels. The area of the shaded region in Figure 8.16 represents the total
number of gallons of gasoline saved over the 10-year period due to the imposi-
tion of the tax. The shaded area can be expressed as the definite integral

$$\int_0^{10} [f(t) - g(t)]\, dt$$

The area between two given curves $y_1 = f(x)$ and $y_2 = g(x)$ bounded on the
left and right, respectively, by the vertical lines $x = a$ and $x = b$ as shown in
Figure 8.17, is given by the equation

$$\boxed{\text{Area} = \int_a^b [f(x) - g(x)]\, dx} \qquad (8.4.1)$$

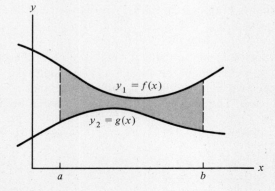

Figure 8.17

where $f(x)$ and $g(x)$ are continuous and $f(x) \geq g(x)$ over the interval $a \leq x \leq b$. This result can be demonstrated by examining Figure 8.18 in which the area beneath each curve and the x axis is shown. The area between the two curves then becomes

$$\text{Area between curves} = A_\text{I} - A_\text{II} = \int_a^b f(x)\, dx - \int_a^b g(x)\, dx$$

$$= \int_a^b [f(x) - g(x)]\, dx$$

When the reverse is true, that is, $g(x) \geq f(x)$ over the interval $a \leq x \leq b$, then

$$\boxed{\text{Area between curves} = \int_a^b [g(x) - f(x)]\, dx} \qquad\qquad \textbf{(8.4.2)}$$

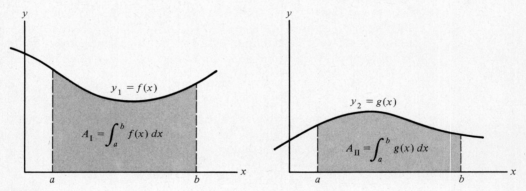

Figure 8.18

Example 1 Find the area of the region located between the two curves

$$y_1 = f(x) = 8 - x^2 \qquad \text{and} \qquad y_2 = g(x) = x + 1$$

that is bounded on the left by the vertical line $x = -1$ and on the right by the vertical line $x = 2$.

Solution It is advisable to make a rough sketch of the two curves to determine the relative positions of the curves $y_1 = f(x)$ and $y_2 = g(x)$. The following figure shows the two curves where it can be seen that $f(x) > g(x)$ over the interval $-1 \leq x \leq 2$. The area of the shaded region can be determined as follows:

$$\text{Area} = \int_{-1}^2 [(8 - x^2) - (x + 1)]\, dx = \int_{-1}^2 (7 - x^2 - x)\, dx$$

$$= \left(7x - \frac{x^3}{3} - \frac{x^2}{2}\right)\Bigg|_{-1}^2 = \left(14 - \frac{8}{3} - 2\right) - \left(-7 + \frac{1}{3} - \frac{1}{2}\right) = \frac{33}{2}$$

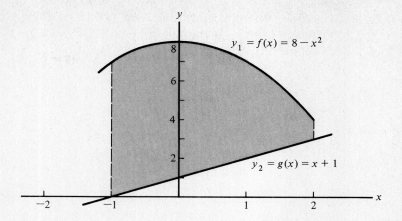

The area between two curves is also useful in making comparisons between two alternative courses of action. This is illustrated in the next example.

Example 2 | The Acme Fish Company is about to take delivery on a fleet of modern fishing trawlers to replace the inefficient and slow boats now in use. The expected annual catch of fish with the new fleet is described by the equation

$$C_1(t) = 25 - \sqrt{t} \qquad 0 \le t \le 16$$

where C_1 is the annual catch in millions of pounds per year and t is the time in years from the present. Expected annual catch with the old fleet is described by the equation

$$C_2(t) = 10e^{-0.12t}$$

What is the increase in the total number of fish caught over the next nine years because of modernization of the fishing fleet?

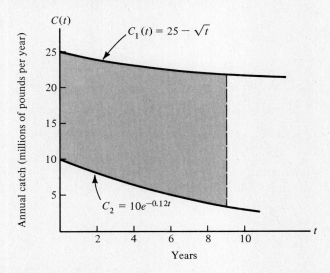

Solution The area of the shaded region in the preceding figure represents the expected additional catch over the next nine years. The additional catch can be written as the definite integral

$$\int_0^9 (25 - \sqrt{t} - 10e^{-0.12t})\, dt$$

Working through the details gives

$$\text{Additional catch} = \left(25t - \frac{2\sqrt{t^3}}{3} + 83.3e^{-0.12t}\right)\Bigg|_0^9$$

$$= 152 \text{ million lb}$$

For many problems, it is necessary to find the x coordinates of the points of intersection of the two curves in order to determine a and b, the upper and lower limits of integration. This situation is illustrated in the next example.

Example 3 Find the area between the two curves

$$y_1 = f(x) = 4 - x^2 \qquad \text{and} \qquad y_2 = g(x) = -x + 2$$

Solution The region whose area we wish to find is shown in the following figure.

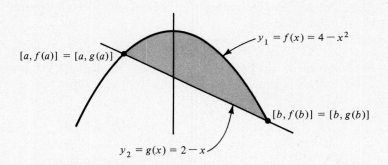

The area in question is given by the definite integral

$$\int_a^b [(4 - x^2) - (2 - x)]\, dx$$

Because a and b represent the x coordinates of the points of intersection, they can be found by setting $y_1 = y_2$ and then solving the resulting equation for x

$$4 - x^2 = -x + 2$$

$$x^2 - x - 2 = 0$$

$$(x - 2)(x + 1) = 0$$

from which we get the solutions

$$x_1 = -1 \qquad x_2 = 2$$

so that the area can be written as

$$\int_{-1}^{2} (2 + x - x^2)\, dx$$

which gives

$$\left(2x + \frac{x^2}{2} - \frac{x^3}{3}\right)\Big|_{-1}^{2} = \left(4 + 2 - \frac{8}{3}\right) - \left(-2 + \frac{1}{2} + \frac{1}{3}\right) = \frac{9}{2}$$

EXERCISE 8.4

Find the area between the curves described in Exercises 1–10.

1. $y_1 = f(x) = 5 - x$ and $y_2 = g(x) = x$ between $x = -1$ and $x = 2$

2. $y_1 = f(x) = \dfrac{8}{x^2}$ and $y_2 = g(x) = \dfrac{x^2}{4}$ between $x = 1$ and $x = 2$

3. $y_1 = f(x) = e^x$ and $y_2 = g(x) = x^3 - 1$ between $x = 0$ and $x = 1$

4. $y_1 = f(x) = 6x - 3x^2$ and $y_2 = -x$ between $x = \frac{1}{2}$ and $x = 2$

5. $y_1 = f(x) = e^x$ and $y_2 = g(x) = \dfrac{1}{x}$ between $x = 1$ and $x = 2$

6. $y_1 = f(x) = 2x + 3$ and $y_2 = g(x) = x^2$ between the points of intersection

7. $y_1 = f(x) = x^2$ and $y_2 = g(x) = x^3$ between the points of intersection

8. $y_1 = f(x) = 1 - x^2$ and $y_2 = g(x) = x^2 - 1$ between the points of intersection

9. $y_1 = f(x) = \sqrt{x}$ and $y_2 = g(x) = x^2$ between the points of intersection

10. $y_1 = f(x) = 5 + 6x - x^2$ and $y_2 = g(x) = x^2 + 5$ between the points of intersection

11. The Clean Cut Razor Blade Company plans to allocate a large part of its advertising budget to promote sales of its new Triple Trac Razor. According to a market analysis, annual sales as a function of time with and without the increase in promotional expenditures are given respectively by the equations

$$S_1(t) = 15e^{0.05t} \qquad \text{and} \qquad S_2(t) = 12 + 0.6t$$

where S is the annual sales in millions of dollars and t is the time in years. Over the next five years, what will be the expected increase in total sales because of the additional promotion?

12. The minimum hourly wage $W(t)$ as a function of time can be approximated by the equation

$$W(t) = 0.25 + 0.06t$$

where t is the time in years from 1938 when the minimum wage law went into effect, and $W(t)$ is expressed in dollars per hour. During this period, the average manufacturing hourly wage $\overline{W}(t)$ as a function of time can be approximated by the equation

$$\overline{W}(t) = 0.50 + 0.125t$$

During the period 1938–1978, how much more money has a worker who received the average manufacturing hourly wage earned than a worker who received only the minimum hourly rate? Assume that both worked 52 forty-hour weeks each year.

13. The Arnold Life Insurance Company has installed a computerized control system in its main office building to control heating and air conditioning and to monitor lighting and equipment operation. Annual energy use during the next 15 years is expected to be described by the equation

$$E_1(t) = 30e^{0.02t}   \quad 0 \le t \le 15$$

whereas annual energy consumption without the control system had been increasing according to the equation

$$E_2(t) = 40e^{0.03t}$$

where E_1 and E_2 are expressed in millions of kilowatt hours of electricity per year and t is the time in years measured from the present.

a. During the next 10 years, what is the expected savings in kilowatt hours of electricity because of the installation of the computerized control system?

b. The unit cost of a kilowatt hour of electricity C, in dollars per kilowatt hour, is expected to increase yearly for the next 10 years according to the equation

$$C(t) = 0.05e^{0.06t}$$

On the basis of this assumption, what will be the total savings in dollars over the next 10 years?

14. The Acme Press has two printing presses, one a small, older unit for which the marginal cost function has the form

$$C_1'(x) = 0.50 + 0.01x$$

and a new, large, high-speed unit for which the marginal cost function is

$$C_2'(x) = 0.30 + 0.002x$$

Set-up costs for the larger press are $200 higher than those for the smaller press. For what size jobs should the small press be used and for which size should the large press be used, assuming that the cost of the job is the only factor to consider?

8.5 Additional Applications of the Definite Integral

There are many other applications of the definite integral in addition to those described in Sections 8.1–8.4. Some of the cases presented in this section will require the use of the following formula:

$$\int te^{at}\, dt = \frac{e^{at}}{a^2}(at - 1) \tag{8.5.1}$$

The techniques for deriving this formula will be presented in Chapter nine. However, you can verify its accuracy by differentiating the right-hand side of Equation 8.5.1 to obtain the expression te^{at}.

APPROXIMATING THE AMOUNT OF AN ANNUITY

When interest is compounded continuously, it was shown in Chapter 5 that a deposit or principal of P dollars grows to an amount A according to Equation 5.3.4

$$A(t) = Pe^{rt}$$

where r represents the quoted annual interest rate and t is the time in years. In obtaining this result, it was assumed that no withdrawals or additional deposits were made. Because this assumption is highly restrictive, more general cases will now be considered.

An *annuity* is defined as a set of equal payments made at equal time intervals. For example, if money is borrowed to finance either a home, an auto, or an education, the repayment process generally takes the form of equal monthly payments for a specified length of time. This set of equal monthly payments forms an annuity. Social security payments and pensions are other examples of annuities.

To keep matters simple, suppose that $100 is deposited in a savings account at the beginning of each year for six years; the annual rate of interest is r. The initial deposit ($t = 0$) is in the account for six years and grows to $100e^{6r}$. The second deposit ($t = 1$) is in the account for 5 years ($6 - 1$) and grows to $100e^{(6-1)r} = 100e^{5r}$. Proceeding in this way, we get for the total amount $S(6)$ in the account at the end of six years

$$S(6) = \underset{\substack{\uparrow \\ t = 0 \\ \text{1st deposit}}}{100e^{(6-0)r}} + \underset{\substack{\uparrow \\ t = 1}}{100e^{(6-1)r}} + \underset{\substack{\uparrow \\ t = 2}}{100e^{(6-2)r}} + \cdots$$

$$+ \underset{\substack{\uparrow \\ t = k}}{100e^{(6-k)r}} + \cdots + \underset{\substack{\uparrow \\ t = 5 \\ \text{last deposit}}}{100e^{(6-5)r}}$$

$$= 100e^{6r} + 100e^{5r} + \cdots + 100e^{(6-k)r} + \cdots + 100e^{r} \qquad \textbf{(8.5.2)}$$

Each term is shown in the graph of Figure 8.19.

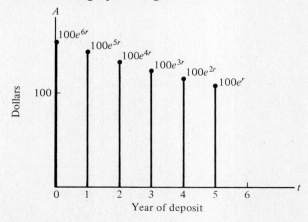

Figure 8.19

The total amount in the account can also be regarded as the sum of the areas of the six rectangles shown in Figure 8.20 because the base of each rectangle equals 1. Superimposed on the figure is the graph of the equation

$$A(t) = 100e^{(6-t)r}$$

The area under the curve between $t = 0$ and $t = 6$ is almost equal to the area of the six rectangles. Therefore, the area under the curve can be used to approximate $S(6)$, the total amount or sum in the account at the end of six years, that is

$$S(6) \approx \int_0^6 100e^{(6-t)r}\, dt = \frac{-100}{r}\, e^{(6-t)r}\, \Big|_0^6$$

$$= \frac{100}{r}\, (e^{6r} - 1) \tag{8.5.3}$$

For example, suppose the annual rate of interest is 5 percent, then

$$\int_0^6 100e^{0.05(6-t)}\, dt = \frac{100}{0.05}\, (e^{0.30} - 1) = 2000(1.3499 - 1) = \$699.80$$

whereas the exact value given by Equation 8.5.2 becomes

$$100e^{0.30} + 100e^{0.25} + \cdots + 100e^{0.05} = \$717.36$$

The generalized form of this procedure takes the form

$$S(T) \approx \int_0^T Re^{r(T-t)}\, dt = \frac{R}{r}\, (e^{rT} - 1) \tag{8.5.4}$$

where R is the size of each payment, r is the interest rate, T is the duration of the annuity, and S is the total value of the annuity.

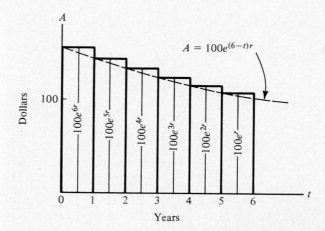

Figure 8.20

The approximation given by Equation 8.5.4 improves as the number of payments increases and/or the time interval between payments decreases.

Example 1 | A person deposits $50 per month in a savings account paying 6 percent per annum compounded continuously. How much is in the account at the end of 10 years?

Solution | Because the equal payments are spaced one month apart, the time t is expressed in months and r must be expressed as the monthly interest rate. Using Equation 8.5.4 with $R = 50$, $r = 0.06/12 = 0.005$, and $T = 12(10) = 120$ months, the total amount in the account becomes

$$S \approx \int_0^{120} 50e^{0.005(120-t)}\, dt = \frac{50}{0.005}(e^{0.60} - 1)$$

$$= 10{,}000(1.8221 - 1) = \$8221$$

PRESENT VALUE OF AN ANNUITY

The present value of a single amount A due t years in the future is given by Equation 5.3.5

$$P = Ae^{-rt}$$

when interest is compounded continuously. Suppose we want to determine the present value of six equal annual payments of $100, the first to begin one year from now; again r is the annual interest rate. The sum of the present values, that is, the present value of the annuity, equals

$$\underset{\substack{\uparrow \\ \text{1st payment}}}{100e^{-r}} + \underset{\substack{\uparrow \\ \text{2nd payment}}}{100e^{-2r}} + \cdots + \underset{\substack{\uparrow \\ \text{6th \& last payment}}}{100e^{-6r}}$$

This sum can be represented geometrically as the sum of the areas of the six rectangles shown in Figure 8.21; also shown is the graph of the equation

$$P(t) = 100e^{-rt}$$

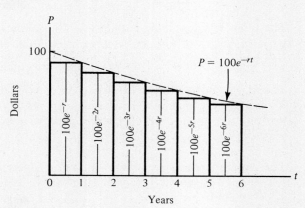

Figure 8.21

The area under the curve between $t = 0$ and $t = 6$ is approximately equal to the sum of the areas of the six rectangles; therefore, it can be used to approximate the present value of the six equal payments

$$\text{Present value} \approx \int_0^6 100e^{-rt}\, dt = \frac{100}{r}(1 - e^{-6r})$$

If $r = 0.05$, we get

$$\text{Present value} \approx \frac{100}{0.05}(1 - e^{0.30}) = \$518.36$$

whereas the sum of the present values of the six equal payments is

$$100e^{-0.05} + 100e^{-0.10} + \cdots + 100e^{-0.30} = \$505.51$$

The generalization of this procedure is described by the equation

$$\text{Present value} \approx \int_0^T Re^{-rt}\, dt = \frac{R}{r}(1 - e^{-rT}) \qquad\qquad \textbf{(8.5.5)}$$

where R represents the amount of each payment, T is the duration of the annuity, and r is the interest rate.

Example 2 An executive wants to establish a fund from which she will withdraw $5000 annually for the next 20 years. If the annual interest rate is 8 percent, how much money should she place in the fund?

Solution We want to find the present value of a series of 20 equal payments, each equal to $5000. Using Equation 8.5.5, we get

$$\text{Present value} \approx \int_0^{20} 5000e^{-0.08t}\, dt = \frac{5000}{0.08}(1 - e^{-1.6})$$
$$= 62{,}500(1 - 0.2019) = \$49{,}881.25$$

The interest rate r in Equation 8.5.5 should be expressed in units compatible with the length of time between successive payments, as demonstrated in the following example.

Example 3 The Massachusetts State Lottery pays $1,000 per week for 20 years to each of its grand prize winners. How much money should the state deposit with an insurance company, which pays $7\frac{1}{2}$ percent annually, to meet the weekly payments for 20 years?

Solution The equally spaced payments constitute an annuity for which $R = 1000$, $T = 52(20) = 1040$, and $r = \frac{0.075}{52}$ (we will leave it in this form). Using Equation 8.5.5, we get

$$\text{Present value} \approx \int_0^{1040} 1000 e^{-(0.075t/52)} \, dt = \frac{(1000)(52)}{0.075} \{ 1 - e^{-[(0.075)(1040)]/52} \}$$

$$\approx 693,331(1 - e^{-1.5}) = \$538,650$$

DISCOUNTED CASH FLOWS

In most situations requiring financial analysis, the cash flows are not uniform as they are in the case of annuities. The definite integral can still be employed to obtain the present value of the set of uneven cash flows by replacing the constant factor R in Equation 8.5.5 with $R(t)$, which expresses the cash flows as a function of time, yielding

$$\text{Present value} = \int_0^T R(t) e^{-rt} \, dt \qquad\qquad \textbf{(8.5.6)}$$

Example 4 | Monthly production at a Yukon oil well is expected to follow the equation

$$B(t) = 3000 \qquad 0 \le t \le 360$$

where t is the time from the present expressed in months and $B(t)$ is the monthly output in barrels per month. If the price of a barrel of crude oil is expected to increase according to the equation

$$p(t) = 13 + 0.10t \qquad 0 \le t \le 360$$

where p is given in dollars, find the present value of the revenue generated during the next 30 years from the sale of oil pumped from the well if the owner expects a 15 percent annual return on his investments.

Solution | The revenue generated each month from the sale of the crude oil is

$$R(t) = B(t)p(t) = 39,000 + 300t$$

Letting $r = 0.15/12 = 0.0125$, the present value of the monthly revenues is given by the equation

$$\text{Present value} = \int_0^{360} (39,000 + 300t) e^{-0.0125t} \, dt$$

$$= \int_0^{360} 39,000 e^{-0.0125} \, dt + \int_0^{360} 300t e^{-0.0125t} \, dt$$

The evaluation can be carried out term by term, noting that Equation 8.5.1 is needed to evaluate the second term. We then get

$$\int_0^{360} 39,000 e^{-0.0125t} \, dt = \frac{39,000}{0.0125} [1 - e^{-(360)(0.0125)}]$$

$$= 3,120,000(1 - e^{-4.5}) = 3,085,368$$

$$\int_0^{360} 300te^{-0.0125t} \, dt = 300 \left[\frac{e^{-0.0125t}}{(0.0125)^2} (-0.0125t - 1) \right] \Bigg|_0^{360}$$
$$= 300(-391.04 + 6400)$$
$$= 1,802,689$$

The present value then becomes

Present value $= 3,085,368 + 1,802,689$
$= \$4,888,057$

CONSUMERS' SURPLUS

As noted previously, a demand equation is used to describe the relationship between the unit price p and the quantity q of a product. The curve of a demand equation generally has a shape similar to that shown in Figure 8.22. Suppose the market price of a product is p_0 and the quantity sold is q_0 units. The area of the shaded rectangular region $p_0 q_0$ in Figure 8.22 represents the revenue generated from the sale of q_0 units. The curve itself shows that many consumers would have been willing to purchase the product at a price higher than the market price p_0. Therefore, this group "saves" money by not having to purchase the product at a price higher than p_0. The amount of money "not spent" by this group is called the *consumers' surplus* and is defined as

$$\text{Consumers' surplus} = \int_0^{q_0} p(q) \, dq - p_0 q_0 \qquad\qquad (8.5.7)$$

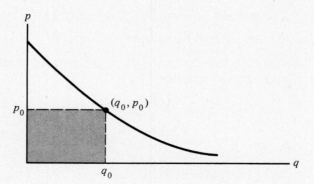

Figure 8.22

Geometrically, this quantity is shown as the area of the shaded region in Figure 8.23.

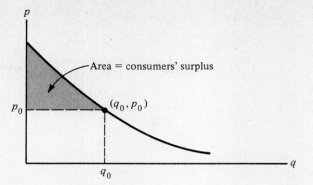

Figure 8.23

Example 5 | The Gremlin Company has developed a new ultrasensitive watch that not only gives the time and date, but forecasts the weather and measures the wearer's pulse rate. The marketing research department has found that the relationship between the unit price and the quantity sold is described by the equation

$$p(q) = -0.5q + 10$$

where q is given in thousands of units and p is expressed in hundreds of dollars. What is the consumers' surplus if the company prices the watch at $400?

Solution | The demand curve is shown in the following figure where the area of the shaded region, which represents consumer surplus, is given by the expression

$$\int_0^{12} (10 - 0.5q) \, dq - (12)(4)$$

Evaluating this yields

$$\text{Consumers' surplus} = \left(10q - \frac{0.5q^2}{2}\right)\Big|_0^{12} - 48$$

$$= 120 - 36 - 48 = \$36 \text{ hundred thousand}$$

$$= \$3.6 \text{ million}$$

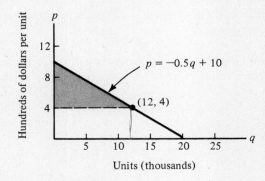

Example 6 | Weekly demand for instant coffee is given by the equation

$$p(q) = 30e^{-0.10q}$$

where p is the price per pound in dollars and q is the amount of coffee sold in millions of pounds. What is the consumers' surplus when 20 million lb are sold?

Solution | The unit price p when 20 million lb are sold becomes

$$p(20) = 30e^{-(0.10)(20)}$$
$$= \$4.06/\text{lb}$$

The demand equation is plotted on the following graph. The consumers' surplus, represented by the shaded region, can be found by evaluating

$$\int_0^{20} 30e^{-0.10q} \, dq - (20)(4.06)$$

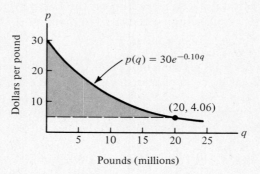

Pounds (millions)

Working through the details, we get

$$\text{Consumers' surplus} = \frac{30}{-0.10} e^{-0.10q} \Big|_0^{20} - 81.20$$
$$= -300(e^{-2} - 1) - 81.20$$
$$= \$178.2 \text{ million}$$

Many of the problems relating to consumers' surplus require the use of both differential and integral calculus in obtaining a solution, as shown in the next example.

Example 7 | Suppose the Gremlin Company (see Example 5) prices the watch to yield maximum revenue. What is the consumers' surplus under those conditions?

Solution | The first matter that must be dealt with is determining the quantity and hence the price at which the company's revenue will be maximized. The company's revenue $R(q)$ is given by the equation

$$R(q) = p(q)q$$
$$= (10 - 0.5q)q = 10q - 0.5q^2$$

Using the optimization techniques developed in Chapter 4, we get first for $R'(q)$

$$R'(q) = 10 - q$$

Next setting $R'(q)$ equal to 0 and solving for q gives

$$10 - q = 0$$

or

$$q = 10 \qquad p(10) = 5$$

Because $R''(q) = -1$ for all values of q, the revenue is maximized when $q = 10$. The consumers' surplus can now be found by evaluating

$$\int_0^{10} (10 - 0.5q)dq - (10)(5)$$

Working through the details gives us

$$\text{Consumers' surplus} = \left(10q - \frac{0.5q^2}{2}\right)\Bigg|_0^{10} - 50$$
$$= \$25 \text{ hundred thousand}$$
$$= \$2.5 \text{ million}$$

EXERCISE 8.5

1. A family deposits $100 per month into a savings account paying 6 percent per annum, compounded continuously. Use Equation 8.5.4 to obtain an approximation for the total amount in the account at the end of 15 years.

2. A worker and his employer each contribute $10 per week to a pension plan for which the annual rate of return is 8 percent, compounded continuously. If the worker should retire after 25 years, how much money will be credited to his account?

3. Another worker in the same pension plan (Exercise 2) wants to have $30,000 at the end of 20 years. Using Equation 8.5.4, determine how much money he should contribute each week to the pension fund.

4. The Family Finance Company charges 12 percent annual interest compounded continuously on automobile loans. If a woman borrows $3000 to purchase an automobile, use Equation 8.5.5 to determine the monthly payments if the term of the loan is four years.

5. A retired college professor is notified that the dollar value of her pension is $100,000. If the insurance company that supervises the pension fund pays an 8 percent annual rate of return, what monthly payment should the professor receive over the next 20 years if the dollar value of her pension fund is zero at the end of this period?

6. The annual output of the Acme Copper Mine as a function of time is expected to follow the equation

$$A(t) = 10 + 0.05t$$

where t is the time in years and $A(t)$ is the annual output in thousands of tons per year. If the per-ton price of copper is expected to remain constant at \$1500, what is the present value of all the copper mined during the next four years if the company seeks a 12 percent annual return on investment?

7. A sporting goods company has developed a new jogging shoe that has received favorable comments in the consumer review sections of the major sports magazines. A market research study has shown that the relation between q, the quantity sold, and p, the unit price, is described by the equation

$$p(q) = 75 - 10q$$

where p is expressed as dollars per pair and q as millions of pairs.

  a. If the company prices the shoes at \$40 per pair, what is the consumers' surplus?
  b. If the company decides to price the shoes to maximize its revenue, what is the consumers' surplus?

8. If the company producing instant coffee (see Example 6) prices the product to maximize revenue, what is the consumers' surplus?

9

Techniques of Integration

The functions whose antiderivatives or integrals were sought in Chapter 8 were deliberately restricted to those of a very simple type. This was done so that the fundamental ideas of the definite and indefinite integrals could be studied without getting bogged down in the process of finding an antiderivative. In most cases, however, the integrals are more complex and require the application of well-known integration techniques: two of the most widely used are the *method of substitution* and *integration by parts,* to be presented in Sections 9.1 and 9.2, respectively. When none of the available techniques is effective, *tables of integrals* can often be used to find an antiderivative; their use will be demonstrated in Section 9.3.

In addition to these cases, many situations arise where a definite integral cannot be evaluated because it may be extremely difficult or even impossible to express an antiderivative in terms of simple functions. Approximation methods such as *Simpson's rule* or the *trapezoidal rule* can be employed to evaluate the definite integral to any desired degree of accuracy; these approximation techniques will be studied in Section 9.4. The final section, 9.5, will be devoted to examining those definite integrals for which one or both limits of integation are infinite. Such integrals belong to a group known as *improper integrals*.

❧ 9.1 Integration by Substitution

Integration by *substitution* is a trial and error method that attempts to define a variable u in terms of x so that the given integral

$$\int f(x)\ dx$$

can be rewritten in terms of u and du as

$$\int g(u)\ du$$

where the antiderivative of $g(u)$ is easier to find than the antiderivative of $f(x)$.

To see how this method works, suppose we want to find

$$\int 2x(x^2 + 1)^6\ dx$$

In its present form, the integral does not fall into one of the simple forms we studied in Chapter 8. Noting that the factor $2x$ is the first derivative of the base $(x^2 + 1)$, let us introduce a new variable u defined as

$$u = x^2 + 1$$

Because the differential du is

$$du = 2x\, dx$$

the original integral can be rewritten in terms of u and du as follows

$$\int 2x(x^2 + 1)^6\, dx = \int (x^2 + 1)^6(2x\, dx) = \int u^6\, du$$

The integral on the right-hand side can be found using Equation 8.1.4, so we can write

$$\int u^6\, du = \frac{u^7}{7} + C$$

Rewriting the equation in terms of x, we can conclude that

$$\int 2x(x^2 + 1)^6\, dx = \frac{(x^2 + 1)^7}{7} + C$$

As usual, we can check our answer by differentiating the function $[(x^2 + 1)^7]/7$ and obtaining $2x(x^2 + 1)^6$, the original integrand.

Many of the integrals for which integration by substitution is an effective technique will assume one of the following forms when u is defined appropriately:

$$\int u^n\, du = \frac{u^{n+1}}{n + 1} + C \tag{9.1.1}$$

$$\int e^u\, du = e^u + C \tag{9.1.2}$$

$$\int \frac{1}{u}\, du = \ln |u| + C \tag{9.1.3}$$

The following examples illustrate situations in which the integration can be carried out using one of the forms given in Equations 9.1.1–9.1.3.

Example 1   Find

$$\int (2x + 3)(x^2 + 3x - 5)^4\, dx$$

Solution | Noting that $(2x + 3)$ equals the first derivative of the base $(x^2 + 3x - 5)$, we define u as

$$u = x^2 + 3x - 5$$

with du becoming

$$du = (2x + 3)\, dx$$

The integral in terms of u now becomes

$$\int u^4\, du$$

which can be integrated according to Equation 9.1.1 as

$$\int u^4\, du = \frac{u^5}{5} + C$$

Now expressing the relationship in terms of x, we get

$$\int (2x + 3)(x^2 + 3x - 5)\, dx = \frac{(x^2 + 3x - 5)^5}{5} + C$$

Again, the expression on the right-hand side can be differentiated with respect to x to show that it is an antiderivative of the integrand $(2x + 3)(x^2 + 3x - 5)^4$.

Example 2 | Find

$$\int x^2(x^3 + 1)^5\, dx$$

Solution | Again, if we define u as

$$u = x^3 + 1$$

the differential du becomes

$$du = 3x^2\, dx$$

The extra factor, that is, 3, poses no problem because constant factors do not affect the process of integration as Equation 8.1.6 indicates. Writing

$$x^2\, dx = \frac{du}{3}$$

the integral can now be written in terms of u and du

$$\int (x^3 + 1)^5(x^2\, dx) = \int u^5 \left(\frac{du}{3}\right) = \frac{1}{3} \int u^5\, du = \frac{u^6}{18} + C$$

Therefore, we conclude that

$$\int x^2(x^3 + 1)^5\, dx = \frac{(x^3 + 1)^6}{18} + C$$

Example 3 | Find

$$\int e^{2x+5} \, dx$$

Solution | Anticipating that this integral might assume a form like that given in Equation 9.1.2, we define u as

$$u = 2x + 5$$

$$du = 2dx$$

or

$$dx = \frac{du}{2}$$

The integral now can be written as

$$\int e^u \left(\frac{du}{2}\right) = \frac{1}{2} \int e^u \, du = \frac{1}{2} e^u + C$$

Substituting $(2x + 5)$ for u gives

$$\int e^{2x+5} \, dx = \frac{1}{2} e^{2x+5} + C$$

Example 4 | Find

$$\int \frac{6x^2 + 14}{x^3 + 7x - 1} \, dx$$

Solution | Noting that the numerator $(6x^2 + 14)$ equals 2 times the first derivative of the denominator $(x^3 + 7x - 1)$, we define u as follows:

$$u = x^3 + 7x - 1$$

$$du = (3x^2 + 7) \, dx$$

or

$$(6x^2 + 14) \, dx = 2du$$

The integral can now be written as

$$\int \frac{6x^2 + 14}{x^3 + 7x - 1} \, dx = \int 2 \frac{du}{u} = 2 \int \frac{du}{u}$$

$$= 2 \ln |u| + C$$

$$= 2 \ln |x^3 + 7x - 1| + C$$

The method of substitution, like most techniques of integration, is a trial and error process that can be applied to situations other than those described by Equations 9.1.1–9.1.3. Success in using this method sometimes requires flexibility in order to carry out the integration.

Example 5 | Find

$$\int x\sqrt{x + 1}\, dx$$

Solution | If we let $u = x + 1$, we quickly find that $du = dx$, which tells us that du is not a multiple of $x\, dx$. At this point the integral can be written as

$$\int x\sqrt{u}\, du$$

Since our ultimate goal is to express the integral in terms of u and du, we want to express the factor x in terms of u if possible. This can be done by noting that

$$x = u - 1$$

so that the integral can be written in terms of u and du as

$$\int (u - 1)\sqrt{u}\, du = \int (u^{3/2} - u^{1/2})\, du$$

$$= \int u^{3/2}\, du - \int u^{1/2}\, du$$

$$= \frac{2u^{5/2}}{5} - \frac{2u^{3/2}}{3} + C$$

$$= \frac{2(x + 1)^{5/2}}{5} - \frac{2(x + 1)^{3/2}}{3} + C$$

The results of this process can be checked by differentiating, that is

$$\frac{d}{dx}\left[\frac{2(x + 1)^{5/2}}{5} - \frac{2(x + 1)^{3/2}}{3} + C\right] = x\sqrt{x + 1}$$

This problem also illustrates one of the features that causes integration to be more difficult than differentiation, namely the necessity to reverse or "unscramble" the algebraic steps that were carried out in simplifying the expression obtained from the differentiation process.

When using the method of substitution to evaluate definite integrals, the process can be carried out by employing either of the following procedures:

1. After defining u, carry out the integration and express the result in terms of x. Then evaluate the definite integral according to the fundamental theorem of calculus, or
2. After defining u, rewrite both the integrand and limits of integration in terms of the variable u. Next, carry out the integration, leaving your

answer expressed in terms of the variable u. Finally evaluate the definite integral according to the fundamental theorem of calculus.

The two alternative methods are illustrated in the next example.

Example 6 | Evaluate

$$\int_0^1 \frac{2x}{(x^2 + 1)^2} \, dx$$

Solution | Letting $u = x^2 + 1$, $du = 2x \, dx$, we can proceed in either of the following ways:

1. $\int \frac{2x}{(x^2 + 1)^2} \, dx = \int \frac{du}{u^2} = \int u^{-2} \, du$

    $$= \frac{u^{-1}}{-1} = \frac{-1}{x^2 + 1}$$

Now, making use of the fundamental theorem of calculus, we have

$$\int_0^1 \frac{2x}{(x^2 + 1)^2} \, dx = \frac{-1}{x^2 + 1} \bigg|_0^1 = -\frac{1}{2} + 1 = \frac{1}{2}$$

2. Noting that $u = 1$ when $x = 0$ and $u = 2$ when $x = 1$, the definite integral can also be written as

$$\int_0^1 \frac{2x}{(x^2 + 1)^2} \, dx = \int_1^2 \frac{du}{u^2}$$

Note the change in the limits of integration in this equation. The integral on the right-hand side can be found directly:

$$\int_1^2 \frac{du}{u^2} = \frac{-1}{u} \bigg|_1^2 = -\frac{1}{2} + 1 = \frac{1}{2}$$

EXERCISE 9.1

In Exercises 1–25, find the given integrals, using the method of substitution where necessary.

1. $\int (x - 5)^3 \, dx$ 2. $\int \sqrt{x + 2} \, dx$

3. $\int e^{x+2} \, dx$ 4. $\int \frac{1}{x + 6} \, dx$

5. $\int 2(2x + 1)^5 \, dx$ 6. $\int 5e^{(5x-1)} \, dx$

7. $\int \frac{4}{4x + 3} \, dx$ 8. $\int \sqrt{4x - 5} \, dx$

9. $\displaystyle\int x(x^2 - 1)^4 \, dx$

10. $\displaystyle\int xe^{x^2} \, dx$

11. $\displaystyle\int xe^{-x^2} \, dx$

12. $\displaystyle\int (x + 2)(x^2 + 4x - 3)^3 \, dx$

13. $\displaystyle\int \frac{5x}{(3x^2 + 7)^2} \, dx$

14. $\displaystyle\int \frac{x}{x^2 + 1} \, dx$

15. $\displaystyle\int x^5(x^6 + 2) \, dx$

16. $\displaystyle\int \frac{1}{x^2} \sqrt{1 + x^{-1}} \, dx$

17. $\displaystyle\int xe^{1-x^2} \, dx$

18. $\displaystyle\int (3x^2 - 4x + 1)(x^3 - 2x^2 + x - 7)^2 \, dx$

19. $\displaystyle\int \frac{e^x}{2 + e^x} \, dx$

20. $\displaystyle\int (x + \sqrt{x - 1}) \, dx$

21. $\displaystyle\int \left(xe^{x^2} - \frac{2x}{x^2 + 1}\right) dx$

22. $\displaystyle\int \left[x(x^2 + 3) - \frac{3}{(x + 5)^4}\right] dx$

23. $\displaystyle\int \frac{e^{1/x}}{x^2} \, dx$

24. $\displaystyle\int \frac{2x}{1 + x} \, dx$

25. $\displaystyle\int (x + 1)\sqrt{x + 3} \, dx$

Evaluate the definite integrals in Exercises 26–33.

26. $\displaystyle\int_0^2 x(x^2 - 2)^3 \, dx$

27. $\displaystyle\int_1^2 xe^{x^2} \, dx$

28. $\displaystyle\int_0^2 (2x + 3)(x^2 + 3x - 5)^2 \, dx$

29. $\displaystyle\int_1^3 \frac{2}{x + 1} \, dx$

30. $\displaystyle\int_2^{\sqrt{12}} x\sqrt{x^2 - 3} \, dx$

31. $\displaystyle\int_0^1 \frac{x}{1 + 3x^2} \, dx$

32. $\displaystyle\int_2^4 \frac{x}{1 + x} \, dx$

33. $\displaystyle\int_0^1 \frac{e^x}{2 + e^x} \, dx$

9.2 Integration by Parts

Another technique, *integration by parts,* is often useful in carrying out the integration process. This method is based, in principle, on reversing the steps contained in the product rule (Equation 3.4.1)

$$\frac{d}{dx}(uv) = u\frac{dv}{dx} + v\frac{du}{dx}$$

Writing this equation as

$$u\frac{dv}{dx} = \frac{d}{dx}(uv) - v\frac{du}{dx}$$

and integrating both sides gives

$$\int u \frac{dv}{dx}\,dx = \int \frac{d}{dx}\,(uv)\,dx - \int v \frac{du}{dx}\,dx$$

The first term on the right-hand side can be written in a simpler form because an antiderivative of the derivative of a function equals the function, that is,

$$\int \frac{d}{dx} f(x)\,dx = f(x)$$

Or in this case

$$\int \frac{d}{dx}\,(uv)\,dx = uv$$

Now it is possible to write

$$\int u \frac{dv}{dx}\,dx = uv - \int v \frac{du}{dx}\,dx \tag{9.2.1}$$

Although Equation 9.2.1 contains the essential features of the integration by parts technique, a simpler and more widely used version can be derived if Equation 9.2.1 is written in terms of the differentials du and dv

$$du = \frac{du}{dx}\,dx \qquad dv = \frac{dv}{dx}\,dx$$

giving

$$\boxed{\int u\,dv = uv - \int v\,du} \tag{9.2.2}$$

The integral we want to find is shown on the left-hand side; it is the responsibility of the person solving the problem to select the expressions defining u and dv. Based on this selection, the right-hand side is then written with the aim of making the integral on the right-hand side less complex and easier to integrate than the integral on the left-hand side. The process will be demonstrated in the following examples, at which time the preceding remarks will become more meaningful.

Example 1 | Find

$$\int xe^x\,dx$$

Solution | First we want to write the integral in the form

$$\int u\,dv$$

by defining u and dv. In this case, there are only two choices

    a. $u = x$ $dv = e^x \, dx$

    b. $u = e^x$ $dv = x \, dx$

Let us begin by selecting choice a. and carrying it through according to Equation 9.2.2.

a. $u = x$ $dv = e^x \, dx$

    Next find du and v

       $du = dx$ $v = e^x$

    Following the procedure described in Equation 9.2.2, we have

$$\underbrace{\int x}_{u} \underbrace{e^x \, dx}_{dv} = \underbrace{x}_{u} \underbrace{e^x}_{v} - \int \underbrace{e^x}_{v} \underbrace{dx}_{du}$$

    The integral on the right-hand side can be found easily

$$\int e^x \, dx = e^x + C$$

    so that we can now write

$$\int x e^x \, dx = x e^x - e^x + C$$

    As usual, this result can be checked by differentiating the right-hand side.

b. Suppose we had defined u and dv as

    $u = e^x$ and $dv = x \, dx$

    The quantities du and v become

       $du = e^x \, dx$ $v = \dfrac{x^2}{2}$

    Again proceeding as described by Equation 9.2.2, we get

$$\int \underbrace{e^x}_{u} \underbrace{x \, dx}_{dv} = \underbrace{e^x}_{u} \underbrace{\left(\frac{x^2}{2}\right)}_{v} - \int \underbrace{\frac{x^2}{2}}_{v} \underbrace{(e^x \, dx)}_{du}$$

Although formally correct, this choice of u and dv generates an integral on the right-hand side that is more complex than that found on the left-hand side. When this happens, it is usually an indication that u and dv should be redefined.

Example 2 | Find

$$\int \ln x \, dx$$

Solution | The choice of u and dv in this case is straightforward

$$u = \ln x \qquad dv = dx$$

$$du = \frac{1}{x} dx \qquad v = x$$

Integrating according to Equation 9.2.2 gives

$$\int \ln x \, dx = x \ln x - \int x \frac{1}{x} dx$$

$$= x \ln x - \int dx$$

yielding the result

$$\int \ln x \, dx = x \ln x - x + C$$

Question: Why would we not select the following for u and dv?

$$u = 1 \qquad dv = \ln x \, dx$$

Often integration by parts must be carried out more than once in order to find the given integral. The following example is an illustration.

Example 3 | Find

$$\int x^2 e^{-x} \, dx$$

Solution | If we let $u = x^2$, $dv = e^{-x} \, dx$

$$du = 2x \, dx \qquad v = -e^{-x}$$

Using Equation 9.2.2, we get

$$\int x^2 e^{-x} \, dx = x^2(-e^{-x}) - \int (-e^{-x})2x \, dx$$

$$\tag{9.2.3}$$

$$= -x^2 e^{-x} + 2 \int x e^{-x} \, dx$$

The integral on the right-hand side can be found by applying integration by parts once more; letting

$$u = x \qquad dv = e^{-x} \, dx$$

$$du = dx \qquad v = -e^{-x}$$

We get

$$2 \int x e^{-x} \, dx = 2 \left[-x e^{-x} + \int e^{-x} \, dx \right] = -2x e^{-x} - 2e^{-x} + C$$

Substituting this result in Equation 9.2.3 gives

$$\int x^2 e^{-x} \, dx = -x^2 e^{-x} - 2x e^{-x} - 2e^{-x} + C$$

EXERCISE 9.2

In Exercises 1–12, find the indefinite integrals using integration by parts.

1. $\int xe^{2x}\,dx$

2. $\int xe^{-3x}\,dx$

3. $\int x^2 e^{4x}\,dx$

4. $\int x^2 e^x\,dx$

5. $\int \ln 3x\,dx$

6. $\int x\ln x\,dx$

7. $\int \sqrt{x}\ln x\,dx$

8. $\int x^2 \ln x\,dx$

9. $\int \ln(x+1)\,dx$ HINT: First use substitution with $w = x + 1$.

10. $\int x^3 e^{4x}\,dx$

11. $\int (x+1)e^{-x}\,dx$

12. $\int x\ln(x+1)\,dx$ HINT: Again use substitution with $w = x + 1$.

Evaluate the definite integrals in Exercises 13–18.

13. $\int_{-1}^{1} xe^{-x}\,dx$

14. $\int_{1}^{2} \ln x\,dx$

15. $\int_{2}^{3} x\ln x^2\,dx$

16. $\int_{0}^{2} \ln(x+1)\,dx$

17. $\int_{0}^{1} x^2 e^x\,dx$

18. $\int_{0}^{2} xe^{-x}\,dx$

19. Find the area of the region between the curve

$$y = f(x) = xe^x$$

and the x axis, bounded on the left by the vertical line $x = 0$ and on the right by the vertical line $x = 1$.

20. Find the area of the region between the curve

$$y = f(x) = \ln x$$

and the x axis, bounded on the left by the vertical line $x = 1$ and on the right by the vertical line $x = e$.

🦢 9.3 Tables of Integrals

When none of the integration techniques presented so far produces satisfactory results, a table of integrals can often be used to find the desired integral. Table D at the end of the book represents a short version of such a table. More extensive tables can be found in more advanced textbooks on calculus or in mathematical handbooks.

In using a table of integrals, the objective is to match the integrand in the problem with the appropriate integrand in the table. The corresponding formula, also given in the table, yields the integral. We should point out in advance that you may have to modify the given integrand and/or use the technique of substitution in order to match one of the forms given in the table. The following examples are intended both to illustrate the use of the tables and to demonstrate some of the devices you may have to employ to carry out the matching.

Example 1 | Find

$$\int \frac{x}{(x + 3)^2}\, dx$$

Solution | Inspection of the integration formulas in Table D reveals that formula 16

$$\int \frac{x}{(ax + b)^2}\, dx = \frac{b}{a^2(ax + b)} + \frac{1}{a^2} \ln |ax + b| + C$$

with $a = 1$ and $b = 3$ is the appropriate formula in this case. So we can write

$$\int \frac{x}{(x + 3)^2}\, dx = \frac{3}{x + 3} + \ln |x + 3| + C$$

Most of the situations we will encounter will require more ingenuity and algebraic manipulation than that required in the preceding example.

Example 2 | Find

$$\int \frac{3}{8 - 2x^2}\, dx$$

Solution | This integral can be put into a form that enables formula 11 in Table D to be applied. Rewriting the integral as

$$\frac{3}{2} \int \frac{1}{4 - x^2}\, dx$$

we can use formula 11 directly with $a = 2$. We then get the following:

$$\int \frac{3}{8 - 2x^2}\, dx = \frac{3}{2} \int \frac{1}{2^2 - x^2}\, dx$$

$$= \frac{3}{2} \left(\frac{1}{4} \ln \left| \frac{2 + x}{2 - x} \right| \right) + C$$

$$= \frac{3}{8} \ln \left| \frac{2 + x}{2 - x} \right| + C$$

Example 3 | Find

$$\int \frac{x^3}{\sqrt{x^2 + 2}}\, dx$$

Solution | A quick check of the integration formulas indicates that none of the integrands shown matches that of our problem. However, if we make the substitution

$$u = x^2 \quad \text{and} \quad du = 2x\,dx$$

the integral can be expressed in terms of the variable u as follows:

$$\int \frac{x^3}{\sqrt{x^2 + 2}}\,dx = \int \frac{x^2(2x\,dx)}{2\sqrt{x^2 + 2}} = \frac{1}{2}\int \frac{u}{\sqrt{u + 2}}\,du$$

Comparing the integral in this form with those given in Table D, we see that formula 15 with $a = 1$ and $b = 2$ can be used. From this, we get

$$\frac{1}{2}\int \frac{u}{\sqrt{u + 2}}\,du = \frac{(u - 4)}{3}\sqrt{u + 2} + C$$

Now, substituting x^2 for u finally gives

$$\int \frac{x^3}{\sqrt{x^2 + 2}}\,dx = \frac{(x^2 - 4)}{3}\sqrt{x^2 + 2} + C$$

Formulas 6 and 9 in Table D are called *recursion formulas*. For any permissible value of n, the given integral is expressed in terms of a second integral whose integrand is identical to that in the original integral except for the replacement of n by $n - 1$. The process of integration is repeated as often as necessary until the last integral is found. The next example illustrates this technique.

Example 4 | Find

$$\int (\ln x)^3\,dx$$

Solution | Using formula 9 from Table D with $n = 3$, we have

$$\int (\ln x)^3\,dx = x(\ln x)^3 - 3\int (\ln x)^2\,dx$$

The integral on the right-hand side is found by the same method; that is, applying formula 9 with $n = 2$ this time

$$\int (\ln x)^2\,dx = x(\ln x)^2 - 2\int \ln x\,dx$$

Putting together the results obtained so far, we have

$$\int (\ln x)^3\,dx = x(\ln x)^3 - 3x(\ln x)^2 + 6\int \ln x\,dx$$

The integral on the right-hand side can be found by either applying formula 9 again, this time with $n = 1$, or applying formula 7. In either case, the final result becomes

$$\int (\ln x)^3\,dx = x(\ln x)^3 - 3x(\ln x)^2 + 6x \ln x - 6x + C$$

EXERCISE 9.3

Use the integration formulas in Table D to find the integrals in Exercises 1–18.

1. $\int x^3 \ln x \, dx$

2. $\int \dfrac{5}{2x(x+1)} \, dx$

3. $\int \dfrac{1}{\sqrt{9x^2 + 9}} \, dx$

4. $\int \dfrac{7}{x^2 - 3} \, dx$

5. $\int \dfrac{x}{(4x+1)^2} \, dx$

6. $\int \sqrt{25x^2 + 8} \, dx$

7. $\int \dfrac{3x}{5x + 2} \, dx$

8. $\int \dfrac{6}{2x^2 + x} \, dx$

9. $\int \dfrac{8}{3x^2 \sqrt{4 - 9x^2}} \, dx$

10. $\int x^3 e^{2x} \, dx$

11. $\int 5^x \, dx$

12. $\int \dfrac{4}{3x(2x + 7)^2} \, dx$

13. $\int \dfrac{x}{x^4 - 1} \, dx$

14. $\int \dfrac{x^2}{\sqrt{x^6 - 9}} \, dx$

15. $\int \dfrac{1}{x(x^2 + 3)^2} \, dx$ HINT: Multiply numerator and denominator by x.

16. $\int e^x \sqrt{e^{2x} + 1} \, dx$ HINT: Use substitution with $u = e^x$.

17. $\int \dfrac{x^3}{\sqrt{x^2 + 5}} \, dx$

18. $\int [\ln(2x - 1)]^3 \, dx$

Evaluate the definite integrals in Problems 19–21.

19. $\int_0^1 \dfrac{1}{4 - x^2} \, dx$

20. $\int_1^6 \dfrac{x}{\sqrt{x + 3}} \, dx$

21. $\int_0^1 \dfrac{3x}{2x + 1} \, dx$

22. Show that formula 6 in Table D can be obtained by using integration by parts with

$u = x^n \qquad dv = e^{ax} \, dx$

23. Show that formula 8 in Table D can be obtained by integration by parts with

$u = \ln x \qquad dv = x^n \, dx$

24. Show that formula 9 in Table D can be obtained by integration by parts with

$u = (\ln x)^n \qquad dv = dx$

25. Show that formula 14 in Table D can be obtained by means of the substitution

$u = ax + b \qquad \text{with } du = a \, dx$

26. Show that formula 15 in Table D can be obtained by means of the substitution

$u = ax + b \qquad \text{with } du = a \, dx$

9.4 Numerical Integration

There are many occasions when the definite integral $\int_a^b f(x)\,dx$ cannot be found by using the fundamental theorem of calculus. This occurs most often when the antiderivative of $f(x)$ cannot be expressed in terms of elementary functions such as those that have been studied in this text. On such occasions, the definite integral can be approximated to any desired degree of accuracy by numerical techniques, two of which, the *trapezoidal* rule and *Simpson's* rule, will be demonstrated in this section.

TRAPEZOIDAL RULE

In order to explain the rationale behind the trapezoidal rule more effectively, the function $y = f(x)$ will be kept positive everywhere over the closed interval $a \le x \le b$, as shown in Figure 9.1. When employing the trapezoidal rule, the first step is to subdivide the closed interval $a \le x \le b$ into n equal subintervals each of whose widths equals h, where

$$h = \frac{b - a}{n}$$

The x coordinates of the endpoints of the subintervals are

$$x_0 = a \qquad x_1 = a + h \qquad x_2 = a + 2h, \ldots$$
$$x_k = a + kh, \ldots \qquad x_n = a + nh = b$$

Next, the y coordinate associated with each x_k is determined. The resulting set of points is shown in Figure 9.1. Finally, the curve $y = f(x)$ is replaced by a set of straight line segments connecting adjacent pairs of points as shown in Figure 9.2. The area under the curve $y = f(x)$ from $x = a$ to $x = b$ is then approximated by the sum of the areas of the n trapezoids, shown in Figure 9.2. The analytical result is stated as the *trapezoidal rule*:

TRAPEZOIDAL RULE

If $y = f(x)$ represents a continuous function over the closed interval $a \le x \le b$, then

$$\int_a^b f(x)\,dx \approx \frac{h}{2}\left[y_0 + 2y_1 + 2y_2 + \cdots + 2y_k + \cdots + y_n\right] \qquad \textbf{(9.4.1)}$$

---

where n equals the number of equal subintervals over $[a, b]$. The approximation improves as the number n of subintervals increases.

Equation 9.4.1 can be obtained by examining more closely the first two trapezoids on the left in Figure 9.2, as shown in Figure 9.3. The area of each

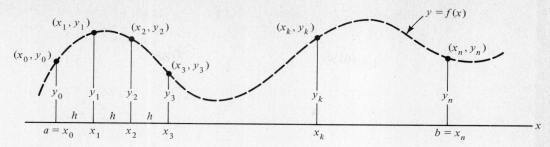

Figure 9.1

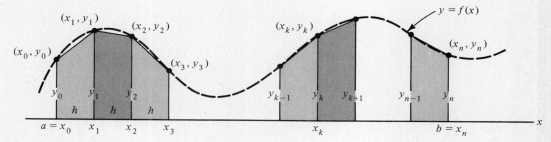

Figure 9.2

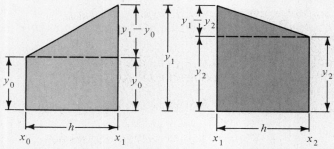

Figure 9.3

trapezoid can be found by adding the area of the rectangular base to that of the accompanying right triangle. The area of the first trapezoid equals

$$\underbrace{hy_0}_{\substack{\text{Area of} \\ \text{rectangle}}} + \underbrace{\frac{h}{2}(y_1 - y_0)}_{\substack{\text{Area of} \\ \text{triangle}}} = \frac{h}{2}(y_0 + y_1)$$

Likewise, the area of the second trapezoid equals

$$hy_2 + \frac{h}{2}(y_1 - y_2) = \frac{h}{2}(y_1 + y_2)$$

The sum of the areas of the first two trapezoids then equals

$$\frac{h}{2}(y_0 + 2y_1 + y_2)$$

If the area of the third trapezoid $(h/2)(y_2 + y_3)$ is added, the sum of the areas of the first three trapezoids becomes

$$\frac{h}{2} (y_0 + 2y_1 + 2y_2 + y_3)$$

If we continue in this manner until the area of the last or nth trapezoid is included, we get Equation 9.4.1 for the area of the n trapezoids.

In the first example, the trapezoidal rule will be demonstrated for a case in which the fundamental theorem of calculus can also be used so that the difference between the two methods can be determined.

Example 1 | Using the trapezoidal rule, find an approximate value for the following definite integral:

$$\int_1^3 x^2 \, dx \qquad \text{with } n = 4$$

Solution | The length h of each subinterval is found first

$$h = \frac{b - a}{n} = \frac{3 - 1}{4} = 0.50$$

The sum within the brackets in Equation 9.4.1 can be found with the aid of a table such as Table 9.1 where the last column contains each term in the sum. Calculations are carried out to four decimal places. Using Equation 9.4.1, we get

$$\int_1^3 x^2 \, dx \approx \frac{(0.50)}{2} (35) = 8.7500$$

Using the fundamental theorem of calculus, we get

$$\int_1^3 x^2 \, dx = \frac{x^3}{3} \Big|_1^3 = \frac{27}{3} - \frac{1}{3} = 8.6667$$

Table 9.1

k	x_k	$y_k = (x_k)^2$	COEFFICIENT	TERM
0	1.0	1.0000	1	1.0000
1	1.5	2.2500	2	4.5000
2	2.0	4.0000	2	8.0000
3	2.5	6.2500	2	12.5000
4	3.0	9.0000	1	9.0000

$$35.000 = \text{Sum}$$

Example 2 | Using the trapezoidal rule, find an approximate value of

$$\int_0^1 \sqrt{2 + x^2} \, dx \qquad \text{with } n = 10$$

Solution | The length of each subinterval is given by

$$h = \frac{b - a}{n} = \frac{1 - 0}{10} = 0.10$$

Again, we set up a table to handle the sum in the brackets of Equation 9.4.1.

Table 9.2

k	x_k	$y_k = \sqrt{2 + (x_k)^2}$	COEFFICIENT	TERM
0	0.0000	1.4142	1	1.4142
1	0.1000	1.4177	2	2.8354
2	0.2000	1.4283	2	2.8566
3	0.3000	1.4457	2	2.8914
4	0.4000	1.4697	2	2.9394
5	0.5000	1.5000	2	3.0000
6	0.6000	1.5362	2	3.0724
7	0.7000	1.5780	2	3.1560
8	0.8000	1.6248	2	3.2496
9	0.9000	1.6763	2	3.3526
10	1.0000	1.7321	1	1.7321

$$30.4997 = \text{Sum}$$

Then, using Equation 9.4.1, we obtain the following:

$$\int_0^1 \sqrt{2 + x^2} \, dx \approx \frac{0.10}{2} (30.4997) = 1.5250$$

The definite integral $\int_0^1 \sqrt{2 + x^2} \, dx$ can be found using formula 19 in Table D

$$\int_0^1 \sqrt{2 + x^2} \, dx = \left[\frac{x}{2} \sqrt{x^2 + 2} + \ln |x + \sqrt{x^2 + 2|} \right]_0^1 = 1.5245$$

SIMPSON'S RULE

Simpson's rule is a technique in which the curve $y = f(x)$ is replaced by a set of adjacent parabolic segments over the closed interval $a \le x \le b$. The sum of the areas beneath these parabolic segments serves as the approximate value of $\int_a^b f(x) \, dx$.

First, the closed interval $a \le x \le b$ is subdivided into an *even* number n of equal subintervals, each of whose widths h equals

$$h = \frac{b - a}{n}$$

Using Figure 9.4 as reference, a parabola is drawn through the three successive points (x_0, y_0), (x_1, y_1), and (x_2, y_2). A second parabola is constructed through the next three points (x_2, y_2), (x_3, y_3), and (x_4, y_4). This procedure is continued until the curve is covered with parabolic segments as displayed in Figure 9.4.

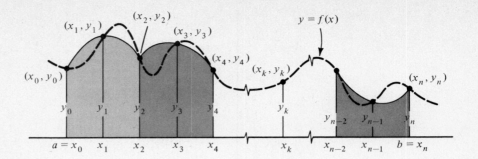

Figure 9.4

When the sum of the areas under the parabolic segments is calculated, the approximation known as *Simpson's rule* emerges*

$$\int_a^b f(x)\ dx \approx \frac{h}{3}\,[y_0 + 4y_1 + 2y_2 + 4y_3$$
$$+ 2y_4 + \cdots + 4y_{n-1} + y_n]$$

(9.4.2)

As with the trapezoidal rule, Example 3 illustrates Simpson's rule for a case in which the fundamental theorem of calculus can also be used.

Example 3 Using Simpson's rule, find an approximate value of

$$\int_1^4 \sqrt{x}\ dx \qquad \text{with } n = 6$$

Solution The length h of each subinterval equals

$$h = \frac{4 - 1}{6} = 0.50$$

Again, Table 9.3 will be set up to evaluate each term within the brackets in

Table 9.3

k	x_k	$y_k = \sqrt{x_k}$	COEFFICIENT	TERM
0	1.0000	1.0000	1	1.0000
1	1.5000	1.2247	4	4.8988
2	2.0000	1.4142	2	2.8284
3	2.5000	1.5811	4	6.3244
4	3.0000	1.7321	2	3.4642
5	3.5000	1.8708	4	7.4832
6	4.0000	2.0000	1	2.0000

27.9990 = Sum

* The derivation of Simpson's rule can be found in S. I. Grossman, *Calculus,* New York: Academic Press, 1977, pp. 428–432.

Equation 9.4.2. Then, according to Simpson's rule (Equation 9.4.2), we get

$$\int_1^4 \sqrt{x}\, dx \approx \frac{0.50}{3}\, (27.990) = 4.6665$$

Using the fundamental theorem of calculus, we obtain

$$\int_1^4 \sqrt{x}\, dx = \frac{2}{3}\, (x^{3/2}) \,\Big|_1^4 = \frac{2}{3}\, (8 - 1) = 4.6667$$

Even for this crude approximation, the error (0.0002) is extremely small.

Example 4   Using Simpson's rule, find an approximate value for

$$\int_1^2 \frac{1}{1 + x^2}\, dx \qquad \text{with } n = 10$$

Solution   First, the length h of each subinterval is found

$$h = \frac{2 - 1}{10} = 0.10$$

We proceed as before by completing Table 9.4. Then, using Simpson's rule (Equation 9.4.2), we get

$$\int_1^2 \frac{1}{1 + x^2}\, dx \approx \frac{0.10}{3}\, (9.6506) = 0.3217$$

Table 9.4

k	x_k	$y_k = \dfrac{1}{1 + (x_k)^2}$	COEFFICIENT	TERM
0	1.00	0.5000	1	0.5000
1	1.10	0.4525	4	1.8100
2	1.20	0.4090	2	0.8180
3	1.30	0.3717	4	1.4868
4	1.40	0.3378	2	0.6756
5	1.50	0.3077	4	1.2308
6	1.60	0.2809	2	0.5618
7	1.70	0.2571	4	1.0284
8	1.80	0.2358	2	0.4716
9	1.90	0.2169	4	0.8676
10	2.00	0.2000	1	0.2000

9.6506 = Sum

EXERCISE 9.4

In Exercises 1–8, find an approximate value for the given definite integral using both the trapezoidal and Simpson's rules. Find the errors associated with these methods by evaluating each integral by means of the fundamental theorem of calculus.

1. $\displaystyle\int_0^2 x^2\, dx \qquad n = 4$

2. $\displaystyle\int_1^2 x^3\, dx \qquad n = 8$

3. $\displaystyle\int_0^1 e^x \, dx \qquad n = 4$ **4.** $\displaystyle\int_1^4 \frac{1}{x} \, dx \qquad n = 6$

5. $\displaystyle\int_3^4 \sqrt{x} \, dx \qquad n = 4$ **6.** $\displaystyle\int_1^3 \frac{1}{x^2} \, dx \qquad n = 8$

7. $\displaystyle\int_0^1 \frac{x}{1 + x^2} \, dx \qquad n = 4$ **8.** $\displaystyle\int_0^1 x^2 \sqrt{1 + x^3} \, dx \qquad n = 4$

In Exercises 9–14, find an approximate value for the given definite integral using both the trapezoidal and Simpson's rules.

9. $\displaystyle\int_0^1 \frac{1}{1 + x^2} \, dx \qquad n = 4$ **10.** $\displaystyle\int_0^1 e^{x^2} \, dx \qquad n = 10$

11. $\displaystyle\int_1^4 \frac{2}{1 + \sqrt{x}} \, dx \qquad n = 6$ **12.** $\displaystyle\int_1^3 \frac{e^x}{x} \, dx \qquad n = 4$

13. $\displaystyle\int_1^2 \frac{\sqrt{1 + x}}{x} \, dx \qquad n = 8$ **14.** $\displaystyle\int_1^3 \frac{5}{1 + x^3} \, dx \qquad n = 8$

9.5 Improper Integrals

In studying and working with the definite integral $\int_a^b f(x) \, dx$, we have assumed, until now, that (1) $f(x)$ is continuous over the closed interval $a \le x \le b$, and (2) both a and b are finite. On the other hand, the integral $\int_a^b f(x) \, dx$ is called *improper* if

    a. The integrand $y = f(x)$ is unbounded at one or more values of x in the closed interval $a \le x \le b$, or

    b. One or more of the limits of integration become infinite.

In this section, we will study improper integrals of type b, which will assume one of the following forms:

$$\int_a^\infty f(x) \, dx \qquad \int_{-\infty}^b f(x) \, dx \qquad \int_{-\infty}^\infty f(x) \, dx$$

Integrals of these types arise when we attempt to calculate the area under a curve, such as $y = f(x) = 1/x^2$, to the right of the vertical line $x = 1$, as shown in Figure 9.5. For a situation such as this, we want to (1) determine whether the area in question is finite, and (2) to evaluate the area, if it is finite. To deal with this problem, we use the following procedure:

    A. The right boundary is set equal to some finite value, say $x = b$, and the area between $x = 1$ and $x = b$ is found by evaluating the integral

$$\int_1^b \frac{1}{x^2} \, dx$$

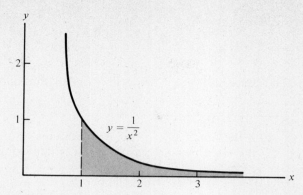

Figure 9.5

B. Next, the limit of the definite integral as $b \to \infty$ is determined using the methods developed in Section 4.4, that is, we look for

$$\lim_{b \to \infty} \int_1^b \frac{1}{x^2}\, dx$$

This procedure is shown graphically in Figure 9.6.

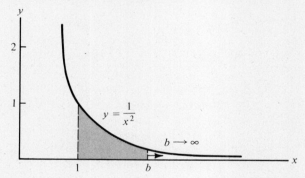

Figure 9.6

The process just described enables us to find the *improper integral* $\int_1^\infty (1/x^2)\, dx$, defined as

$$\int_1^\infty \frac{1}{x^2}\, dx = \lim_{b \to \infty} \int_1^b \frac{1}{x^2}\, dx$$

Following the steps described in A and B yields the following:

A. $\displaystyle \int_1^b \frac{1}{x^2}\, dx = \frac{-1}{x}\,\Big|_1^b = 1 - \frac{1}{b}$

B. $\displaystyle \lim_{b \to \infty} \int_1^b \frac{1}{x^2}\, dx = \lim_{b \to \infty} \left(1 - \frac{1}{b} \right)$

As b is assigned values that increase without limit, the quantity $\dfrac{1}{b}$ approaches 0, that is, $\dfrac{1}{b} \to 0$ as $b \to \infty$. Thus we get

$$\lim_{b \to \infty} \left(1 - \frac{1}{b} \right) = 1$$

so now we can write

$$\int_1^\infty \frac{1}{x^2}\, dx = 1$$

The generalization of this process takes the following form:

I. If the function $y = f(x)$ is continuous on the interval $a \le x < \infty$, then
    the improper integral $\int_a^\infty f(x)\, dx$ is defined as

$$\int_a^\infty f(x)\, dx = \lim_{b \to \infty} \int_a^b f(x)\, dx \qquad\qquad \textbf{(9.5.1)}$$

If the limit exists, the integral is said to be *convergent;* if it does not, it
is called *divergent.*

In the same way, it is possible to define the improper integrals

$$\int_{-\infty}^b f(x)\, dx \quad \text{and} \qquad \int_{-\infty}^\infty f(x)\, dx$$

II. If the function $y = f(x)$ is continuous over the interval $-\infty < x \le b$,
     the improper integral $\int_{-\infty}^b f(x)\, dx$ is defined as

$$\int_{-\infty}^b f(x)\, dx = \lim_{a \to -\infty} \int_a^b f(x)\, dx \qquad\qquad \textbf{(9.5.2)}$$

III. If the function $y = f(x)$ is continuous over the interval $-\infty < x < \infty$,
      then the improper integral $\int_{-\infty}^\infty f(x)\, dx$ is defined as

$$\int_{-\infty}^\infty f(x)\, dx = \lim_{a \to -\infty} \int_a^0 f(x)\, dx + \lim_{b \to \infty} \int_0^b f(x)\, dx \qquad \textbf{(9.5.3)}$$

NOTE: In order for $\int_{-\infty}^\infty f(x)\, dx$ to be convergent, both limits in Equation 9.5.3 must exist.

Example 1   Find $\int_1^\infty (1/\sqrt{x^3})\, dx$, if the improper integral exists.

Solution   Geometrically, we want to determine whether the area beneath the curve $y = f(x) = 1/\sqrt{x^3}$ to the right of the vertical line $x = 1$ is finite, and, if so, to evaluate it. The area in question is shown in the following figure. According to Equation 9.5.1, we first find $\int_1^b (1/\sqrt{x^3})\,dx$. Working this through gives

$$\int_1^b \frac{1}{\sqrt{x^3}}\,dx = \int_1^b x^{-3/2}\,dx = -2x^{-1/2}\,\bigg|_1^b = 2 - \frac{2}{\sqrt{b}}$$

Next, we find the limit of this result as $b \to \infty$, obtaining

$$\int_1^\infty \frac{1}{\sqrt{x^3}}\,dx = \lim_{b \to \infty}\left(2 - \frac{2}{\sqrt{b}}\right)$$

As b is assigned values that increase without limit, the quantity $\dfrac{2}{\sqrt{b}}$ approaches 0, so we get

$$\lim_{b \to \infty}\left(2 - \frac{2}{\sqrt{b}}\right) = 2$$

Using this result, we can write

$$\int_1^\infty \frac{1}{\sqrt{x^3}}\,dx = 2$$

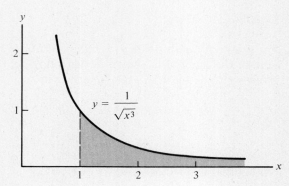

$$y = \frac{1}{\sqrt{x^3}}$$

Example 2   Find $\int_1^\infty (1/\sqrt{x})\,dx$, if it exists.

Solution   Again, we seek to determine whether the area under the curve $y = f(x) = 1/\sqrt{x}$ to the right of the line $x = 1$ is finite, and to evaluate the area if it is finite. The area of the shaded region in the figure at the top of the next page represents the integral $\displaystyle\int_1^\infty \frac{1}{\sqrt{x}}\,dx$. Proceeding as we did in Example 1, we first evaluate $\int_1^b (1/\sqrt{x})\,dx$

$$\int_1^b \frac{1}{\sqrt{x}}\,dx = \int_1^b x^{-1/2}\,dx = 2x^{1/2}\,\bigg|_1^b = 2\sqrt{b} - 2$$

Next, taking the limit as $b \to \infty$, we have

$$\int_1^\infty \frac{1}{\sqrt{x}}\,dx = \lim_{b \to \infty}(2\sqrt{b} - 2)$$

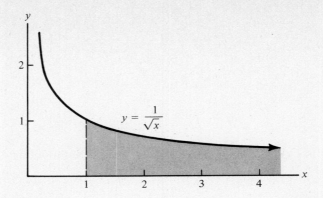

As b is assigned values that increase indefinitely, the quantity $(2\sqrt{b} - 2)$ also increases, although more slowly than b, without limit; that is, $(2\sqrt{b} - 2) \to \infty$ *as* $b \to \infty$. Thus the limit does not exist, so we conclude that $\int_1^\infty \frac{1}{\sqrt{x}} \, dx$ is divergent.

Example 3 | Evaluate $\int_{-\infty}^0 e^x \, dx$.

Solution | The graph of the function $y = f(x) = e^x$ together with the shaded region representing the integral is shown in the following figure. Following the procedure given in Equation 9.5.2, we have

$$\int_{-\infty}^0 e^x \, dx = \lim_{a \to -\infty} \int_a^0 e^x \, dx = \lim_{a \to -\infty} e^x \Big|_a^0$$
$$= \lim_{a \to -\infty} (1 - e^a)$$

Noting that $e^a \to 0$ as $a \to -\infty$, we obtain

$$\int_{-\infty}^0 e^x \, dx = 1$$

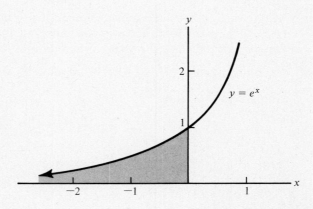

Example 4 | Evaluate $\int_{-\infty}^{\infty} xe^{-x^2}\, dx$.

Solution | The graph of the function $y = xe^{-x^2}$ is shown in the next figure. Proceeding according to Equation 9.5.3, we have

$$\int_{-\infty}^{\infty} xe^{-x^2}\, dx = \lim_{a \to -\infty} \int_{a}^{0} xe^{-x^2}\, dx + \lim_{b \to \infty} \int_{0}^{b} xe^{-x^2}\, dx$$

$$= \lim_{a \to -\infty} \frac{-e^{-x^2}}{2} \bigg|_{a}^{0} + \lim_{b \to \infty} \frac{-e^{-x^2}}{2} \bigg|_{0}^{b}$$

$$= \lim_{a \to -\infty} \left(-\frac{1}{2} + \frac{e^{-a^2}}{2} \right) + \lim_{b \to \infty} \left(\frac{1}{2} - \frac{e^{-b^2}}{2} \right)$$

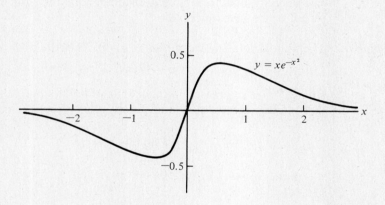

Noting that

$$\frac{e^{-a^2}}{2} = \frac{1}{2e^{a^2}} \to 0 \qquad \text{as } a \to -\infty$$

and

$$\frac{e^{-b^2}}{2} = \frac{1}{2e^{b^2}} \to 0 \qquad \text{as } b \to \infty$$

we have

$$\int_{-\infty}^{\infty} xe^{-x^2}\, dx = -\frac{1}{2} + \frac{1}{2} = 0$$

The improper integral is very useful in calculating the present value of a stream of cash flows that is anticipated to continue indefinitely. The present value of the cash flows is obtained by letting the upper limit T in Equation 8.5.6 become infinite so that we have

$$\text{Present value} = \int_0^\infty R(t)e^{-rt}\,dt \qquad\qquad \textbf{(9.5.4)}$$

where $R(t)$ is the cash flow in dollars per unit time and r is the rate of return.

Example 5 | Annual production at the Wildcat Coal Mine is expected to follow the equation

$$M(t) = 50e^{-0.03t}$$

where t is the time from the present expressed in years and $M(t)$ is the annual output in thousands of tons per year. If the per-ton price of coal remains constant at \$15 per ton, what is the present value of the revenue generated from the sale of *all* the coal to be mined, assuming that the company expects a 15 percent return on its investments?

Solution | The annual revenue from mining and selling the coal is given by the equation

$$R(t) = M(t)p(t) = (50e^{-0.03t})(15) = 750e^{-0.03t}$$

where $R(t)$ is expressed in thousands of dollars per year. The present value of the revenue obtained from all future sales of coal is

$$\text{Present value} = \int_0^\infty R(t)e^{-rt}\,dt = \int_0^\infty 750e^{-0.03t}e^{-0.15t}\,dt$$

$$= \int_0^\infty 750e^{-0.18t}\,dt = \lim_{b\to\infty}\int_0^b 750e^{-0.18t}\,dt$$

$$= \lim_{b\to\infty}\left[\frac{750}{-0.18}(e^{-0.18b}-1)\right]$$

$$= \$4167 \text{ thousand}$$

$$= \$4.167 \text{ million}$$

where we made use of the result $e^{-0.18b} = \dfrac{1}{e^{0.18b}} \to 0$ as $b \to \infty$.

EXERCISE 9.5

In Exercises 1–16, evaluate the improper integrals when they exist.

1. $\displaystyle\int_3^\infty \frac{2}{x^2}\,dx$ \qquad\qquad **2.** $\displaystyle\int_2^\infty \frac{1}{(x-1)^2}\,dx$

3. $\displaystyle\int_1^\infty \frac{4}{x^3}\,dx$ \qquad\qquad **4.** $\displaystyle\int_2^\infty x\,dx$

5. $\displaystyle\int_{-\infty}^1 e^{2x}\,dx$ \qquad\qquad **6.** $\displaystyle\int_1^\infty \frac{dx}{x}$

7. $\displaystyle\int_{-\infty}^{-1} \frac{2}{\sqrt[3]{x^4}}\, dx$

8. $\displaystyle\int_{-\infty}^{-1} \frac{2}{\sqrt[3]{x^2}}\, dx$

9. $\displaystyle\int_{0}^{\infty} e^{-2x}\, dx$

10. $\displaystyle\int_{-\infty}^{1} \frac{dx}{(x-2)^2}$

11. $\displaystyle\int_{-\infty}^{0} \frac{x}{(x^2+2)^2}\, dx$

12. $\displaystyle\int_{-\infty}^{\infty} x^2\, dx$

13. $\displaystyle\int_{0}^{\infty} \frac{x}{x^2+1}\, dx$

14. $\displaystyle\int_{1}^{\infty} e^{1-x}\, dx$

15. $\displaystyle\int_{-\infty}^{\infty} xe^{-3x^2}\, dx$

16. $\displaystyle\int_{0}^{\infty} xe^{-x}\, dx$

17. In Example 5, suppose the per-ton price of coal increases according to the equation

$$p(t) = 15 + t$$

What is the present value of all future revenues obtained from mining and selling the coal?

10

Multivariable Calculus

Our attention has been focused exclusively on functions of a single independent variable. However functions of two or more independent variables arise far more frequently in applications. For example, most companies derive their revenue from selling many different products so that the revenue R is a function of the number of each type of product sold. If a retailer sells n different products whose unit prices are $p_1, p_2, \ldots, p_n$, and if $x_1, x_2, \ldots, x_n$ represent the corresponding number of each type of product sold, then the revenue function $R(x_1, x_2, \ldots, x_n)$ can be written as

$$R(x_1, x_2, \ldots, x_n) = p_1 x_1 + p_2 x_2 + \cdots + p_n x_n$$

Mathematically, little is lost by restricting our study to functions of only two variables; the methods developed for functions of two variables can be extended easily to functions of three or more variables. Section 10.1 will be devoted to studying functions of two variables and their graphs. As with functions of a single variable, techniques for describing rates of change of functions of two or more variables are desirable and useful. The first *partial derivatives,* presented in Section 10.2, are the most widely used methods to accomplish this. Optimization is also a very important consideration in dealing with functions of two or more variables; this topic will be studied in Sections 10.3 and 10.4. Finally, the optimization techniques will be used to find the equation of the line of best fit to a set of experimental data; the procedure known as the *method of least squares* will be presented in Section 10.5.

❧ 10.1 Functions of Two Variables and their Graphs

Suppose, for simplicity, that a retailer sells only two different types of cameras, one an inexpensive instamatic that sells for $20 and the second a 35-mm camera that retails for $125. If x represents the number of instamatics sold and y the number of 35-mm cameras sold, the revenue R can be expressed mathematically as a function of x and y by the equation

$$R(x, y) = 20x + 125y$$

For each pair of values assigned to the *independent* variables x and y, the dependent variable R assumes a single value. In this situation, R is said to be a *function* of the two independent variables x and y.

DEFINITION

In general, the variable z is said to be a *function* of the *two independent* variables x and y, written as

$$z = f(x, y)$$

if for each ordered pair of real numbers (x, y), for which $f(x, y)$ is defined, the *dependent variable* z is assigned a *unique* value.

Example 1   If the function $z = f(x, y)$ is defined by the equation

$$z = f(x, y) = 3x^2 - 2y^2 + xy$$

find each of the following:

a. $f(1, 4)$ b. $f(4, 1)$ c. $f(a, b)$ d. $f(x + h, y)$ e. $f(x, y + h)$

Solution   The value assigned to the dependent variable z is found by substituting the given values of x and y into the equation that defines the function, so we get

a. $z = f(1, 4) = 3(1)^2 - 2(4)^2 + 1(4) = -25$
b. $z = f(4, 1) = 3(4)^2 - 2(1)^2 + 4 = 50$
c. $z = f(a, b) = 3a^2 - 2b^2 + ab$
d. $z = f(x + h, y) = 3(x + h)^2 - 2y^2 + (x + h)y$
$$= 3x^2 + 6xh + 3h^2 - 2y^2 + xy + yh$$
e. $z = f(x, y + h) = 3x^2 - 2(y + h)^2 + x(y + h)$
$$= 3x^2 - 2y^2 - 4yh - 2h^2 + xy + xh$$

Example 2   Suppose that the unit cost C of manufacturing a refrigerator is dependent on the amount of steel W, in pounds, and the number of man-hours of labor L according to the following equation

$$C(W, L) = 20 + W + 6L$$

What is the cost of a refrigerator that weighs 125 lb and requires 10 man-hours of labor to build?

Solution   The unit cost of each refrigerator is found by setting $W = 125$, $L = 10$, and evaluating. We get

$$C(125, 10) = 20 + 125 + 6(10) = \$205$$

THREE-DIMENSIONAL COORDINATE SYSTEMS

In order to plot the graph of a function of two variables $z = f(x, y)$, it is necessary to have a coordinate system whose points are represented by ordered triples (x, y, z) of real numbers. A three-dimensional coordinate system can be

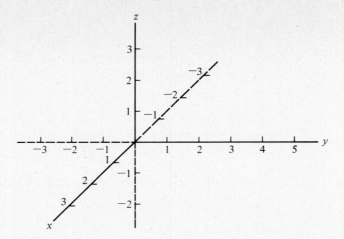

Figure 10.1

formed by setting three number lines, usually denoted as the x, y, and z axes, at right angles to one another and aligning them so that the common point of intersection, called the origin, corresponds to the number 0 on all three axes. The customary alignment of three axes is shown in Figure 10.1. The three planes formed by each pair of coordinate axes are called the xy, xz, and yz planes, portions of which are shown in Figure 10.2.

Each ordered triple (x, y, z) of real numbers can be represented as a point P in a three-dimensional coordinate system and vice versa; that is, each point can be represented as an ordered triple of numbers. The first element x of an ordered triple (x, y, z) is called the x coordinate and represents the directed distance of the point P from the yz plane. In the same way, the second and third elements of the ordered triple are called the y and z coordinates and represent the directed distances of P from the xz and xy planes, respectively. Figure 10.3 shows the point P corresponding to the ordered triple $(2, 4, 5)$. The xz, yz, and xy planes divide space into eight regions known as *octants*. The x, y, and z coordinates of all points in the first octant are positive. Points on the x, y, and z

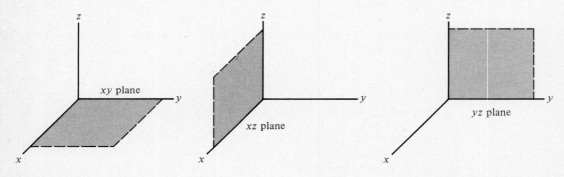

Figure 10.2

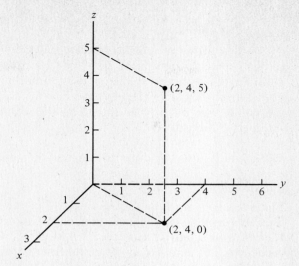

Figure 10.3

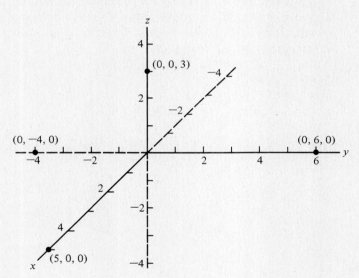

Figure 10.4

axes are characterized by the fact that the coordinates of the other two variables equal 0. A selected group of points is shown in Figure 10.4.

As in the two-dimensional case, the graph of a function of two variables is the set of all points (x, y, z) satisfying the equation $z = f(x, y)$, which defines the function. Sketching the graph of a function of two variables is generally difficult and time consuming because the graph usually is a three-dimensional surface, whereas the graph of a function of one variable is generally a two-dimensional curve. However, there is one type of function whose graph can be sketched easily, the linear function

$$z = f(x, y) = ax + by + c \qquad a, b, \text{ and } c \text{ are constant} \tag{10.1.1}$$

The graph of a linear function is a plane in the three-dimensional coordinate system.

Noting that a plane is determined uniquely by three noncollinear points, a partial representation of its graph can be constructed by locating, whenever possible, the points where the plane intersects the three coordinate axes. The point where the plane intersects the x axis is called the x intercept of the plane; the points where the plane intersects the y and z axes are called the y and z intercepts, respectively.

Example 3 | Plot a graph of the function whose equation is

$$z = f(x, y) = -4x - 2y + 8$$

Solution | The intercepts are found by setting two of the three variables equal to 0 and then solving the resulting equation for the third variable. This procedure yields the following set of ordered triples: (0, 0, 8), (0, 4, 0), and (2, 0, 0). These points are shown in the following figure together with the shaded surface that represents that portion of the plane located in the first octant. The three lines shown with their equations on the graph result from the intersection of the given plane with the xy, yz, and xz planes. These lines are called *traces* of the given plane and are very useful in graphing.

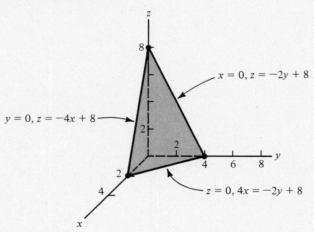

If one variable is missing from Equation 10.1.1, its graph is a plane parallel to the axis represented by the missing variable, as demonstrated in the following example.

Example 4 | Plot a graph of the equation

$$z = f(x, y) = \frac{-2y}{3} + 2 \qquad \text{or} \qquad 2y + 3z = 6$$

Solution | The graph is found by plotting the graph of the equation in the xz plane ($x = 0$) and then extending the line indefinitely parallel to the x axis, as shown in the next figure, to generate the graph of the plane.

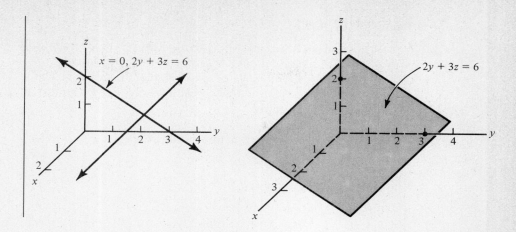

If two variables are missing from Equation 10.1.1, its graph is a plane parallel to the coordinate plane represented by the missing variables, as shown by the following example.

Example 5 | Plot a graph of the function represented by the equation

$$y = 4$$

Solution | The equation $y = 4$ is satisfied by all ordered triples having the form $(x, 4, z)$, that is, the set of all points whose y-coordinate equals 4. This set of points lies in the plane parallel to the xz coordinate plane whose y intercept equals 4, as shown in the following figure.

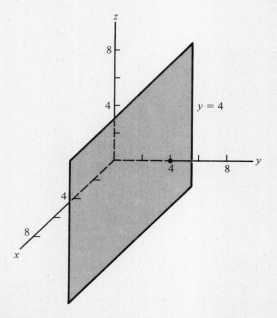

LEVEL CURVES

Functions of two variables can also be represented graphically by means of *level curves*. Level curves are projections of the surface $z = f(x, y)$ onto the xy plane for selected values of the dependent variable z. The level curves are generally easier to plot and when used wisely, can be almost as revealing as a three-dimensional sketch.

Example 6   Plot level curves for the function

$$z = f(x, y) = -2x - 4y + 8$$

for the following values of z:

$$z = 0, 2, 4$$

Solution   A three-dimensional sketch of the graph is shown in Example 3. When z is held fixed, the graph of the resulting equation in x and y is a straight line in the xy plane. Therefore, we have the following equations to graph:

$$z = 0 \qquad 0 = -2x - 4y + 8$$

$$z = 2 \qquad 2 = -2x - 4y + 8$$

$$z = 4 \qquad 4 = -2x - 4y + 8$$

The graph of the functions is shown in part a of the following figure where the curves for $z = 0, 2, 4$ are shown. The corresponding level curves are shown in part b of the same figure.

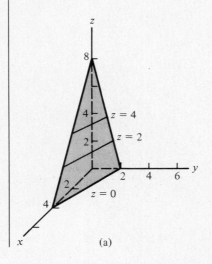

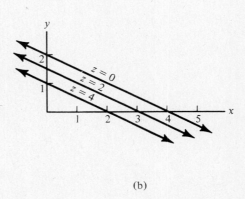

(a)   (b)

Example 7   Plot level curves for the function

$$z = f(x, y) = x^2 + y^2$$

for the following values of z:

$$z = 0, 1, 4, 9, 16$$

Solution   The graph of the function is shown at the top in the following figure. The level curves for the given values of z form a set of concentric circles about the origin $(0, 0)$ in the x, y plane.

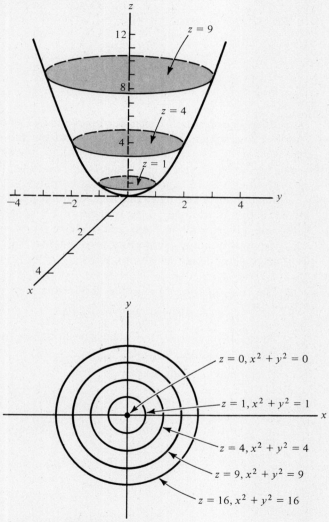

Example 8   Show the level curves for the function

$$z = f(x, y) = \frac{1}{xy} \quad \text{when} \quad z = \frac{1}{4}, \frac{1}{2}, 1, 2$$

Solution   The level curves are equilateral hyperbolas, displayed in the following figure.

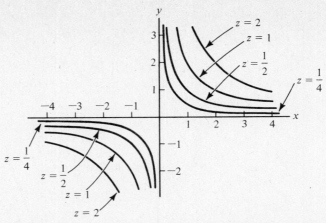

Level curves can sometimes be useful in evaluating the allocation of available resources among competing demands as shown in the following example.

Example 9 a. The developers of a new shampoo, Fair Hair, have decided to promote the product via TV and magazine advertising. If a 30-second TV spot costs $1,000 and a full-page magazine spread $4,000, what is the monthly cost function in terms of x, the number of TV spots, and y, the number of magazine advertisements? Assume that fixed costs amount to $3,000 per month.

b. If monthly advertising costs have been set at $63,000, what does the corresponding level curve look like?

Solution a. The cost function $C(x, y)$ has the form

$$C(x, y) = 1000x + 4000y + 3000$$

b. The level curve for $C = 63,000$ is described by the equation $63,000 = 1000x + 4000y + 3000$, or

$$60,000 = 1000x + 4000y$$

and its graph is shown in the next figure. The points on the line represent the various combinations of TV and magazine promotions that satisfy the $63,000 total cost requirement. The next step would then be to select the most effective combination.

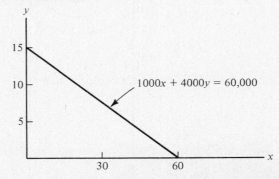

EXERCISE 10.1

1. If $z = f(x, y) = x^2 - 5y + 3$, find each of the following:

  a. $f(2, 1)$ **b.** $f(-1, 3)$ **c.** $f(2, 4)$ **d.** $f(2 + h, 1)$

  e. $f(2 + h, 1) - f(2, 1)$ **f.** $\dfrac{f(2 + h, 1) - f(2, 1)}{h}$

  g. $\lim\limits_{h \to 0} \dfrac{f(2 + h, 1) - f(2, 1)}{h}$

2. If $z = f(x, y) = (x + 2y)/(x + 3)$, find each of the following:

  a. $f(0, 1)$ **b.** $f(-2, 0)$ **c.** $f(2, 1)$ **d.** $f(2, 1 + k)$

3. If $z = f(x, y) = 2x\sqrt{y} - 3y\sqrt{x} + 1$, find each of the following:

  a. $f(1, 4)$ **b.** $f(4, 9)$ **c.** $f(0, 16)$ **d.** $f(\frac{1}{4}, \frac{16}{9})$

4. If $z = x \ln y - y \ln x$, find each of the following:

  a. $f(e, e)$ **b.** $f(1, 1)$ **c.** $f(1, e)$ **d.** $f(e, 1)$

5. If $z = f(x, y) = ye^x + xe^{-y}$, find each of the following:

  a. $f(0, 2)$ **b.** $f(2, 0)$ **c.** $f(1, 1)$ **d.** $f(a, -b)$

On the following coordinate system, plot the points in Exercises 6–12.

6. $(0, 3, 0)$ **7.** $(2, 0, 0)$

8. $(5, 0, 0)$ **9.** $(1, 4, 0)$

10. $(0, 3, 2)$ **11.** $(1, 4, 5)$

12. $(-5, 2, 3)$

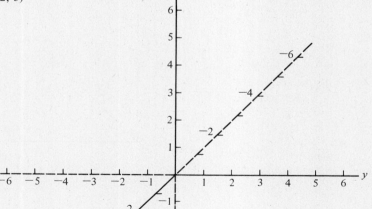

Plot graphs of the planes in Exercises 13–14 on the following coordinate systems.

13. $z = -3x - 6y + 9$

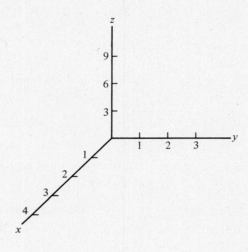

14. $z = 2x - 4y + 6$

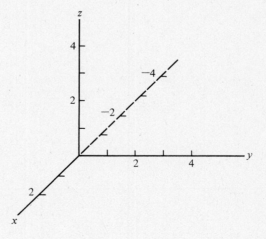

Sketch the level curves for the functions in Exercises 15–17 for the given values of z.

15. $z = f(x, y) = -x^2 + y$ $z = -1, 0, 1, 2, 3$

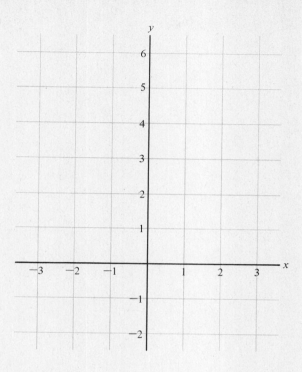

16. $z = f(x, y) = \dfrac{1}{x^2 y}$ $z = \dfrac{1}{4}, \dfrac{1}{2}, 1, 2$

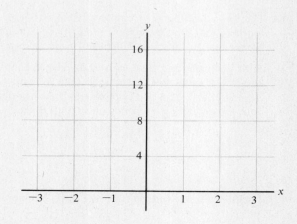

17. $z = f(x, y) = \sqrt{xy}$ $z = 0, 1, 2, 3$

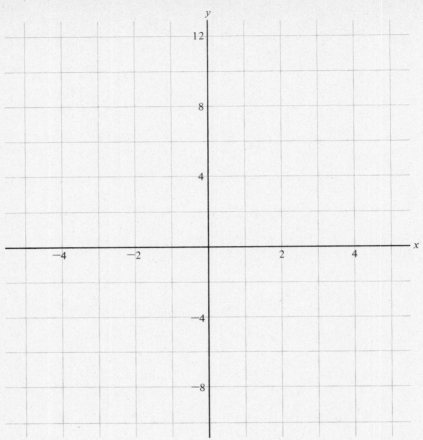

18. The Tasty Hamburger Company sells hamburgers for 75¢ each and cheeseburgers for 85¢ each. Find the company's revenue R as a function of x, the number of hamburgers sold, and y, the number of cheeseburgers sold.

19. Both Schmaltz and Bash breweries have recently introduced their own brands of low-calorie beer. Let p_1 and q_1 denote the unit price and quantity of beer sold by Schmaltz, and let p_2 and q_2 represent the unit price and quantity sold by Bash. Suppose the quantity q_1 depends upon both p_1 and p_2 as described by the equation

$$q_1 = f(p_1, p_2) = 10 - p_1 + 0.5p_2$$

while the dependence of q_2 upon p_1 and p_2 is given by the equation

$$q_2 = g(p_1, p_2) = 8 + 0.4p_1 - p_2$$

Find the revenue functions R_1 and R_2 for each brewery in terms of the unit prices p_1 and p_2.

20. The Fine Furniture Company manufactures two types of student desks: one is completely assembled and finished; the other unassembled, unfinished, and sold in the form of a kit. The number of each kind that can be sold monthly is a function not only of its own unit price but also of the unit price of the other. Let q_1 represent the number of finished desks and q_2 the number of kits; let p_1 and p_2 represent the unit prices of each. The dependence of q_1 and q_2 upon the unit prices p_1 and p_2 is described by the equations

$$q_1 = 500 - 5p_1 + 2p_2$$

$$q_2 = 400 + 2p_1 - 4p_2$$

Find the monthly revenue function $R(p_1, p_2)$.

21. The Office Remodeling Company charges \$2/ft² to carpet a floor, \$1/ft² to panel a wall, and \$1.50/ft² to install an acoustical ceiling. Find an equation describing the cost C of completely remodeling a rectangular office whose length is L, width is W, and height is H. What is the cost of remodeling an office whose dimensions are 20 ft by 15 ft by 8 ft?

22. The cost C of making a long distance phone call is dependent on the length t of the call and the distance d between the stations. Suppose the cost of an out of state long distance call can be described by the equation

$$C(d, t) = \sqrt[4]{d}(1 + 3t) \qquad 16 \le d \le 625$$

where d is the distance in miles and t is the length of the call in hours. What is the cost of a two-hour phone call to a station 256 miles away?

10.2 Partial Derivatives

When dealing with functions of two variables, the concept of rate of change is not defined uniquely as it is in the case of functions of one variable. However, the rate of change of the function can be determined if only one variable is permitted to change while the other is kept constant. When the variable y is held constant, the function $z = f(x, y)$ can be treated as a function of the variable x alone. The first derivative with respect to the variable x then can be found by using methods developed in Chapter 3. This derivative, called the *first partial derivative of f with respect to x* is written

$$f_x(x, y) \qquad \text{or} \qquad \frac{\partial z}{\partial x}$$

For example, if $z = f(x, y) = x^3 + 2xy + y^4$, the quantity $f_x(x, y)$ is found by treating y as a constant, and differentiating with respect to x, giving

$$f_x(x, y) = 3x^2 + 2y$$

In the same way, the first partial derivative of $f(x, y)$ with respect to y is found by keeping the variable x constant while y is permitted to change. This derivative is written

$$f_y(x, y) \qquad \text{or} \qquad \frac{\partial z}{\partial y}$$

For the function $z = f(x, y) = x^3 + 2xy + y^4$, we have

$$f_y(x, y) = 2x + 4y^3$$

The first partial derivatives $f_x(x, y)$ and $f_y(x, y)$ are defined as follows:

$$f_x(x, y) = \frac{\partial z}{\partial x} = \lim_{h \to 0} \frac{f(x + h, y) - f(x, y)}{h}$$

(10.2.1)

$$f_y(x, y) = \frac{\partial z}{\partial y} = \lim_{k \to 0} \frac{f(x, y + k) - f(x, y)}{k}$$

(10.2.2)

The definitions will not be needed to find the first partial derivatives; the methods developed in Chapter 3 are sufficient to enable us to find $f_x(x, y)$ and $f_y(x, y)$. It is essential that y be treated as a constant when finding $f_x(x, y)$ and x be treated as a constant when finding $f_y(x, y)$.

Example 1   Find the first partial derivatives $f_x(x, y)$ and $f_y(x, y)$ for the function

$$z = f(x, y) = x^3 - 6xy + 2y^2$$

Solution   The quantity $f_x(x, y)$ is found by differentiating each term with respect to x while simultaneously keeping y constant, yielding

$$\frac{\partial z}{\partial x} = f_x(x, y) = 3x^2 - 6y$$

When finding $f_y(x, y)$, the roles of x and y are reversed, so we get

$$\frac{\partial z}{\partial y} = f_y(x, y) = -6x + 4y$$

Example 2   Find $f_x(x, y)$ and $f_y(x, y)$ for the function

$$z = f(x, y) = y \ln x - xe^{2y}$$

Solution   Differentiating each term with respect to x while keeping y constant gives

$$\frac{\partial z}{\partial x} = f_x(x, y) = y \left(\frac{1}{x}\right) - (1)e^{2y} = \frac{y}{x} - e^{2y}$$

The first derivative $f_y(x, y)$ can be found by differentiating with respect to y while keeping x constant; we get

$$\frac{\partial z}{\partial y} = f_y(x, y) = (1) \ln x - xe^{2y}(2) = \ln x - 2xe^{2y}$$

Let us now turn our attention to the question of the geometrical interpretation of the first partial derivatives $f_x(x, y)$ and $f_y(x, y)$. In particular, let us examine $f_x(x, y)$ and $f_y(x, y)$ for the function

$$z = f(x, y) = 9 - x^2 - y^2$$

at the point $(1, 2, 4)$. To see what the partial derivative $f_x(1, 2)$, also written

$$\frac{\partial z}{\partial x}\bigg|_{(1,2)}$$

represents geometrically, let us consider Figure 10.5, which shows the graph of the function together with a portion of the plane $y = 2$ (remember y is kept constant when finding $f_x(x, y)$). The intersection of the surface $z = 9 - x^2 - y^2$ with the plane $y = 2$ generates a curve called the *trace* of the surface in the plane. The equation of the trace is

$$z = f(x, 2) = 9 - x^2 - 2^2 = 5 - x^2 \qquad y = 2$$

The partial derivative of z with respect to x has the form

$$f_x(x, 2) = \frac{\partial z}{\partial x}\bigg|_{y=2} = -2x$$

and represents the *slope* of the line tangent to the trace. The slope of the line tangent to the trace at the point $(1, 2, 4)$, also shown in Figure 10.5 (a), equals

$$f_x(1, 2) = \frac{\partial z}{\partial x}\bigg|_{(1,2)} = -2(1) = -2$$

In the same way, $f_y(1, 2)$ represents the slope of the line tangent to the trace that results from the intersection of the surface $z = f(x, y) = 9 - x^2 - y^2$ with

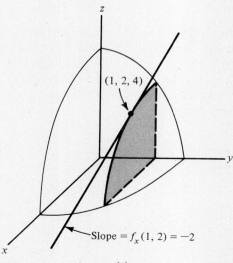

Figure 10.5 (a)   (a)

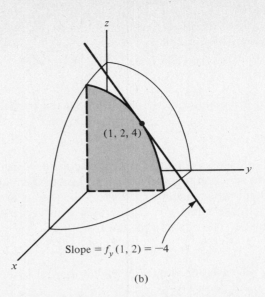

Figure 10.5 (b)   (b)

the plane $x = 1$, shown in Figure 10.5 (b). The relation between the y and z coordinates along the trace is described by the equation

$$z = f(1, y) = 9 - 1^2 - y^2 = 8 - y^2$$

The partial derivative $f_y(1, y)$ takes the form

$$f_y(1, y) = \frac{\partial z}{\partial y}\bigg|_{x=1} = -2y$$

so that the slope of the line tangent to the trace at $(1, 2, 4)$ becomes

$$f_y(1, 2) = \frac{\partial z}{\partial y}\bigg|_{(1,2)} = -2(2) = -4$$

This tangent line is also shown in Figure 10.5(b). The two perpendicular lines shown in Figures 10.5 (a) and 10.5 (b) define the plane tangent to the surface $z = f(x, y) = 9 - x^2 - y^2$ at the point $(1, 2, 4)$, shown in Figure 10.6.

Each of the first partial derivatives $f_x(x, y)$ and $f_y(x, y)$ can be differentiated with respect to the variables x and y giving rise to four second partial derivatives as follows:

$$\frac{\partial}{\partial x}\left(\frac{\partial z}{\partial x}\right) = \frac{\partial^2 z}{\partial x^2} = f_{xx}$$

$$\frac{\partial}{\partial y}\left(\frac{\partial z}{\partial y}\right) = \frac{\partial^2 z}{\partial y^2} = f_{yy}$$

$$\frac{\partial}{\partial x}\left(\frac{\partial z}{\partial y}\right) = \frac{\partial^2 z}{\partial x \partial y} = f_{yx}$$

$$\frac{\partial}{\partial y}\left(\frac{\partial z}{\partial x}\right) = \frac{\partial^2 z}{\partial y \partial x} = f_{xy}$$

(10.2.3)

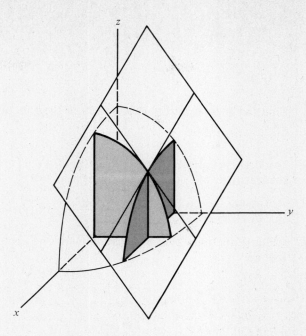

Figure 10.6

Although the order of differentiation is the same for both $\partial^2 z/\partial x \partial y$ and f_{yx}, that is, differentiate first with respect to y and second with respect to x, the order of the variables x and y is not the same for the two expressions. This is because both notations are abbreviations for expressions in which the correct order is displayed explicitly, that is,

$$\frac{\partial^2 z}{\partial x \partial y} = \frac{\partial}{\partial x}\left(\frac{\partial z}{\partial y}\right) \qquad f_{yx} = (f_y)_x$$

Fortunately, it is not necessary to concern yourself with the order in which the differentiation is carried out because $f_{yx} = f_{xy}$ for all the functions we shall encounter as well as for those used in most applications; however, it should be noted that $f_{xy} = f_{yx}$ is not true for all functions.

Example 3 | Find the second partial derivatives of the function

$$z = f(x, y) = x^3 - 6xy + 2y^2$$

given in Example 1.

Solution | Using Equations 10.2.3 and the results from Example 1, we get

$$f_{xx} = \frac{\partial^2 z}{\partial x^2} = \frac{\partial}{\partial x}(3x^2 - 6y) = 6x$$

$$f_{yy} = \frac{\partial^2 z}{\partial y^2} = \frac{\partial}{\partial y}(-6x + 4y) = 4$$

$$f_{yx} = \frac{\partial^2 z}{\partial x \partial y} = \frac{\partial}{\partial x}(-6x + 4y) = -6$$

$$f_{xy} = \frac{\partial^2 z}{\partial y \partial x} = \frac{\partial}{\partial y}(3x^2 - 6y) = -6$$

Example 4   Find the second partial derivatives of the function

$$z = f(x, y) = x^2 y - xy^3$$

Solution   The first partial derivatives $f_x(x, y)$ and $f_y(x, y)$ are found by differentiating with respect to x and y, respectively, giving

$$\frac{\partial z}{\partial x} = f_x(x, y) = 2xy - y^3 \qquad \frac{\partial z}{\partial y} = f_y(x, y) = x^2 - 3xy^2$$

Using Equations 10.2.3, we obtain for the second partial derivatives the following:

$$\frac{\partial^2 z}{\partial x^2} = f_{xx} = \frac{\partial}{\partial x}(2xy - y^3) = 2y$$

$$\frac{\partial^2 z}{\partial y^2} = f_{yy} = \frac{\partial}{\partial y}(x^2 - 3xy^2) = -6xy$$

$$\frac{\partial^2 z}{\partial x \partial y} = \frac{\partial}{\partial x}\left(\frac{\partial x}{\partial y}\right) = f_{yx} = \frac{\partial}{\partial x}(x^2 - 3xy^2) = 2x - 3y^2$$

$$\frac{\partial^2 z}{\partial y \partial x} = \frac{\partial}{\partial y}\left(\frac{\partial z}{\partial x}\right) = f_{xy} = \frac{\partial}{\partial y}(2xy - y^3) = 2x - 3y^2$$

The concept of marginal cost, revenue, and profit can be extended to situations where functions of two or more variables are required, as shown in the following example.

Example 5   A company producing wood-burning stoves finds that its monthly costs C depend on L, the number of man-hours of labor, and W, the number of pounds of steel used, according to the equation

$$C(L, W) = 1500 + 7L + 2W$$

The quantity $\partial C/\partial L = C_L(L, W)$ represents the *marginal cost* of *labor,* while $\partial C/\partial W = C_W(L, W)$ represents the *marginal cost* of *material.* For this case

$$\frac{\partial C}{\partial L} = \$7 \text{ per man-hour}$$

$$\frac{\partial C}{\partial W} = \$2 \text{ per lb of steel}$$

The number of units q that a firm produces in a given length of time, called output or productivity, is a function of both the number of man-hours of labor L

and the number of machine hours K, also called capital, allocated to producing a given product. The relationship between the variables can in many cases be described by an equation of the form

$$q = aL^\alpha K^{1-\alpha} \qquad \text{where } a \text{ and } \alpha \text{ are constants}$$

called the Cobb-Douglas production function. The first derivatives $\partial q/\partial L$ and $\partial q/\partial k$ are called the marginal productivity of labor and capital, respectively. They measure the approximate change in output per unit increase in labor and capital, respectively.

Example 6   The weekly output q of a firm producing skis is described by the equation

$$q = 8L^{0.5}K^{0.5}$$

where L represents the number of man-hours per week and K the number of machine hours per week; q is output in hundreds of units per week. Find the marginal productivity of labor and capital, respectively when $L = 900$ and $K = 400$.

Solution   The marginal productivity of labor and capital are found by taking the first partial derivatives with respect to L and K, respectively, giving

$$\frac{\partial q}{\partial L} = 4 \left(\frac{K}{L}\right)^{0.5} \qquad \frac{\partial q}{\partial K} = 4 \left(\frac{L}{K}\right)^{0.5}$$

When $L = 900$ and $K = 400$, the marginal productivities $\partial q/\partial L$ and $\partial q/\partial K$ become

$$\left.\frac{\partial q}{\partial L}\right|_{(900,\,400)} = 4 \left(\frac{900}{400}\right)^{0.5} = 6 \text{ units/man-hour of labor}$$

$$\left.\frac{\partial q}{\partial K}\right|_{(900,\,400)} = 4 \left(\frac{400}{900}\right)^{0.5} = \frac{8}{3} \text{ units/machine hour}$$

EXERCISE 10.2

Find $f_x(x, y)$ and $f_y(x, y)$ for the functions in Exercises 1–12.

1. $z = f(x, y) = 3x^4 - 6y^3$

2. $z = f(x, y) = x^2y^2 - 5x^3 + 7y$

3. $z = f(x, y) = \sqrt{xy}$

4. $z = f(x, y) = \dfrac{x}{y}$

5. $z = f(x, y) = \dfrac{y}{x}$

6. $z = f(x, y) = ye^x + x \ln y$

7. $z = f(x, y) = (x^2 - y^2)^3$

8. $z = f(x, y) = \dfrac{x - y}{x + y}$

9. $z = f(x, y) = e^{x \cdot y}$

10. $z = f(x, y) = \ln (2x + 3y)$

11. $z = f(x, y) = e^{x+y}$

12. $z = f(x, y) = e^x \ln y$

13. Find the slope of the line tangent to the curve formed by the intersection of the surface $z = f(x, y) = 2x^2 + y^2$ and the plane $x = 1$ at the point $(1, 1, 3)$.

14. Repeat Exercise 13 with the surface $z = f(x, y) = e^{xy}$, the plane $y = 2$, and the point $(0, 2, 1)$.

15. Repeat Exercise 13 with the surface $z = f(x, y) = \sqrt{xy}$, the plane $y = 4$, and the point $(1, 4, 2)$.

16. Repeat Exercise 13 with the surface $z = f(x, y) = \dfrac{x}{y}$, the plane $x = 6$, and the point $(6, -2, -3)$.

Find f_{xx}, f_{yy}, f_{xy}, and f_{yx} for the functions in Exercises 17–22; in each case verify that $f_{xy} = f_{yx}$.

17. $z = f(x, y) = 3x^2 + 2xy - 7y^2$

18. $z = f(x, y) = (x^2 - 3y)^2$

19. $z = f(x, y) = e^x + \ln xy$

20. $z = f(x, y) = xe^{y^2}$

21. $z = f(x, y) = y\sqrt{x}$

22. $z = f(x) = \dfrac{2x}{5y}$

23. Weekly sales for a vendor selling ice cream bars and novelties depend on the temperature T in degrees Fahrenheit and the population density d of his route according to the equation

$$S(T, d) = 20(T - 60)^2\, d^{3/2} \qquad T > 60°$$

where S is expressed in dollars and d in thousands of people per square mile. Assuming that $d = 16$, find the quantity $\partial S / \partial T$ when $T = 70°F$ and when $T = 90°F$.

24. The monthly output of a plant producing fragrances for both men and women is described by the equation

$$q(K, L) = 3K^{0.75}L^{0.25}$$

where K represents the investment in plant and machinery and L is the number of man-hours of labor. Find the marginal productivities of both labor and capital and evaluate each when $K = 16$ and $L = 81$.

10.3 Maxima and Minima

The first and second partial derivatives are very useful in locating and analyzing the shape of a surface at points which are relative maxima and minima. Figure 10.7(a) illustrates a case where the point $[a, b, f(a, b)]$ represents a relative maximum for the surface whose equation is $z = f(x, y)$. A case where $[a, b, f(a, b)]$ represents a relative minimum is shown in Figure 10.7(b). At points where the surface "peaks" or "bottoms out," the plane tangent to the surface is parallel to the xy plane; at such points, the first partial derivatives f_x and f_y equal 0, that is

$$f_x(x, y) = 0 \qquad \text{and} \qquad f_y(x, y) = 0 \tag{10.3.1}$$

Points satisfying Equation 10.3.1 are called *critical points*. In locating relative maxima and minima, the first step is to find the coordinates of all critical points.

After the critical points have been found, further analysis is required be-

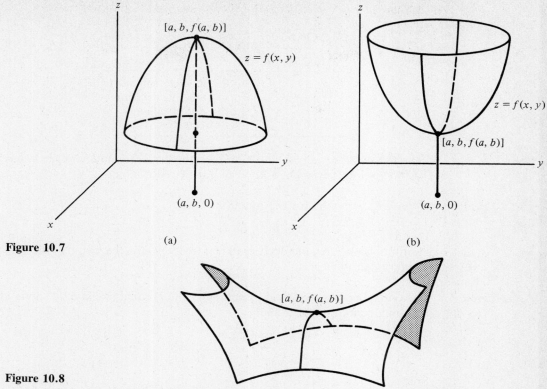

Figure 10.7

(a)   (b)

Figure 10.8

cause all critical points do not represent relative maxima or minima. A critical point may describe a situation such as that shown in Figure 10.8, where the critical point $[a, b, f(a, b)]$ represents a *saddle point* of the surface. The analysis takes the form of a second derivative test similar to that used in testing critical points for functions of a single variable.

SECOND DERIVATIVE TEST

Let $[a, b, f(a, b)]$ be a critical point for the function $z = f(x, y)$, that is, $f_x(a, b) = 0$ and $f_y(a, b) = 0$, and let the quantities A, B, and C be defined as follows:

$$A = f_{xx}(a, b) \qquad B = f_{xy}(a, b) \qquad C = f_{yy}(a, b)$$

The second derivative test takes the following form:

   I. If $AC - B^2 > 0$ and $A > 0$, then $[a, b, f(a, b)]$ is a relative *minimum*.
  II. If $AC - B^2 > 0$ and $A < 0$, then $[a, b, f(a, b)]$ is a relative *maximum*.
 III. If $AC - B^2 < 0$, then $[a, b, f(a, b)]$ is a *saddle point*.
 IV. If $AC - B^2 = 0$, then the test *fails*, that is, no conclusions about the nature of the surface in the vicinity of the critical point can be drawn.

In tabular form, the second derivative test can be written

Table 10.1

$AC - B^2$	A	$[a, b, f(a, b)]$
+	+	Relative minimum
+	−	Relative maximum
−		Saddle point
0		Test fails

The following examples illustrate how the critical points and the behavior of the function in the neighborhood of each is determined.

Example 1 | Find the critical points for the function

$$z = f(x, y) = x^2 + y^2 - 8x + 2y + 7$$

In addition, use the second derivative test to determine the shape of the surface at each critical point.

Solution | a. The critical points are found by setting f_x and f_y equal to 0 and by solving the resulting equations for x and y

$$f_x = 2x - 8 \qquad f_y = 2y + 2$$

$$0 = 2x - 8 \qquad 0 = 2y + 2$$

Solutions:
$$x = 4 \qquad y = -1$$

The z coordinate is calculated next

$$z = (4)^2 + (-1)^2 - 8(4) + 2(-1) + 7 = -10$$

We find that $(4, -1, -10)$ is the only critical point.

b. Next, the second derivative test is applied. It is necessary to first find $f_{xx}(x, y), f_{xy}(x, y)$, and $f_{yy}(x, y)$ in order to calculate A, B, and C

$$f_{xx} = 2 \qquad f_{xy} = 0 \qquad f_{yy} = 2$$

We have in this case $A = 2$, $B = 0$, and $C = 2$. Thus

$$AC - B^2 = (2)(2) - 0 = 4 > 0 \qquad \text{and} \qquad A = 2 > 0$$

so we conclude that the point $(4, -1, -10)$ is a relative minimum.

Problem 1 Find the critical points and determine the shape of the surface in the vicinity of each for the function

$$z = f(x, y) = -x^2 - y^2 - 6x + 4y + 1$$

Answer $(-3, 2, 14)$ Relative maximum

Example 2 | Find the critical points and determine the shape of the surface in the vicinity of each for the function

$$z = f(x, y) = y^2 - x^2 + 1$$

Solution | a. The critical points are found first

$$f_x = -2x \qquad f_y = 2y$$

Setting f_x and f_y equal to 0 gives

$$0 = -2x \qquad 0 = 2y$$

and solving yields the critical point $(0, 0, 1)$.

b. The shape of the surface in the vicinity of the critical point is found by applying the second derivative test

$$f_{xx} = -2 \qquad f_{xy} = 0 \qquad f_{yy} = 2$$

We get $AC - B^2 = -4$, so the point is a saddle point.

Problem 2 Find the critical points and determine the shape of the surface in the vicinity of each for the function

$$z = f(x, y) = y^2 - x^2 + 4x + 3$$

Answer $(2, 0, 7)$ Saddle point

Example 3 | Find the critical points and determine the shape of the surface in the vicinity of each for the function

$$z = f(x, y) = x^2 + y^2 - xy + 3y$$

Solution | a. Again, the critical points are found by setting both f_x and f_y equal to 0 and solving the resulting set of equations

$$f_x = 2x - y \qquad f_y = 2y - x + 3$$

Setting f_x and f_y equal to 0 gives

$$0 = 2x - y \tag{1}$$

$$0 = 2y - x + 3 \tag{2}$$

We are looking for those values of x and y that satisfy both Equations 1 and 2. One method of finding the solutions is to solve Equation 1 for y in terms of x, that is

$$y = 2x \tag{3}$$

and then to substitute $2x$ for y in Equation 2. Carrying this out generates a single equation containing the variable x. Equation 2 now becomes

$$0 = 2(2x) - x + 3 \quad \text{or} \quad 0 = 3x + 3$$

Solving this equation yields the solution

$$x = -1$$

Substituting this result into Equation 3 to find y gives

$$y = -2$$

Finally, the value of z is found to be

$$z = f(-1, -2)$$
$$= (-1)^2 + (-2)^2 - (-1)(-2) + 3(-2) = -3$$

Thus, the point $(-1, -2, -3)$ is the only critical point.

b. The second derivative test is applied to determine the shape of the surface in the vicinity of the critical point. The second partial derivatives f_{xx}, f_{xy}, and f_{yy} have the following forms:

$$f_{xx} = 2 \qquad f_{xy} = -1 \qquad f_{yy} = 2$$

so

$$A = 2 \qquad B = -1 \qquad C = 2$$

Applying the second derivative test, we get

$$AC - B^2 = (2)(2) - 1 = 3 > 0 \qquad \text{and} \qquad A = +2 > 0$$

which leads us to conclude that $(-1, -2, -3)$ is a relative minimum.

Problem 3 Find the critical points and the shape of the surface in the vicinity of each for the function

$$z = f(x, y) = x^2 + y^2 + xy - 6x + 2$$

Answer $(4, -2, -10)$ Relative minimum

These methods can also be used to maximize quantities such as profits or revenue or, on the other hand, to minimize costs as shown in the following examples.

Example 4 The Television Hut Company sells two types of portable color TV sets. The demand functions for each are given by the equations

$$q_1 = 20 - 8p_1 + 2p_2$$
$$q_2 = 15 + 2p_1 - 3p_2$$

where the subscript 1 refers to the less expensive model and the subscript 2 to the more expensive model; p is expressed in hundreds of dollars and q in thousands of sets. Find the revenue function $R(p_1, p_2)$ and determine at what price levels the revenue is maximized.

Solution The revenue R can be expressed as

$$R = q_1 p_1 + q_2 p_2$$

Using the demand equations, the total revenue can be expressed as a function of p_1 and p_2

$$R = (20 - 8p_1 + 2p_2)p_1 + (15 + 2p_1 - 3p_2)p_2$$
$$= 20p_1 - 8p_1^2 + 4p_2 p_1 + 15p_2 - 3p_2^2$$

The values of p_1 and p_2 that maximize R can be found by setting $\partial R/\partial p_1$ and $\partial R/\partial p_2$ both equal to 0 and solving the resulting equations. First, we find the first partial derivatives

$$\frac{\partial R}{\partial p_1} = 20 - 16p_1 + 4p_2$$

$$\frac{\partial R}{\partial p_2} = 15 + 4p_1 - 6p_2$$

Setting each equal to 0 generates the set of equations

$$0 = 20 - 16p_1 + 4p_2 \tag{1}$$
$$0 = 15 + 4p_1 - 6p_2 \tag{2}$$

Solving Equation 1 for p_2 gives

$$p_2 = -5 + 4p_1$$

Substituting the expression $(-5 + 4p_1)$ for p_2 in Equation 2 gives

$$0 = 15 + 4p_1 - 6(-5 + 4p_1)$$
$$0 = 45 - 20p_1$$

From this equation we then get the solution

$$p_1 = 2.25$$

Substituting this value of p_1 into either Equation 1 or 2 gives for p_2

$$p_2 = 4$$

from which we then obtain the revenue R

$$R = 20(2.25) - 8(2.25)^2 + 4(4)(2.25) + 15(4) - 3(4)^2$$
$$= \$52.5 \text{ hundred thousand} = \$5.25 \text{ million}$$

To show that this result represents a maximum revenue, we can use the second derivative test

$$\frac{\partial^2 R}{\partial p_1^2} = -16 \qquad \frac{\partial^2 R}{\partial p_1 \partial p_2} = 4 \qquad \frac{\partial^2 R}{\partial p_2^2} = -6$$

Thus $A = -16$, $B = 4$, and $C = -6$, so we get

$$AC - B^2 = +80 > 0 \qquad \text{and} \qquad A = -16 < 0$$

so we conclude that the revenue is maximized.

Example 5   A home improvement firm has decided to market in kit form an inexpensive rectangular shed that has no floor and one open side. If material for the three sides and the roof costs 25¢/ft², find the dimensions of the shed that minimize the cost of material if the volume of the shed equals 686 ft³.

Solution   A sketch of the shed when assembled is shown in the following figure. Denoting the dimensions of the shed x, y, and w, the cost of materials can be written as

$$C = 25xy + 50yw + 25xw \tag{1}$$

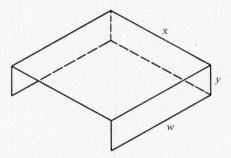

The fact that the volume is fixed at 686 ft³ can be represented by the equation

$$xyw = 686 \tag{2}$$

The restriction expressed by Equation 2 indicates that x, y, and w are not independent variables. We can solve Equation 2 for one of the variables in terms of the other two and substitute this result into Equation 1. Solving for w gives

$$w = \frac{686}{xy}$$

and substituting this expression into Equation 1 gives the cost as a function of x and y

$$C(x, y) = 25xy + 50y\left(\frac{686}{xy}\right) + 25x\left(\frac{686}{xy}\right)$$

$$C(x, y) = 25xy + \frac{(50)(686)}{x} + \frac{(25)(686)}{y}$$

The values of x and y that minimize $C(x, y)$ can be found by solving the equations $C_x = 0$ and $C_y = 0$

$$C_x = 25y - \frac{(50)(686)}{x^2} = 0 \tag{3}$$

$$C_y = 25x - \frac{(25)(686)}{y^2} = 0 \tag{4}$$

First, rewriting Equation 3 as

$$yx^2 - 2(686) = 0 \tag{5}$$

and solving Equation 4 for x in terms of y gives

$$x = \frac{686}{y^2} \tag{6}$$

Substituting this expression for x into Equation 5 gives

$$y \left(\frac{686}{y^2}\right)^2 - 2(686) = 0$$

or

$$686 - 2y^3 = 0 \tag{7}$$

Solving Equation 7 for y gives

$$y = \sqrt[3]{343} = 7$$

Substituting $y = 7$ into Equation 6 yields the solution $x = 14$ and from Equation 2 we obtain $w = 7$. Therefore, the dimensions of the shed that minimize the cost of materials are 14 ft by 7 ft by 7 ft. The minimum cost C equals

$$C = 25(14)(7) + 50(7)(7) + 25(14)(7)$$
$$= 7350\cent = \$73.50$$

Proving that the cost is minimized when the dimensions are 14 ft by 7 ft by 7 ft is left to an exercise.

EXERCISE 10.3

For the functions in Exercises 1–18, find the critical points and determine the shape of the curve in the vicinity of each critical point using the second derivative test.

1. $z = f(x, y) = x^2 + 2y^2 + 1$

2. $z = f(x, y) = 1 - x^2 - 2y^2$

3. $z = f(x, y) = x^2 - y^2 + 3$

4. $z = f(x, y) = x^2 + y^2 - 2x - 4y - 3$

5. $z = f(x, y) = x^2 + y^2 - 8x + 2y + 5$

6. $z = f(x, y) = 2 - 4x + 4y - x^2 - y^2$

7. $z = f(x, y) = x^2 - y^2 + xy + 2x - 9y + 3$

8. $z = f(x, y) = x^2 + y^2 - xy - 3x - y + 2$

9. $z = f(x, y) = x^3 + 3x^2 - y^2 + 2y + 4$

10. $z = f(x, y) = y^3 - 6y^2 + x^2 - 2x + 1$

11. $z = f(x, y) = x^3 + y^3 - 3xy - 5$

12. $z = f(x, y) = x^3 - y^3 - 3xy + 2$

13. $z = f(x, y) = 2x^3 - 3x^2 - 12x + y^2 - 2y + 3$

14. $z = f(x, y) = y^3 - 3y^2 - 9y + x^2 - 2x + 6$

15. $z = f(x, y) = 4 + 6xy - x^3 - y^2$

16. $z = f(x, y) = 2xy + \dfrac{4}{x} - \dfrac{1}{y}$

17. $z = f(x, y) = xy + \dfrac{8}{x} + \dfrac{27}{y}$

18. $z = f(x, y) = e^{x^2 + y^2}$

19. The owner of a sporting goods store carries two kinds of jogging shoes. He pays $10 per pair for the less expensive shoe and $15 per pair for the more expensive shoe. The demand functions for the two types of shoes are described by the equations

$$q_1 = 150 - 10p_1 + 4p_2$$

$$q_2 = 93 + 4p_1 - 5p_2$$

where the subscript 1 refers to the less expensive and 2 to the more expensive shoes, respectively. Set up the profit function $P(p_1, p_2)$ and determine the unit prices p_1 and p_2 that maximize his profit.

20. The length plus the girth of a package mailed through the U.S. Postal Service cannot exceed 84 in. What are the dimensions of the rectangular package which maximize the volume?

NOTE: In the following figure we have length + girth = $l + (2w + 2h) = 84$.

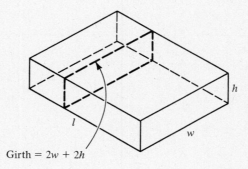

Girth = $2w + 2h$

21. Find three positive numbers x, y, and w whose sum is 45 and whose product xyw is a maximum.

22. Find three positive numbers x, y, and w whose product is 27 and whose sum ($x + y + w$) is a minimum.

23. A builder has a contract to build a one-story rectangular warehouse that will have a volume of 375,000 ft³. Material for the floor costs $2/ft². Material for three of the walls and the roof costs $1/ft². The fourth wall, consisting of sliding doors, costs $3/ft². What are the dimensions of the building that minimize the cost of material? Although building costs are minimized, does this building have some features that might make it difficult to operate as a warehouse?

24. Complete the analysis in Example 5 by showing that the cost of material is minimized for a shed that measures 14 ft by 7 ft by 7 ft.

ఈ 10.4 Lagrange Multipliers

Most maximization or minimization problems require that the optimization be carried out subject to constraints on the independent variables. For example, a utility that burns coal to generate electricity cannot operate its furnaces to generate the maximum energy possible from the coal because too many pollutants would be emitted from its stacks in the process. The company must try to generate the maximum amount of energy consistent with Environmental Protection Agency (EPA) guidelines on pollution control. In the economic realm, all companies have finite resources that serve as constraints on the operations of the firm. It is the function of the firm's management to allocate the available resources so that the company's profits are maximized.

Geometrically, the introduction of constraints is demonstrated in Figure 10.9. Figure 10.9 (a) shows a surface $z = f(x, y)$ whose minimum point is found by the method described in Section 10.3, that is, solving the equations $f_x = 0$ and $f_y = 0$. The minimum in such a situation is said to be unconstrained. If in addition, the variables x and y must satisfy an equation of the form $g(x, y) = 0$, called a constraint, the ordered triple (x, y, z) that satisfies both $z = f(x, y)$ and $g(x, y) = 0$ will lie on a curve such as that shown in Figure 10.9 (b), which forms the outer boundary of the shaded region. The minimum value of z on the curve (the constrained minimum) will generally differ from the minimum point on the surface $z = f(x, y)$.

When constraints were encountered previously in Section 4.2 and Example 5 of Section 10.3, the constraint equation was solved for one of the variables in terms of the remaining variables. Substituting this result into the function to be optimized reduced the total number of variables by one. Another approach, known as the method of Lagrange multipliers, can be used when

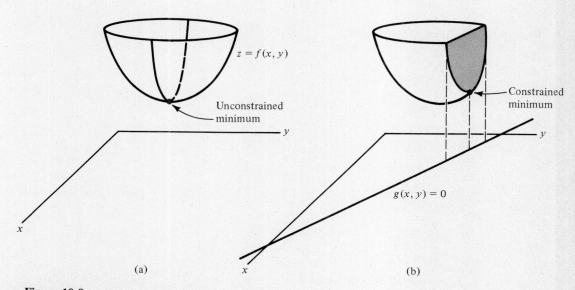

Figure 10.9

the constraint equation is unwieldy or difficult to solve for one variable in terms of the remaining variables. The method works as follows.

Let $f(x, y)$ and $g(x, y)$ be differentiable functions of both x and y. A new function $F(x, y, \lambda)$ is defined as follows:

$$F(x, y, \lambda) = f(x, y) - \lambda g(x, y)$$

where the quantity λ is known as the *Lagrange multiplier*. The values of x and y that optimize $f(x, y)$ subject to the constraint $g(x, y) = 0$ are found by solving the system of equations

$$F_x = 0 \qquad F_y = 0 \qquad F_\lambda = 0$$

or equivalently

$$\boxed{f_x - \lambda g_x = 0} \qquad\qquad\qquad (10.4.1)$$

$$\boxed{f_y - \lambda g_y = 0} \qquad\qquad\qquad (10.4.2)$$

$$\boxed{g(x, y) = 0} \qquad\qquad\qquad (10.4.3)$$

The method is illustrated in the following examples.

Example 1 | Minimize the function

$$z = f(x, y) = x^2 + 3y^2$$

subject to the constraint

$$x + y = 2$$

Solution | The constraint equation $g(x, y) = 0$ is written as

$$g(x, y) = x + y - 2 = 0$$

According to Equations 10.4.1–10.4.3, the system of equations to be solved takes the form

$$f_x - \lambda g_x = 2x - \lambda = 0 \qquad\qquad (1)$$

$$f_y - \lambda g_y = 6y - \lambda = 0 \qquad\qquad (2)$$

$$g(x, y) = x + y - 2 = 0 \qquad\qquad (3)$$

From Equations 1 and 2 we can eliminate the variable λ to yield

$$2x = 6y \quad \text{or} \quad x = 3y$$

Substituting this result into Equation 3 yields for y

$$y = \tfrac{1}{2}$$

From Equation 3, we get for x

$$x = \tfrac{3}{2}$$

and for z

$$z = (\tfrac{3}{2})^2 + 3(\tfrac{1}{2})^2 = 3$$

Problem 1 Maximize the function

$$z = f(x, y) = 4 - x^2 - y^2$$

subject to the constraint

$$x - 2y = 10$$

Answer $(2, -4, -16)$

Although the proof of the Lagrange multiplier method is beyond the scope of this course, the method can be demonstrated graphically for the function $z = f(x, y) = x^2 + 3y^2$ subject to the constraint $x + y = 2$ (Example 1). A number of level curves for $z = f(x, y)$ are shown in Figure 10.10 together with the graph of the constraint equation. The Lagrange multiplier selects from among all the points common to both types of curves, those for which the constraint curve is tangent to the level curves. The optimum values of the function $z = f(x, y)$ are located at these points.

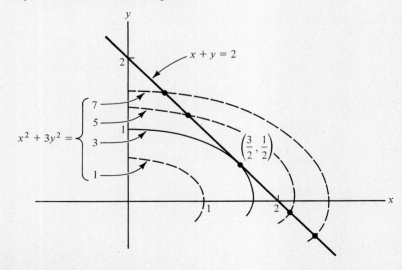

Figure 10.10

Problem 2 Minimize the function $z = f(x, y) = xy$ subject to the constraint

$$g(x, y) = x + 3y - 6 = 0$$

Answer $f(3, 1) = 3$

In addition, draw a graph of both the constraint curve and that portion of the level curve $xy = 3$ in the first quadrant on the accompanying graph.

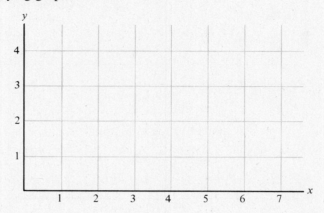

Example 2 Maximize the function

$$z = f(x, y) = xy$$

subject to the constraint

$$x^2 + y^2 = 8$$

Solution Again, the constraint equation is written

$$g(x, y) = x^2 + y^2 - 8 = 0$$

Equations 10.4.1–10.4.3 yield in this case the following:

$$f_x - \lambda g_x = y - 2\lambda x = 0 \tag{1}$$

$$f_y - \lambda g_y = x - 2\lambda y = 0 \tag{2}$$

$$g(x, y) = x^2 + y^2 - 8 = 0 \tag{3}$$

If Equation 1 is solved for λ, we get

$$\lambda = \frac{y}{2x}$$

Substituting this result into Equation 2 gives

$$x - 2\left(\frac{y}{2x}\right)y = 0$$

or

$$x^2 = y^2 \tag{4}$$

Substituting this result into Equation 3 yields the equation

$$2x^2 = 8$$

for which the solutions are

$$x_1 = 2 \quad \text{and} \quad x_2 = -2$$

Equation 4 indicates that for each of these solutions, there are two values of y, that is, $y = \pm 2$, so there are four ordered pairs that satisfy Equations 1–3: (2, 2), (2, −2), (−2, 2), and (−2, −2). The corresponding values of z are

$$z(2, 2) = 4$$

$$z(2, -2) = -4$$

$$z(-2, 2) = -4$$

$$z(-2, -2) = 4$$

The maximum value of $f(x, y)$, that is, 4, occurs at (2, 2, 4) and (−2, −2, 4). The remaining two solutions represent minimum values of $f(x, y)$. These results are shown graphically in Figure 10.11 where various level curves for the function $z = f(x, y)$ are shown together with the constraint curve $g(x, y) = x^2 + y^2 - 8 = 0$. As before, the optimum values of $f(x, y)$ occur where the level curves are tangent to the constraint curve.

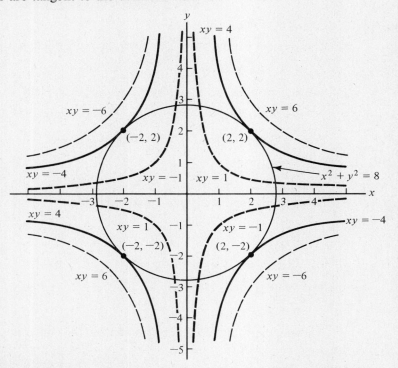

Figure 10.11

These methods can also be used to determine how resources should be allocated to maximize profits.

Example 3   An aircraft company manufactures two kinds of helicopters. The monthly profit function $P(x, y)$, expressed in thousands of dollars, is described by the equation

$$P(x, y) = -x^2 - \frac{y^2}{2} + 60x + 40y + xy - 100$$

where x and y represent the number of units of each type produced. At its present capacity, the company can produce 10 units each month. How many of each type of helicopter should the company produce in order to maximize monthly profits?

Solution   The constraint equation in this case takes the form

$$g(x, y) = x + y - 10 = 0$$

We want to maximize $P(x, y)$ subject to this constraint. When the company finds those values of x and y that maximize P, then resources can be allocated to attain the desired output. According to Equations 10.4.1–10.4.3, we want to solve the system of equations

$$P_x - \lambda g_x = -2x + 60 + y - \lambda = 0 \tag{1}$$

$$P_y - \lambda g_y = -y + 40 + x - \lambda = 0 \tag{2}$$

$$g(x, y) = x + y - 10 = 0 \tag{3}$$

This system can be solved by first solving Equation 1 for λ, obtaining

$$\lambda = -2x + 60 + y$$

and substituting this result into Equation 2 to yield

$$-y + 40 + x - (-2x + 60 + y) = 0 \qquad \text{or} \qquad 3x - 2y = 20$$

Solving this equation for x in terms of y gives

$$x = \frac{20}{3} + \frac{2y}{3}$$

Substituting this result into Equation 3 gives us

$$\left(\frac{20}{3} + \frac{2y}{3}\right) + y - 10 = 0$$

$$20 + 2y + 3y - 30 = 0 \qquad \text{or} \qquad 5y = 10$$

The solution for y is

$$y = 2$$

whereas the value of x can be found from the constraint $x + y - 10 = 0$ to give

$$x = 8$$

We can now calculate $P(8, 2)$ to give

$$P(8, 2) = -64 - 2 + 480 + 80 + 16 - 100 = \$410 \text{ thousand}$$

EXERCISE 10.4

Use the technique of Lagrange multipliers to solve Exercises 1–8.

1. Find the maximum value of the function $z = f(x, y) = xy$ subject to the constraint $x + y = 2$. In addition, draw the level curve corresponding to the maximum value of z together with a sketch of the constraint equation on the following coordinate system.

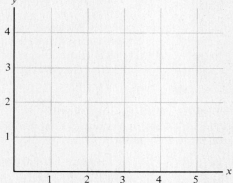

2. Find the maximum and minimum values of the function $z = f(x, y) = x + y$ subject to the constraint $x^2 + y^2 = 2$. Again, draw the level curves for both the minimum and maximum values of z together with a sketch of the constraint equation on the following coordinate system.

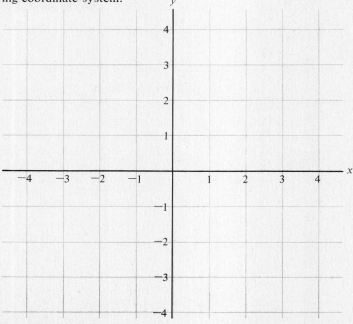

3. Find the maximum value of the function $z = f(x, y) = 3xy + x$ subject to the constraint $x + y = 1$.

4. Find the minimum value of the function $z = f(x, y) = 2x^2 + y^2 - xy$ subject to the constraint $2x + 3y = 14$.

5. Find the maximum value of the function $z = f(x, y) = 2x^2 - y^2 + xy$ subject to the constraint $y + 2x = 4$.

6. Find the maximum value of the function $z = f(x, y) = 2x + 3y - x^2 - y^2$ subject to the constraint $2x + y = 6$.

7. Find the minimum value of the function $z = f(x, y) = e^{x^2+y^2+xy}$ subject to the constraint $x + y = 2$.

8. A steel company produces two grades of steel at one of its plants. The better grade steel sells for $1200 per hundred tons, whereas the standard grade sells for $900 per hundred tons. Daily plant output is currently 148 hundred tons. The daily joint cost function C is described by the equation

$$C(q_1, q_2) = 2q_1^2 + q_2^2 - q_1 q_2$$

where q_1 and q_2 represent the daily outputs in hundreds of tons of the better and standard grades, respectively. Find the amount of each that should be produced to maximize daily profits.

9. Using the method of Lagrange multipliers, solve problem 2 in Exercise 4.2.

10. Using the method of Lagrange multipliers, solve problem 3 in Exercise 4.2.

11. Using the method of Lagrange multipliers, solve problem 4 in Exercise 4.2.

12. Using the method of Lagrange multipliers, solve problem 6 in Exercise 4.2.

13. Using the method of Lagrange multipliers, solve problem 7 in Exercise 4.2.

14. Using the method of Lagrange multipliers, solve problem 8 in Exercise 4.2.

10.5 Least Squares Method

Until now, we have purposely avoided discussing how the functions describing the relationships between two variables are determined. In many cases, they are determined by finding what is called a "curve of best fit" to experimental data. To see what this means, suppose the cost accountant of a firm has gathered the data, shown in Table 10.2 which give the company's total cost C for

Table 10.2

ORDER SIZE (thousands) x	TOTAL COST (thousands of $) C
30	40
40	45
25	33
45	54
35	47

each of five different orders for the same item where x represents the number of items sold. For planning purposes, the company's management would like to develop a method to determine the total cost of each order as a function of order size x. The data are shown graphically in Figure 10.12 (a). Inspection of the graph indicates that the underlying relationship between C and x could be approximated reasonably well by a linear relationship of the form

$$C = mx + b$$

It is possible to sketch many lines that visually represent reasonable approximations to the underlying relationship; two such lines are shown in Figure 10.12 (b). However, a technique known as the *least squares method* enables us to determine the line of best fit. To see how the method works, let us suppose that we are given a set of n points such as those shown in Figure 10.13 for which a linear relationship seems to exist.

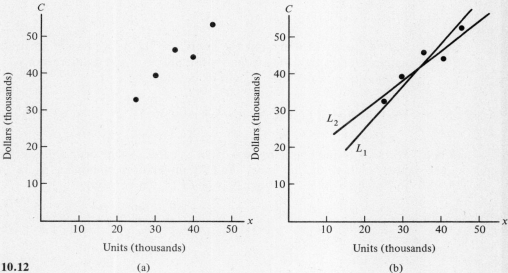

Figure 10.12   (a)   (b)

The method of least squares locates the line for which the sum of the squares of the vertical distances between the data points and the line is minimized. Using the slope-intercept formula $y = mx + b$, we will now show how the method of least squares enables us to find m and b. The square of the vertical distance from (x_1, y_1) to the line equals $(mx_1 + b - y_1)^2$. The sum of the squares S of the distances between all n points and the line given by the equation

$$S = (mx_1 + b - y_1)^2 + (mx_2 + b - y_2)^2$$
$$+ \cdots + (mx_n + b - y_n)^2 \tag{10.5.1}$$

The next step is to find those values of m and b that minimize S. This is accomplished by setting both $\partial S/\partial m$ and $\partial S/\partial b$ equal to 0 and then solving the re-

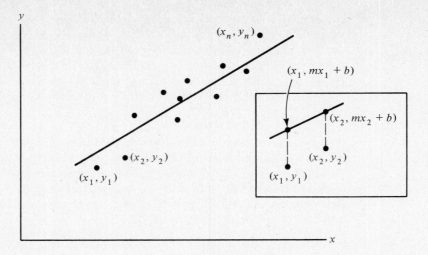

Figure 10.13

sulting system of equations for m and b. This procedure is illustrated in Example 1 following which formulas for finding m and b in terms of x_1 x_2, . . . , x_n and y_1, y_2, y_3, . . . , y_n will be developed.

Example 1 | Find the line of best fit for the points $(0, 6)$, $(1, 4)$, $(4, 3)$, $(5, 1)$

Solution | We want to minimize the sum of the squares of the vertical distances from the four points to the unknown line L as shown in the following figure. According to Equation 10.5.1, the sum S is given by the equation

$$S = (0m + b - 6)^2 + (1m + b - 4)^2 + (4m + b - 3)^2 + (5m + b - 1)^2$$
$$= (b - 6)^2 + (m + b - 4)^2 + (4m + b - 3)^2 + (5m + b - 1)^2$$

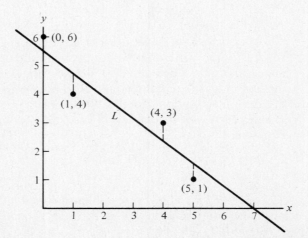

The values of m and b that minimize S are determined by finding $\partial S/\partial m$ and $\partial S/\partial b$ and setting each equal to 0

$$\frac{\partial S}{\partial m} = 2(m + b - 4) + 2(4m + b - 3)(4) + 2(5m + b - 1)(5)$$

$$= 2m + 2b - 8 + 32m + 8b - 24 + 50m + 10b - 10$$

$$= 84m + 20b - 42 = 0$$

$$\frac{\partial S}{\partial b} = 2(b - 6) + 2(m + b - 4) + 2(4m + b - 3) + 2(5m + b - 1)$$

$$= 2b - 12 + 2m + 2b - 8 + 8m + 2b - 6 + 10m + 2b - 2$$

$$= 20m + 8b - 28 = 0$$

The following system of equations must be solved to find m and b

$$42m + 10b - 21 = 0 \qquad\qquad\qquad\qquad\qquad\qquad\qquad (1)$$

$$5m + 2b - 7 = 0 \qquad\qquad\qquad\qquad\qquad\qquad\qquad (2)$$

The solution for m can be obtained if we multiply each term in Equation 2 by 5, thereby making the coefficients of b identical for both equations. We now have the equivalent system

$$42m + 10b - 21 = 0 \qquad\qquad\qquad\qquad\qquad\qquad\qquad (3)$$

$$25m + 10b - 35 = 0 \qquad\qquad\qquad\qquad\qquad\qquad\qquad (4)$$

Subtracting Equation 4 from 3 gives us a single equation in m

$$17m + 14 = 0 \qquad \text{or} \qquad m = -\tfrac{14}{17} = -0.82$$

Substituting this result into Equation 3 or 4 and solving for b yields

$$b = 5.56$$

The equation of the least squares line has the form

$$\boxed{y = -0.82x + 5.56}$$

It is not necessary to follow the procedure described in Example 1 each time the least squares line is desired. Formulas are available to enable you to find m and b in terms of $x_1, x_2, \ldots, x_n$ and $y_1, y_2, \ldots, y_n$ and n. The formulas are generalizations of the procedure described in Example 1 and are obtained by differentiating S in Equation 10.5.1 with respect to both m and b and setting each equal to 0

$$\frac{\partial S}{\partial m} = 2x_1(mx_1 + b - y_1) + 2x_2(mx_2 + b - y_2)$$

$$+ \cdots + 2x_n(mx_n + b - y_n)$$

Noting that m and b are the only variables on the right-hand side, we can group like terms together

$$\frac{\partial S}{\partial m} = 2m(x_1^2 + x_2^2 + \cdots + x_n^2) + 2b(x_1 + x_2 + \cdots + x_n)$$
$$- 2(x_1y_1 + x_2y_2 + \cdots + x_ny_n) = 0$$

The second equation required in the analysis is found by setting $\partial S/\partial b$ equal to 0

$$\frac{\partial S}{\partial b} = 2(mx_1 + b - y_1) + 2(mx_2 + b - y_2) + \cdots + 2(mx_n + b - y_n) = 0$$

Grouping all terms that contain m as a factor and doing the same for those that contain b as a factor give

$$(2mx_1 + 2mx_2 + \cdots + 2mx_n) + (2b + 2b + \cdots + 2b)$$
$$- (2y_1 + 2y_2 + \cdots + 2y_n) = 0$$

$$2m(x_1 + x_2 + \cdots + x_n) + 2nb - 2(y_1 + y_2 + \cdots + y_n) = 0$$

Now we have the system

$$(x_1 + x_2 + \cdots + x_n)m + nb = y_1 + y_2 + \cdots + y_n \qquad \textbf{(10.5.2)}$$

$$(x_1^2 + x_2^2 + \cdots + x_n^2)m + (x_1 + x_2 + \cdots + x_n)b$$
$$= x_1y_1 + x_2y_2 + \cdots + x_ny_n \quad \textbf{(10.5.3)}$$

To avoid becoming immersed in too much detail when solving the system for m and b, Equations 10.5.2 and 10.5.3 will be written in terms of four quantities A, B, C, and D defined as follows:

$$A = x_1 + x_2 + \cdots + x_n$$

$$B = y_1 + y_2 + \cdots + y_n$$

$$C = x_1^2 + x_2^2 + \cdots + x_n^2$$

$$D = x_1y_1 + x_2y_2 + \cdots + x_ny_n$$

Equations 10.5.2 and 10.5.3 can now be written as

$$Am + nb = B \qquad \textbf{(10.5.4)}$$

$$Cm + Ab = D \qquad \textbf{(10.5.5)}$$

We can eliminate the variable b in Equations 10.5.4 and 10.5.5 by multiplying each term in Equation 10.5.4 by A and each term in Equation 10.5.5 by n to produce the equivalent system

$$A^2m + nAb = AB \qquad \textbf{(10.5.6)}$$

$$nCm + nAb = nD \qquad \textbf{(10.5.7)}$$

Subtracting Equation 10.5.6 from 10.5.7 gives

$$(nC - A^2)m = nD - AB$$

Solving for m gives

$$m = \frac{nD - AB}{nC - A^2}$$

or

$$m = \frac{n(x_1 y_1 + x_2 y_2 + \cdots + x_n y_n) - (x_1 + x_2 + \cdots + x_n)(y_1 + y_2 + \cdots + y_n)}{n(x_1^2 + x_2^2 + \cdots + x_n^2) - (x_1 + x_2 + \cdots + x_n)^2} \qquad \textbf{(10.5.8)}$$

Similarly, we find that the y intercept b is given by the formula

$$b = \frac{(y_1 + y_2 + y_3 + \cdots + y_n)(x_1^2 + x_2^2 + \cdots + x_n^2) - (x_1 + x_2 + \cdots + x_n)(x_1 y_1 + x_2 y_2 + \cdots + x_n y_n)}{n(x_1^2 + x_2^2 + \cdots + x_n^2) - (x_1 + x_2 + \cdots + x_n)^2}$$

$$\textbf{(10.5.9)}$$

Example 2 Find the equation of the least squares line for the points $(0, 3)$, $(1, 5)$, $(2, 4)$, $(3, 3)$, and $(4, 7)$, using Equations 10.5.8 and 10.5.9.

Solution It is advisable to set up a table such as the following so that the addition can be carried out more quickly. The quantities m and b can now be found using Equations 10.5.8 and 10.5.9

$$m = \frac{(5)(50) - (10)(22)}{5(30) - (10)^2} = 0.60 \qquad b = \frac{(22)(30) - (10)(50)}{5(30) - (10)^2} = 3.20$$

The line $y = 0.60x + 3.20$ and the data points are shown in the following figure.

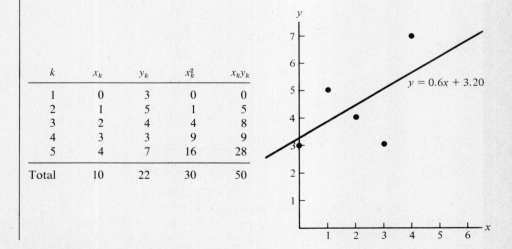

k	x_k	y_k	x_k^2	$x_k y_k$
1	0	3	0	0
2	1	5	1	5
3	2	4	4	8
4	3	3	9	9
5	4	7	16	28
Total	10	22	30	50

Example 3 | Using Equations 10.5.8 and 10.5.9, find the least squares line for the cost-accounting data shown in Table 10.2.

Solution | The total cost data represent the y coordinates for each ordered pair in the table. Constructing the following table enables us to carry out the calculations. Using Equations 10.5.8 and 10.5.9, we get

$$m = \frac{5(7900) - (175)(219)}{5(6375) - (175)^2} = \frac{1175}{1250} = 0.94$$

$$b = \frac{(219)(6375) - (175)(7900)}{5(6375) - (175)^2} = \frac{13625}{1250} = 10.90$$

k	x_k	C_k	x_k^2	$x_k C_k$
1	30	40	900	1200
2	40	45	1600	1800
3	25	33	625	825
4	45	54	2025	2430
5	35	47	1225	1645
Totals	175	219	6375	7900

So the cost function takes the form

$$C = 0.94x + 10.90$$

The line and the data points are shown in the next figure.

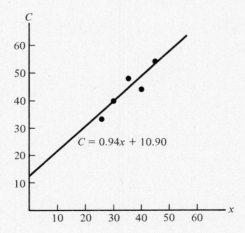

Before leaving this subject, a word of caution is in order. Although Equations 10.5.8 and 10.5.9 enable you to fit a least squares line to any set of data, it should not be inferred that a linear equation always represents the best relation-

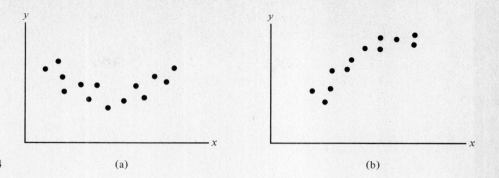

Figure 10.14   (a)   (b)

ship between the two variables. For example, the points shown in Figures 10.14 (a) and (b) indicate that a nonlinear relationship between the variables seems to exist, and therefore the equations developed in this section are not applicable. Once a particular functional relationship has been discovered or assumed, methods similar to those just studied will enable you to obtain a curve of best fit.

EXERCISE 10.5

Using the technique described in Example 1 of this section, find the equation of the line of best fit for the sets of points in Exercises 1–3. In addition, plot the points together with the line on the accompanying coordinate system.

1. $\{(2, 1), (3, 3), (1, 2), (5, 4)\}$

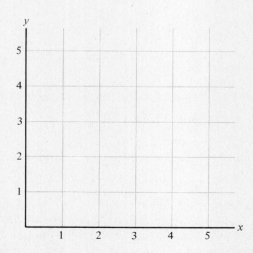

2. {(0, 5), (3, 4),(5, 2), (2, 3)}

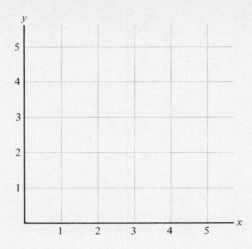

3. {(0, 1), (2, 2), (5, 4), (3, 3)}

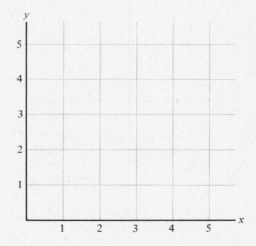

Find the equation of the line of best fit for the sets of points in Exercises 4–10.

4. {(1, 3), (3, 4), (5, 5), (6, 4)}

5. {(2, 4), (3, 3), (5, 2), (7, 1)}

6. {(1, −2), (3, 0), (4, 1), (5, 4)}

7. {(1, 5), (2, 3), (3, 1), (5, 0)}

8. {(2, 4), (3, 2), (4, 0), (6, −1)}

9. {(1, 5), (2, 6), (3, 3), (4, 1), (5, 0)}

10. {(0, −2), (2, 0), (3, 1), (4, 3), (5, 4)}

11. Demand and supply equations can be approximated in many cases by determining the line of best fit to a set of data. Suppose a market research study provides the following data on the unit price and quantity for a new smooth-writing pen. Find the demand equation that represents the line of best fit for this data.

QUANTITY (millions)	PRICE/UNIT
4	$2.00
3	2.40
2	2.75
1	3.00

12. As a machine ages, more money must be spent annually on maintenance and repair. The following data represent annual maintenance and repair costs for six pumps, identical in all respects except for age, in a sewage treatment plant.

| AGE (years) | ANNUAL COSTS |
x	y
5	$150
4	160
6	175
2	140
9	180
1	125

   a. Find the equation for the line of best fit.
   b. If the company purchases (secondhand) a seventh pump that is eight years old, how much money should be budgeted for repair and maintenance of the pump?

13. The amount of money a family of four spends on entertainment is related to the family's income. The following data represent the result of a study of six different families.

| ANNUAL INCOME (1000s) | ENTERTAINMENT EXPENDITURES (1000s) |
x	y
$13	$1.0
20	1.5
22	1.9
30	3.0
35	2.7
15	0.7

   a. Find the equation for the line of best fit.
   b. How much money does a typical family of four with an annual income of $25,000 spend on entertainment?

14. Monthly sales of a popular soft drink are affected by the temperature. The accompanying table shows monthly sales data for six consecutive months together with the average daytime temperature for the corresponding month. Find the equation of the line of best fit.

TEMPERATURE (°F)	MONTHLY SALES (1000s)
55	10
62	12
69	16
78	20
75	17
70	16
66	14

Appendixes
and
Tables

APPENDIX A 🦃 SETS AND SET ALGEBRA

The concept of a *set* is basic to the study of every area of mathematics. When the word set is used, it refers to a *well-defined collection* or *group* of objects. The following are examples of sets:

1. The set of automobiles produced by General Motors during the month of February.
2. The set of baseball bats used by Ted Williams during his major league career.
3. The set of integers greater than 5.
4. The set of living Americans who fought in the War of 1812.

The phrase "well-defined" indicates that we must be able to determine with certainty whether or not a given object belongs to the set under study. For example, the group of all dishonest building inspectors in the United States does not constitute a set because the word "dishonest" is not well-defined, that is, there is no universal standard by which to gauge the virtue of honesty.

Sets under discussion will be denoted by capital letters such as A, B, C, X, Y, and Z. The individual members or *elements* of a set will be denoted by lower-case letters such as a, b, c, x, y, and z. There are two common ways of defining sets:

I. Listing the individual elements, separated by commas, within braces; for example

$$A = \{0, 1, 2\}$$

$$B = \{1, 2, 3, 4, \ldots, 98, 99, 100\}$$

$$C = \{\text{G. Washington, J. Adams,} \ldots, \text{G. Ford, J. Carter}\}$$

The elements of the set A are the integers 0, 1, and 2. The elements of the set B are the positive integers from 1 to 100, inclusive. The elements of the set C are the presidents of the United States.

II. Set-building notation in which the common property or condition shared by all the elements of the set is stated in words. For example, set C defined previously can also be defined as

$$C = \{x | x \text{ is/was president of the United States}\}$$

where x, called a *variable,* represents an arbitrary member of the set. The vertical bar | is a short-hand way to denote the phrase "has the property that" and to the right of the bar is stated in words the common property shared by all elements of the set.

Set B is defined in set-builder notation as

$$B = \{x | x \text{ is a positive integer between 1 and 100 inclusive}\}$$

NOTE: There are many sets for which no discernible common property exists; in those cases, the set-builder notation is not applicable. For example, $\{1, ?, \delta\}$

436

Membership in a set is denoted by the symbol $\in$, which means "is a member of" or "belongs to". For example, if B is the set

$B = \{-3, 0, 2\}$

$0 \in B \qquad 2 \in B \qquad -3 \in B$

Nonmembership is denoted by $\notin$, which means "does not belong to"; for example

$7 \notin B, \frac{1}{2} \notin B$

Two sets, A and B, are said to be *equal*, written as

$A = B$

if every element of the set A belongs to set B and every element of the set B belongs to set A.

Example 1 | Suppose $A = \{a, e, i, o, u\}$, $B = \{i, e, u, a, o\}$. Because every element of set A is an element of set B and vice versa, the two sets are equal.

Example 2 | Suppose $B = \{a, b, c, d, e\}$, $A = \{d, e, a, b\}$. Because the element c belongs to B, but does not belong to A, the two sets are not equal

$c \in B \qquad$ but $c \notin A \qquad$ so $A \neq B$

An important set is the *empty* or *null* set, denoted by $\varnothing$. As its name implies, the empty set has no elements and is written as

$\varnothing = \{ \quad \}$

The following are examples of empty or null sets: (1) set of athletes who have run 3-minute miles and (2) set of integers that are divisible by 0.

When every element of the set A is also an element of a second set B, we say that A is a *subset* of B. For example, the set of registered voters in the United States is a subset of the set of U.S. citizens.

DEFINITION

If every element of the set A also belongs to the set B, then A is called a subset of B, written as

$A \subseteq B$

where $\subseteq$ means "is a subset of." If A is not a subset of B, we write $A \not\subseteq B$.

Example 3 | Let $A = \{k, l, m, n\}$ and $B = \{m, r, l, n, k\}$

a. Is A a subset of B? b. Is B a subset of A?

Solution | a. Since every element of set A also belongs to set B, then $A \subseteq B$.
b. There is an element of B, namely r, which does not belong to A, so we conclude that $B \not\subseteq A$.

Every set is a subset of itself; for example, $A \subseteq A$, and $B \subseteq B$. In addition, the empty set $\emptyset$ is a subset of every set, that is, $\emptyset \subseteq A$.

In Example 3, we saw that A is a subset of B, but B is not a subset of A. When this occurs, we say that set A is a *proper* subset of set B.

DEFINITION

If a set A is a subset of the set B, that is, $A \subseteq B$, and B has at least one element that is not an element of A, then A is called a proper subset of B, written as

$A \subset B$

where $\subset$ means "is a proper subset of." If A is not a proper subset of B, we write $A \not\subset B$.

Example 4 | Let $A = \{w, x, y\}$.

a. List all the subsets.   b. List all the proper subsets.

Solution | a. $\{w\}, \{x\}, \{y\}, \{w, x\}, \{w, y\}, \{x, y\}, \emptyset, A$
b. $\{w\}, \{x\}, \{y\}, \{w, x\}, \{w, y\}, \{x, y\}, \emptyset$

Just as there are operations such as addition, subtraction, multiplication, and division with real numbers, so are there operations involving sets, called the *union* and the *intersection*.

DEFINITION

Given the two sets A and B, the *union* of A and B, written $A \cup B$, is the set consisting of elements that belong to set A and/or to set B.

Example 5 | If $A = \{a, e, i\}$, $B = \{a, b, c\}$, $C = \{c, d, e, i\}$ find

a. $A \cup B$ b. $A \cup C$ c. $B \cup C$

Solution | a. $A \cup B = \{a, b, c, e, i\}$ b. $A \cup C = \{a, c, d, e, i\}$
c. $B \cup C = \{a, b, c, d, e, i\}$

DEFINITION

Given two sets A and B, the *intersection* of A and B, written $A \cap B$, is the set consisting of elements that belong to both set A and to set B.

Example 6 | Let A, B, and C be defined as in Example 6. Find each of the following:

a. $A \cap B$ b. $B \cap C$ c. $C \cap A$

Solution | a. $A \cap B = \{a\}$ b. $B \cap C = \{c\}$ c. $A \cap C = \{e, i\}$

If two sets A and B have no elements in common, that is

$A \cap B = \varnothing$

then A and B are said to be *disjoint*.

EXERCISE A

1. Which of the following statements are true and which are false:

   a. $4 \in \{2, 4, 6, 8\}$
   b. $4 \subset \{2, 4, 6, 8\}$
   c. $\{4\} \in \{2, 4, 6, 8\}$
   d. $\{4\} \subset \{2, 4, 6, 8\}$
   e. $\{6, 8, 9, k\} = \{9, 8, k, 6\}$
   f. $\{1, 2\} \not\subset \{0, 2, 5, 7\}$
   g. $\{1, 2\} \not\subseteq \{2, 1\}$
   h. $\{0\} = \varnothing$

2. List the elements of the following sets:

   a. $\{x | x$ is an even integer less than 100 that is divisible by 5 and 4$\}$
   b. $\{x | x$ is a letter in the name of the largest U.S. city west of the Rockies$\}$
   c. $\{x | x$ is a month having less than 30 days$\}$
   d. $\{x | x$ is a New England State$\}$
   e. $\{x | x$ is a U.S. president who served more than three terms in office$\}$

3. **a.** List all the subsets of the set $A = \{?, !, \$\}$.
   b. List all the proper subsets of the set $A = \{?, !, \$\}$.

In Exercises 4–12, let $A = \{0, 1, 2, 3, 4\}$, $B = \{0, 2, 4\}$, $C = \{3, 5\}$. Find each of the following sets:

4. $A \cup B$ 5. $B \cap C$

6. $A \cap (B \cup C)$ 7. $(A \cap C) \cup (A \cap B)$

8. $C \cup \varnothing$ 9. $C \cap \varnothing$

10. $B \cup (B \cap C)$ 11. $(A \cup C) \cap (B \cup C)$

12. $(A \cup C) \cap (C \cap B)$

APPENDIX B 🐦 REVIEW OF ALGEBRA

We will not attempt to review all areas of algebra but will concentrate on those you are likely to encounter repeatedly in your reading and in solving problems.

EXPONENTS

Integral Exponents   When a number is to be multiplied by itself repeatedly, exponents are used to represent the repeated multiplication, for example,

$5 \cdot 5 \cdot 5 \cdot 5 = 5^4$ $t \cdot t \cdot t \cdot t \cdot t = t^5$

in general

$$\underbrace{b \cdot b \cdot b \ldots b}_{n \text{ factors}} = b^n$$

In the expression b^n, b is referred to as the *base* and n as the *exponent*.

The exponent *zero* and *negative integral* exponents are defined as follows:

$$b^0 = 1 \qquad \text{provided that } b \neq 0$$

Examples | $\quad 3^0 = 1 \quad x^{4-4} = x^0 = 1 \quad (x \neq 0)$

$$b^{-n} = \frac{1}{b^n} \qquad b \neq 0$$

Examples | $\quad 5^{-3} = \frac{1}{5^3} = \frac{1}{125} \qquad \left(\frac{2}{3}\right)^{-2} = \frac{1}{(\frac{2}{3})^2} = \frac{1}{(\frac{4}{9})} = \frac{9}{4}$

$$2x^{-6} = 2\frac{1}{x^6} = \frac{2}{x^6} \qquad (x \neq 0)$$

Working with expressions that contain exponents can be made easier if advantage is taken of the following properties of exponents:

> a. $b^n \cdot b^m = b^{m+n}$

Examples | $\quad x^3 \cdot x^7 = x^{7+3} = x^{10} \qquad 5^4 \cdot 5^3 = 5^{4+3} = 5^7$

$$t^7 \cdot t^{-4} = t^{7+(-4)} = t^3$$

> b. $\dfrac{b^m}{b^n} = b^{m-n} = \dfrac{1}{b^{n-m}}$

Examples | $\quad \dfrac{y^8}{y^5} = y^{8-5} = y^3 \qquad \dfrac{6^7}{6^3} = 6^{7-3} = 6^4 = 1296$

$$\frac{(x+y)^4}{(x+y)^9} = \frac{1}{(x+y)^{9-4}} = \frac{1}{(x+y)^5}$$

> c. $(b^m)^n = b^{m \cdot n}$

Examples | $\quad (4^3)^5 = 4^3 \cdot 4^3 \cdot 4^3 \cdot 4^3 \cdot 4^3 = 4^{3+3+3+3+3} = 4^{3 \cdot 5} = 4^{15}$

$$(x^4)^6 = x^{4\cdot6} = x^{24} \qquad (y^{-2})^4 = y^{(-2)\cdot(4)} = y^{-8} = \frac{1}{y^8}$$

d. $(a \cdot b)^n = a^n \cdot b^n$

Examples | $(2 \cdot 3)^4 = 2^4 \cdot 3^4 = 16 \cdot 81 = 1296 \qquad (x^2 \cdot y)^3 = (x^2)^3(y)^3 = x^6y^3$

e. $\left(\dfrac{a}{b}\right)^n = \dfrac{a^n}{b^n}$

Examples | $\left(\dfrac{2}{5}\right)^3 = \dfrac{2^3}{5^3} = \dfrac{8}{125} \qquad \left(\dfrac{x^3}{y^4}\right)^2 = \dfrac{(x^3)^2}{(y^4)^2} = \dfrac{x^6}{y^8}$

Rational Exponents Before dealing with rational exponents, it is necessary to define what is meant by the nth root of a real number.

DEFINITION

If b and c are real numbers such that $c^n = b$, where n is a positive integer, then c is called an nth root of b.

For example

$4^2 = 16$; therefore, 4 is a square (second) root of 16
$(-4)^2 = 16$; therefore, -4 is also a square root of 16
$2^3 = 8$; therefore, 2 is a cube (third) root of 8
$3^4 = 81$; therefore, 3 is a fourth root of 81
$(-3)^4 = 81$; therefore, -3 is also a fourth root of 81

Because 16 has two real square roots and 81 has two real fourth roots, we introduce the concept of the *principal nth root* to distinguish between the two roots in such cases.

DEFINITION

If n is a positive integer greater than 1, the principal nth root of b, denoted as $\sqrt[n]{b}$, is defined as follows:

If $b > 0$, $\sqrt[n]{b}$ is the positive nth root of b
If $b < 0$ and n is odd, $\sqrt[n]{b}$ is the negative nth root of b
$\sqrt[n]{0} = 0$

NOTE: The principal square root of b is written $\sqrt{b}$, where $n = 2$ is understood.

Examples | $\sqrt{16} = 4$ $\sqrt[3]{125} = 5$ $\sqrt[5]{-32} = -2$ $-\sqrt[4]{81} = -3$

DEFINITION

The quantity $b^{1/n}$, where n is a positive integer, is defined as follows:

$b^{1/n} = \sqrt[n]{b}$ provided that $\sqrt[n]{b}$ is a real number

---

Examples | $8^{1/3} = \sqrt[3]{8} = 2$ $64^{1/2} = \sqrt{64} = 8$ $(-32)^{1/5} = \sqrt[5]{-32} = -2$

DEFINITION

The quantity $b^{m/n}$, where m and n are positive integers, is defined as follows:

$b^{m/n} = (\sqrt[n]{b})^m$ provided that $\sqrt[n]{b}$ is a real number

---

Examples | $25^{3/2} = (\sqrt{25})^3 = 5^3 = 125$ $x^{2/5} = (\sqrt[5]{x})^2$

Properties a–e, which were defined for positive integral exponents, also hold for rational exponents.

Examples | $8^{2/3} \cdot 8^{1/3} = 8^{(2/3)+(1/3)} = 8^1 = 8$

$\dfrac{x^{3/5}}{x^{2/5}} = x^{(3/5)-(2/5)} = x^{1/5}$

$(x^{2/3} \cdot y^{1/3})^3 = (x^{2/3})^3 \cdot (y^{1/3})^3 = x^2 y$

FACTORING

The words *terms* and *factors* are used both to describe the way in which a mathematical expression is constructed and also to identify components within the expression. Quantities combined via the operations of *addition* or *subtraction* are called *terms;* those combined via the operation of *multiplication* are called *factors.* When an expression is written as a product of two or more quantities, it is said to be *factored.* The distributive law, written in the form

$$ab + ac = a(b + c)$$

enables us to write an expression containing two or more terms as a single term that has two or more factors; for example

$$3x + 9 = 3(x + 3)$$

$$x^2 - 5x = x(x - 5)$$

There are a number of frequently occurring types of expressions whose factors are well known. Some of them are listed next.

DIFFERENCE OF TWO SQUARES

$$(ax)^2 - b^2 = (ax + b)(ax - b)$$

Examples

$$x^2 - 16 = (x + 4)(x - 4) \qquad 9x^2 - 25 = (3x + 5)(3x - 5)$$

$$\frac{4x^2}{9} - 49y^2 = \left(\frac{2x}{3} + 7y\right)\left(\frac{2x}{3} - 7y\right)$$

TRINOMIAL THAT IS A PERFECT SQUARE

$$(ax)^2 \pm 2abx + b^2 = (ax \pm b)^2$$

Examples

$$9x^2 + 24x + 16 = (3x + 4)^2$$

$$36x^2 - 60xy + 25y^2 = (6x - 5y)^2$$

TRINOMIAL THAT IS NOT A PERFECT SQUARE (factored by trial and error)

$$acx^2 + (ad + bc)x + bd = (ax + b)(cx + d)$$

Examples

$$x^2 + 5x - 6 = (x + 6)(x - 1)$$

$$2x^2 + 11x + 12 = (2x + 3)(x + 4)$$

SUM AND DIFFERENCE OF TWO CUBES

$$x^3 + y^3 = (x + y)(x^2 - xy + y^2)$$

$$x^3 - y^3 = (x - y)(x^2 + xy + y^2)$$

Examples

$$x^3 + 1 = x^3 + 1^3 = (x + 1)(x^2 - x + 1)$$

$$x^3 - 8 = x^3 - 2^3 = (x - 2)(x^2 + 2x + 4)$$

$$8x^3 + 27y^3 = (2x)^3 + (3y)^3 = (2x + 3y)(4x^2 - 6xy + 9y^2)$$

RATIONAL EXPRESSIONS

Algebraic expressions such as

$$\frac{x}{x + 2} \qquad \frac{x^2 + 3x}{x^2 - 4x + 1} \qquad \frac{1 - x}{4 - x^3}$$

are called *rational expressions*. Because division by zero is not defined, we will assume that the denominators of all rational expressions are not equal to zero. The methods for simplifying and combining rational expressions are identical to those for numerical fractions in arithmetic.

Reducing Fractions   Reducing or simplifying a fraction is a process in which all factors common to both the numerator and denominator are canceled or removed. This process is based on the following property:

$$\frac{ac}{bc} = \frac{a}{b}$$

Examples

$$\frac{3x}{5x^2} = \frac{3x}{5xx} = \frac{3}{5x}$$

$$\frac{x^2 - 1}{x^2 + 2x + 1} = \frac{(x + 1)(x - 1)}{(x + 1)^2} = \frac{x - 1}{x + 1}$$

Multiplying Fractions   The multiplication of two fractions is carried out by multiplying the numerators and denominators separately, that is,

$$\frac{a}{b} \cdot \frac{c}{d} = \frac{ac}{bd}$$

and then simplifying the resulting expression if possible

Examples

$$\frac{x}{2} \cdot \frac{3}{x + 1} = \frac{3x}{2x + 2}$$

$$\frac{x + 2}{5} \cdot \frac{10}{x^2 + 4x + 4} = \frac{10(x + 2)}{5(x^2 + 4x + 4)} = \frac{10(x + 2)}{5(x + 2)^2} = \frac{2}{x + 2}$$

Dividing Fractions   Dividing the fraction a/b by the fraction c/d is carried out by multiplying a/b by d/c, the *reciprocal* of c/d, that is,

$$\frac{a}{b} \div \frac{c}{d} = \frac{a}{b} \cdot \frac{d}{c} = \frac{ad}{bc}$$

Again, the result should be simplified as far as possible.

Examples

$$\frac{3}{x} \div \frac{2}{x - 1} = \frac{3}{x} \cdot \frac{x - 1}{2} = \frac{3x - 3}{2x}$$

$$\frac{x^2 - 4}{4} \div \frac{2x + 4}{3} = \frac{x^2 - 4}{x} \cdot \frac{3}{2x + 4} = \frac{3(x + 2)(x - 2)}{2x(x + 2)} = \frac{3x - 6}{2x}$$

Adding Fractions   a. The denominators of the two fractions are identical

$$\frac{a}{c} + \frac{b}{c} = \frac{a + b}{c}$$

Examples

$$\frac{2}{x + 1} + \frac{x}{x + 1} = \frac{2 + x}{x + 1}$$

$$\frac{3x}{x - 2} - \frac{6}{x - 2} = \frac{3x - 6}{x - 2} = \frac{3(x - 2)}{x - 2} = 3 \qquad x \neq 2$$

b. The denominators of the two fractions are not identical: The numerator and denominator of each fraction are multiplied by factors that will cause the denominators of the two fractions to become equal

$$\frac{a}{b} + \frac{c}{d} = \frac{ad}{bd} + \frac{cb}{db} = \frac{ad + cb}{bd}$$

Examples

$$\frac{4}{x} + \frac{2x}{x+1} = \frac{4(x+1)}{x(x+1)} + \frac{2x^2}{x(x+1)} = \frac{2(x^2 + 2x + 2)}{x(x+1)}$$

The expression $x(x + 1)$ is called the *least common denominator* of the original fractions.

SOLVING EQUATIONS IN ONE VARIABLE

Solving an equation is an exercise in finding those values of the variable that make the given equation a true statement. The solution of the equation

$$x + 3 = 0$$

is

$$x = -3$$

because

$$-3 + 3 = 0$$

First-degree (Linear) Equations   A first-degree equation in one variable is one that can be written in the form

$$ax + b = 0$$

Example | Solve the equation

$$5x + 7 = 3x + 13$$

Solution | Combine all terms containing x on one side of the equation and the remaining terms on the other by subtracting the expression $(3x + 7)$ from both sides giving

$$5x - 3x = 13 - 7$$
$$2x = 6$$

Next, divide both sides by the coefficient of x, giving

$$x = 3$$

Example | Solve the equation

$$\frac{2}{x} + \frac{4}{3} = \frac{1}{x} - \frac{2}{3}$$

Solution | Multiply both sides of the equation by the lowest common denominator $3x$, giving

$$3x\left(\frac{2}{x}\right) + 3x\left(\frac{4}{3}\right) = 3x\left(\frac{1}{x}\right) - 3x\left(\frac{2}{3}\right)$$

from which we get

$$6 + 4x = 3 - 2x$$

Proceeding as in the previous example, we get

$$6x = -3$$

or

$$x = -\tfrac{1}{2}$$

Nonlinear Equations   Before delving into this case, let us note a seemingly trivial but nevertheless very important principle used in finding solutions to many nonlinear equations. If a and b are two real numbers, and if

$$a \cdot b = 0$$

then either $a = 0$, or $b = 0$, or both a and b equal 0. This principle can be applied to solve equations when bringing all terms onto one side of the equation yields an expression that can be factored. Setting each factor equal to 0 generates additional equations that are generally easier to solve than the original equation

Example   Solve the equation

$$x^2 = 2x$$

Solution   First subtract $2x$ from both sides yielding

$$x^2 - 2x = 0$$

Factoring the left-hand side gives

$$x(x - 2) = 0$$

Setting each factor equal to 0 and solving gives

$$x_1 = 0 \qquad \text{and} \qquad x_2 = 2$$

as the two solutions.

Example   Solve the equation

$$\frac{3x}{x + 2} = \frac{x + 4}{x}$$

Solution   Multiply both sides by the lowest common denominator $x(x + 2)$, yielding the equation

$$3x^2 = x^2 + 6x + 8$$

Next, subtract $x^2 + 6x + 8$ from both sides, giving

$$2x^2 - 6x - 8 = 0$$

$$2(x - 4)(x + 1) = 0$$

Setting each factor equal to 0 gives the solution

$$x_1 = 4 \qquad x_2 = -1$$

QUADRATIC EQUATIONS

A standard quadratic equation is one that can be written in the form

$$ax^2 + bx + c = 0 \qquad a, b, \text{ and } c \text{ are constants and } a \neq 0$$

When they exist, solutions of quadratic equations can be obtained by either

1. Factoring the left-hand side, if possible, and setting each factor equal to 0, or
2. Using the *quadratic formula*

$$x = \frac{-b \pm \sqrt{b^2 - 4ac}}{2a}$$

Example | Solve the equation

$$x^2 - 6x - 7 = 0$$

Solution | The expression $(x^2 - 6x - 7)$ can be factored, so the equation can be written as

$$(x - 7)(x + 1) = 0$$

Setting each factor equal to zero yields the solutions

$$x_1 = 7 \qquad x_2 = -1$$

Example | Solve the equation

$$2x^2 + 3x - 1 = 0$$

Solution | Because the left-hand side cannot be factored, we use the quadratic formula to obtain the solution

$$x = \frac{-3 \pm \sqrt{9 - 4(2)(-1)}}{4} = \frac{-3 \pm \sqrt{17}}{4}$$

INTERVALS

The set of real numbers can be represented graphically by the set of points on a line, called the real number line, where each point on the line corresponds to a real number. Points to the right of 0 represent the positive numbers $(x > 0)$, while points to the left of 0 represent the negative numbers $(x < 0)$.

The set of numbers which satisfies the inequality $a \leq x \leq b$ is called a "closed interval," denoted by $[a, b]$, that is,

$$[a, b] = \{x | a \leq x \leq b\}$$

The following figure illustrates the closed interval $[a, b]$ where the closed circles at the endpoints a and b indicate that both a and b are included in the interval.

Example | The closed interval $[-1, 2]$ is shown in the next figure.

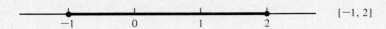

The set of numbers which satisfies the inequality $a < x < b$ is called an "open interval," denoted by (a, b), that is,

$$(a, b) = \{x | a < x < b\}$$

The following figure illustrates the open interval (a, b) where the open circles at the endpoints indicate that both a and b are not included in the interval.

Example | The open interval $(-2, 3)$ is shown in the following figure.

EXERCISE B

Simplify the expressions in Exercises 1–23. Express all answers with positive exponents, assuming that all variables are positive numbers.

1. $x^2 x^3 x$

2. $(\frac{1}{2})^4$

3. $(-\frac{3}{4})^3$

4. $\dfrac{x^7}{x^5}$

5. $3x^{-4}$

6. $(2x^3)^5$

7. $x^{-2}x^5$

8. $\left(\dfrac{x^{-3}}{2}\right)^5$

9. $\dfrac{x^0 + y^0}{2x^0 - y^0}$

10. $(x^{-2}y^3)^{-3}$

11. $(x^{-1} + y^{-1})^{-1}$

12. $\left(\dfrac{2x^2}{y^3}\right)^{-2}$

13. $(\frac{49}{16})^{1/2}$

14. $(\frac{8}{27})^{-4/3}$

15. $x^{1/3}x^{4/3}x^{7/3}$

16. $\dfrac{2x}{4x^{1/3}}$

17. $\left(\dfrac{2x^{2/3}}{y^{4/3}}\right)^3$

18. $\left(\dfrac{x^2}{4y^6}\right)^{1/2}$

19. $\sqrt{64x^2}$

20. $\sqrt[3]{4}\,\sqrt[3]{2}$

21. $\sqrt{3x}\sqrt{27x}$

22. $\dfrac{\sqrt[3]{16x^4}}{\sqrt[3]{2x}}$

23. $\sqrt{\dfrac{25x^6}{9y^{10}}}$

Factor the expressions in Exercises 24–33.

24. $6x^2 - 8x - 14$

25. $x^2 - 16$

26. $3x^3 - 75x$

27. $x^2 - 10x + 25$

28. $x^2 - 10x + 16$

29. $2x^2 - x - 3$

30. $4x^2 + 12xy + 9y^2$

31. $4x^2 + 13xy + 9y^2$

32. $x^3 + 27$

33. $2x^4 - 128x$

In Exercises 34–41, carry out the indicated operations and simplify.

34. $\dfrac{2x^2}{9y} \cdot \dfrac{3y^2}{8x^3}$

35. $\dfrac{x^2 - 4}{2x + 2} \cdot \dfrac{3x^2 + 6x + 3}{x + 2}$

36. $\dfrac{4x}{5y^2} \div \dfrac{2x^3}{15y}$

37. $\dfrac{x^2 + 2x - 3}{x^2 + x - 6} \div \dfrac{x^2 + 4x - 5}{x^2 - 4}$

38. $\dfrac{3y + 2}{y} + \dfrac{2y - 1}{y}$

39. $\dfrac{2}{x + 1} - \dfrac{1}{x}$

40. $\dfrac{3}{1 - x} + \dfrac{3x}{x - 1}$

41. $\dfrac{3}{x^2 - 4} + \dfrac{2x}{x + 2}$

Solve the equations in Exercises 42–52.

42. $5x + 3 = x + 11$

43. $2x - 7 = x + 1$

44. $\dfrac{x}{3} - \dfrac{1}{2} = \dfrac{3x}{2} + \dfrac{1}{3}$

45. $\dfrac{2x}{3} + \dfrac{1}{2} = \dfrac{x}{3} - \dfrac{1}{2}$

46. $\dfrac{4}{x} - 1 = \dfrac{3}{x}$

47. $\dfrac{x}{x - 1} + 5 = \dfrac{3}{x - 1}$

48. $x^2 = 3x$

49. $2x^2 = 5x$

50. $x^2 + x = 2$

51. $x^2 - 3x = 1$

52. $2x^2 - 3x - 2 = 0$

53. $5x^2 + x - 1 = 0$

54. $x^2 - 5x + 4 = 0$

On the number lines in Exercises 55–58, sketch the indicated intervals.

55. $[-1, 1]$

56.

$[-3, 4]$

57. $(1, 3)$

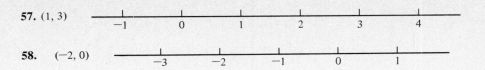

58. $(-2, 0)$

APPENDIX C ❧ PROOFS OF DIFFERENTIATION TECHNIQUES

SIMPLE POWER RULE

If the function to be differentiated has the form

$$y = f(x) = Cx^n \qquad \text{where } n \text{ is a positive integer}$$

its first derivative $f'(x)$ is

$$\frac{dy}{dx} = f'(x) = Cnx^{n-1}$$

This result can be demonstrated by applying the definition

$$\frac{dy}{dx} = f'(x) = \lim_{h \to 0} \frac{f(x + h) - f(x)}{h}$$

to the function $y = f(x) = Cx^n$. We get

$$f'(x) = \lim_{h \to 0} \frac{C(x + h)^n - Cx^n}{h} \tag{C.1}$$

The expression $(x + h)^n$ can be expanded to give

$$(x + h)^n = x^n + nx^{n-1}h + \frac{n(n - 1)}{2}x^{n-2}h^2 + \cdots + nxh^{n-1} + h^n$$

When the expanded version is substituted into Equation C.1, we get

$$\frac{dy}{dx} = f'(x) = \lim_{h \to 0} \frac{Cx^n + Cnx^{n-1}h + \dfrac{Cn(n - 1)}{2}x^{n-2}h^2 + \cdots + Ch^n - Cx^n}{h}$$

$$= \lim_{h \to 0} \left[Cx^{n-1} + \frac{Cn(n - 1)}{2}x^{n-2}h + \cdots + Ch^{n-1} \right] \tag{C.2}$$

Since every term on the right-hand side of Equation C.2, except the first, contains h as a factor, the only term to survive as $h \to 0$ is the first, yielding

$$\frac{dy}{dx} = f'(x) = Cnx^{n-1}$$

PRODUCT RULE

If the function to be differentiated has the form

$$y = f(x) = u(x)v(x)$$

the first derivative $f'(x)$ is

$$\frac{dy}{dx} = f'(x) = u(x)v'(x) + v(x)u'(x)$$

This result can be derived from the definition 3.2.1. Noting that

$$f(x + h) = u(x + h)v(x + h)$$

$f'(x)$ can be written as

$$f'(x) = \lim_{h \to 0} \frac{u(x + h)v(x + h) - u(x)v(x)}{h}$$

The next step in the development is one that initially has the aura of "black magic" about it; however, it does enable us to find the limit. The expression $u(x + h)v(x)$ is added to and subtracted from the numerator in the following manner:

$$f'(x) = \lim_{h \to 0} \frac{u(x + h)v(x + h) - u(x + h)v(x) + u(x + h)v(x) - u(x)v(x)}{h}$$

Next, the right-hand side is rewritten as

$$f'(x) = \lim_{h \to 0} \left\{ u(x + h) \left[\frac{v(x + h) - v(x)}{h} \right] + v(x) \left[\frac{u(x + h) - u(x)}{h} \right] \right\}$$

$$= \lim_{h \to 0} \left\{ u(x + h) \frac{v(x + h) - v(x)}{h} \right\} + \lim_{h \to 0} \left\{ v(x) \frac{u(x + h) - u(x)}{h} \right\}$$

Since

$$\lim_{h \to 0} u(x + h) = u(x) \qquad \lim_{h \to 0} \frac{v(x + h) - v(x)}{h} = v'(x)$$

$$\lim_{h \to 0} \frac{u(x + h) - u(x)}{h} = u'(x)$$

it is possible to write

$$\lim_{h \to 0} \left[u(x + h) \frac{v(x + h) - v(x)}{h} \right] = \lim_{h \to 0} u(x + h) \lim_{h \to 0} \frac{v(x + h) - v(x)}{h} = u(x)v'(x)$$

$$\lim_{h \to 0} \left[v(x) \frac{u(x + h) - u(x)}{h} \right] = \lim_{h \to 0} v(x) \lim_{h \to 0} \frac{u(x + h) - u(x)}{h} = v(x)u'(x)$$

from which we get

$$f'(x) = u(x)v'(x) + v(x)u'(x)$$

QUOTIENT RULE

If the function to be differentiated has the form

$$y = f(x) = \frac{u(x)}{v(x)}$$

its first derivative can be written as

$$\frac{dy}{dx} = f'(x) = \frac{v(x)u'(x) - u(x)v'(x)}{[v(x)]^2}$$

Like the product rule, an operation that looks initially like a "sleight of hand" is needed to generate $f'(x)$. Working from the definition 3.2.1, we get

$$f'(x) = \lim_{h \to 0} \frac{1}{h} \left[\frac{u(x + h)}{v(x + h)} - \frac{u(x)}{v(x)} \right]$$

$$= \lim_{h \to 0} \frac{v(x)u(x + h) - u(x)v(x + h)}{hv(x)v(x + h)}$$

Next, the expression $v(x)u(x)$ is added to and subtracted from the numerator

$$f'(x) = \lim_{h \to 0} \frac{v(x)u(x + h) - v(x)u(x) + v(x)u(x) - u(x)v(x + h)}{hv(x)v(x + h)}$$

Proceeding as we did with the product rule, we can write

$$f'(x) = \lim_{h \to 0} \frac{\left[v(x) \cdot \dfrac{u(x + h) - u(x)}{h} \right] - \left[u(x) \cdot \dfrac{v(x + h) - v(x)}{h} \right]}{v(x) \cdot v(x + h)}$$

$$= \frac{v(x) \cdot \lim_{h \to 0} \dfrac{u(x + h) - u(x)}{h} - u(x) \cdot \lim_{h \to 0} \dfrac{v(x + h) - v(x)}{h}}{v(x) \cdot \lim_{h \to 0} v(x + h)}$$

$$= \frac{v(x)u'(x) - u(x)v'(x)}{[v(x)]^2}$$

GENERAL POWER RULE

If the function to be differentiated has the form

$$y = f(x) = C[u(x)]^n$$

its first derivative can be written as

$$\frac{dy}{dx} = f'(x) = Cn[u(x)]^{n-1}u'(x)$$

This result will be developed from the definition 3.2.1 for the case where the exponent n is restricted to positive integral values

$$\frac{dy}{dx} = f'(x) = \lim_{h \to 0} \frac{C[u(x + h)]^n - C[u(x)]^n}{h}$$

$$= \lim_{h \to 0} \left(C \left[\frac{u(x + h) - u(x)}{h} \right] \right.$$

$$\left. \underbrace{\{[u(x + h)]^{n-1} + [u(x + h)]^{n-2}u(x) + \cdots + [u(x)]^{n-1}\}}_{n \text{ terms}} \right)$$

The expression can be simplified by noting that, as $h \to 0$, the quantity $u(x + h) \to u(x)$ and $[u(x + h) - u(x)]/h \to u'(x)$. We then get

$$= Cn[u(x)]^{n-1}u'(x)$$

APPENDIX D ❧ TWO ADDITIONAL PROOFS

I. Show that $1 + 2 + 3 + 4 + \cdots + k + \cdots + n = [n(n + 1)]/2$. If we represent the sum by the symbol S, we have

$$S = 1 + 2 + 3 + \cdots + k + \cdots + (n - 2) + (n - 1) + n \qquad \textbf{(D.1)}$$

Writing the terms on the right-hand side in reverse order gives

$$S = n + (n - 1) + (n - 2) + \cdots + k + \cdots + 3 + 2 + 1 \qquad \textbf{(D.2)}$$

Adding the corresponding terms on the left- and right-hand sides of Equations D.1 and D.2 gives

$$\begin{aligned} 2S &= (n + 1) + (n - 1 + 2) + (n - 2 + 3) \\ &\quad + \cdots + (3 + n - 2) + (2 + n - 1) + (1 + n) \\ &= (n + 1) + (n + 1) + (n + 1) + \cdots + (n + 1) \\ &\quad + (n + 1) + (n + 1) \\ &= n(n + 1) \end{aligned}$$

Dividing both sides by 2 gives

$$S = \frac{n(n + 1)}{2}$$

II. Show that $1^2 + 2^2 + 3^2 + \cdots + k^2 + \cdots + n^2 = [n(2n + 1)(n + 1)]/6$. The following relationship will prove to be very useful:

$$m^3 - (m - 1)^3 = m^3 - (m^3 - 3m^2 + 3m - 1) = 3m^2 - 3m + 1 \qquad \textbf{(D.3)}$$

First we create the sum

$$(1^3 - 0^3) + (2^3 - 1^3) + (3^3 - 2^3) + \cdots$$
$$+ [(n - 1)^3 - (n - 2)^3] + [n^3 - (n - 1)^3] \qquad \textbf{(D.4)}$$

This sum is nothing more than an exotic way to write n^3, but coupled with Equation D.3, it will enable us to obtain the desired result. If each term in D.4 is replaced by an equivalent term using Equation D.3, we get

$$[3(1)^2 - 3(1) + 1] + [3(2)^2 - 3(2) + 1] + \cdots$$
$$+ [3k^2 - 3k + 1] + \cdots + [3n^2 - 3n + 1] \qquad \textbf{(D.5)}$$

Rearranging the terms in D.5 as

$$[3(1)^2 + 3(2)^2 + \cdots + 3k^2 + \cdots + 3n^2]$$
$$- [3(1) + 3(2) + \cdots + 3k + \cdots + 3n] + \underbrace{(1 + 1 + \cdots + 1)}_{n \text{ terms}}$$

Making use of the result found in part I, we can replace the second term with $[3n(n + 1)]/2$; finally, the resulting expression can be set equal to n^3, yielding

$$3(1^2 + 2^2 + 3^2 + \cdots + n^2) - \frac{3n(n + 1)}{2} + n = n^3$$

Solving this equation for the sum $(1^2 + 2^2 + 3^2 + \cdots + n^2)$ gives

$$1^2 + 2^2 + 3^2 + \cdots + n^2 = \frac{n(2n + 1)(n + 1)}{6}$$

Table A
Values of $(1 + i)^n$

n	(0.5%) 0.005	(1%) 0.01	(1.5%) 0.015	(2%) 0.02	(3%) 0.03	(4%) 0.04	(5%) 0.05	(6%) 0.06	(7%) 0.07	(8%) 0.08	(9%) 0.09	(10%) 0.10	n
1	1.0050	1.0100	1.0150	1.0200	1.0300	1.0400	1.0500	1.0600	1.0700	1.0800	1.0900	1.1000	1
2	1.0100	1.0201	1.0302	1.0404	1.0609	1.0816	1.1025	1.1236	1.1449	1.1664	1.1881	1.2100	2
3	1.0151	1.0303	1.0457	1.0612	1.0927	1.1248	1.1576	1.1910	1.2250	1.2597	1.2950	1.3310	3
4	1.0202	1.0406	1.0614	1.0824	1.1255	1.1698	1.2155	1.2624	1.3107	1.3604	1.4115	1.4641	4
5	1.0253	1.0510	1.0773	1.1040	1.1592	1.2166	1.2762	1.3382	1.4025	1.4693	1.5386	1.6105	5
6	1.0304	1.0615	1.0934	1.1261	1.1940	1.2653	1.3400	1.4185	1.5007	1.5868	1.6771	1.7715	6
7	1.0355	1.0721	1.1098	1.1486	1.2298	1.3159	1.4071	1.5036	1.6057	1.7138	1.8280	1.9487	7
8	1.0407	1.0828	1.1265	1.1716	1.2667	1.3685	1.4774	1.5938	1.7181	1.8509	1.9925	2.1435	8
9	1.0459	1.0936	1.1434	1.1950	1.3047	1.4233	1.5513	1.6894	1.8384	1.9990	2.1718	2.3579	9
10	1.0511	1.1046	1.1605	1.2189	1.3439	1.4802	1.6288	1.7908	1.9671	2.1589	2.3673	2.5937	10
11	1.0564	1.1156	1.1779	1.2433	1.3842	1.5394	1.7103	1.8982	2.1048	2.3316	2.5804	2.8531	11
12	1.0617	1.1268	1.1956	1.2682	1.4257	1.6010	1.7958	2.0121	2.2521	2.5181	2.8126	3.1384	12
13	1.0670	1.1380	1.2136	1.2936	1.4685	1.6650	1.8856	2.1329	2.4098	2.7196	3.0658	3.4522	13
14	1.0723	1.1494	1.2318	1.3194	1.5125	1.7316	1.9799	2.2609	2.5785	2.9371	3.3417	3.7974	14
15	1.0777	1.1609	1.2502	1.3458	1.5579	1.8009	2.0789	2.3965	2.7590	3.1721	3.6424	4.1772	15
16	1.0831	1.1725	1.2690	1.3727	1.6047	1.8729	2.1828	2.5403	2.9521	3.4259	3.9703	4.5949	16
17	1.0885	1.1843	1.2880	1.4002	1.6528	1.9479	2.2920	2.6927	3.1588	3.7000	4.3276	5.0544	17
18	1.0939	1.1961	1.3073	1.4282	1.7024	2.0258	2.4066	2.8543	3.3799	3.9960	4.7171	5.5599	18
19	1.0994	1.2081	1.3270	1.4568	1.7535	2.1068	2.5269	3.0255	3.6165	4.3157	5.1416	6.1159	19
20	1.1049	1.2201	1.3469	1.4859	1.8061	2.1911	2.6532	3.2071	3.8696	4.6609	5.6044	6.7274	20
21	1.1104	1.2323	1.3671	1.5156	1.8602	2.2787	2.7859	3.3995	4.1405	5.0338	6.1088	7.4002	21
22	1.1170	1.2447	1.3876	1.5459	1.9161	2.3699	2.9252	3.6035	4.4304	5.4365	6.6586	8.1402	22
23	1.1216	1.2571	1.4084	1.5768	1.9735	2.4647	3.0715	3.8197	4.7405	5.8714	7.2578	8.9543	23
24	1.1272	1.2697	1.4295	1.6084	2.0327	2.5633	3.2250	4.0489	5.0723	6.3411	7.9110	9.8497	24
25	1.1328	1.2824	1.4509	1.6406	2.0937	2.6658	3.3863	4.2918	5.4274	6.8484	8.6230	10.8347	25
30	1.1614	1.3478	1.5631	1.8114	2.4273	3.2434	4.3219	5.7435	7.6123	10.0627	13.2676	17.4494	30
35	1.1907	1.4165	1.6839	1.9999	2.8138	3.9460	5.5160	7.6861	10.6766	14.7853	20.4140	28.1024	35
40	1.2208	1.4889	1.8140	2.2080	3.2620	4.8010	7.0399	10.2857	14.9744	21.7245	31.4094	45.2592	40

Table B

e^x and e^{-x}

x	e^x	e^{-x}	x	e^x	e^{-x}
0.00	1.0000	1.0000	1.5	4.4817	0.2231
0.01	1.0101	0.9901	1.6	4.9530	0.2019
0.02	1.0202	0.9802	1.7	5.4739	0.1827
0.03	1.0305	0.9705	1.8	6.0496	0.1653
0.04	1.0408	0.9608	1.9	6.6859	0.1496
0.05	1.0513	0.9512	2.0	7.3891	0.1353
0.06	1.0618	0.9418	2.1	8.1662	0.1225
0.07	1.0725	0.9324	2.2	9.0250	0.1108
0.08	1.0833	0.9331	2.3	9.9742	0.1003
0.09	1.0942	0.9139	2.4	11.0230	0.0907
0.10	1.1052	0.9048	2.5	12.182	0.0821
0.11	1.1163	0.8958	2.6	13.464	0.0743
0.12	1.1275	0.8869	2.7	14.880	0.0672
0.13	1.1388	0.8781	2.8	16.445	0.0608
0.14	1.1503	0.8694	2.9	18.174	0.0550
0.15	1.1618	0.8607	3.0	20.086	0.0498
0.16	1.1735	0.8521	3.1	22.198	0.0450
0.17	1.1853	0.8437	3.2	24.533	0.0408
0.18	1.1972	0.8353	3.3	27.113	0.0369
0.19	1.2092	0.8270	3.4	29.964	0.0334
0.20	1.2214	0.8187	3.5	33.115	0.0302
0.21	1.2337	0.8106	3.6	36.598	0.0273
0.22	1.2461	0.8025	3.7	40.447	0.0247
0.23	1.2586	0.7945	3.8	44.701	0.0224
0.24	1.2712	0.7866	3.9	49.402	0.0202
0.25	1.2840	0.7788	4.0	54.598	0.0183
0.30	1.3499	0.7408	4.1	60.340	0.0166
0.35	1.4191	0.7047	4.2	66.686	0.0150
0.40	1.4918	0.6703	4.3	73.700	0.0136
0.45	1.5683	0.6376	4.4	81.451	0.0123
0.50	1.6487	0.6065	4.5	90.017	0.0111
0.55	1.7333	0.5769	4.6	99.484	0.0101
0.60	1.8221	0.5488	4.7	109.950	0.0091
0.65	1.9155	0.5220	4.8	121.510	0.0082
0.70	2.0138	0.4966	4.9	134.290	0.0074
0.75	2.1170	0.4724	5.0	148.41	0.0067
0.80	2.2255	0.4493	5.5	244.69	0.0041
0.85	2.3396	0.4274	6.0	403.43	0.0025
0.90	2.4596	0.4066	6.5	665.14	0.0015
0.95	2.5857	0.3867	7.0	1096.60	0.0009
1.0	2.7183	0.3679	7.5	1808.0	0.0006
1.1	3.0042	0.3329	8.0	2981.0	0.0003
1.2	3.3201	0.3012	8.5	4914.8	0.0002
1.3	3.6693	0.2725	9.0	8103.1	0.0001
1.4	4.0552	0.2466	10.0	22026.0	0.00005

Table C
Natural Logarithms

x	ln x	x	ln x	x	ln x
		4.5	1.5041	9.0	2.1972
0.1	−2.3026	4.6	1.5261	9.1	2.2083
0.2	−1.6094	4.7	1.5476	9.2	2.2192
0.3	−1.2040	4.8	1.5686	9.3	2.2300
0.4	−0.9163	4.9	1.5892	9.4	2.2407
0.5	−0.6931	5.0	1.6094	9.5	2.2513
0.6	−0.5108	5.1	1.6292	9.6	2.2618
0.7	−0.3567	5.2	1.6487	9.7	2.2721
0.8	−0.2231	5.3	1.6677	9.8	2.2824
0.9	−0.1054	5.4	1.6864	9.9	2.2925
1.0	0.0000	5.5	1.7047	10	2.3026
1.1	0.0953	5.6	1.7228	11	2.3979
1.2	0.1823	5.7	1.7405	12	2.4849
1.3	0.2624	5.8	1.7579	13	2.5649
1.4	0.3365	5.9	1.7750	14	2.6391
1.5	0.4055	6.0	1.7918	15	2.7081
1.6	0.4700	6.1	1.8083	16	2.7726
1.7	0.5306	6.2	1.8245	17	2.8332
1.8	0.5878	6.3	1.8405	18	2.8904
1.9	0.6419	6.4	1.8563	19	2.9444
2.0	0.6931	6.5	1.8718	20	2.9957
2.1	0.7419	6.6	1.8871	25	3.2189
2.2	0.7885	6.7	1.9021	30	3.4012
2.3	0.8329	6.8	1.9169	35	3.5553
2.4	0.8755	6.9	1.9315	40	3.6889
2.5	0.9163	7.0	1.9459	45	3.8067
2.6	0.9555	7.1	1.9601	50	3.9120
2.7	0.9933	7.2	1.9741	55	4.0073
2.8	1.0296	7.3	1.9879	60	4.0943
2.9	1.0647	7.4	2.0015	65	4.1744
3.0	1.0986	7.5	2.0149	70	4.2485
3.1	1.1314	7.6	2.0281	75	4.3175
3.2	1.1632	7.7	2.0412	80	4.3820
3.3	1.1939	7.8	2.0541	85	4.4427
3.4	1.2238	7.9	2.0669	90	4.4998
3.5	1.2528	8.0	2.0794	100	4.6052
3.6	1.2809	8.1	2.0919	110	4.7005
3.7	1.3083	8.2	2.1041	120	4.7875
3.8	1.3350	8.3	2.1163	130	4.8676
3.9	1.3610	8.4	2.1282	140	4.9416
4.0	1.3863	8.5	2.1401	150	5.0106
4.1	1.4110	8.6	2.1518	160	5.0752
4.2	1.4351	8.7	2.1633	170	5.1358
4.3	1.4586	8.8	2.1748	180	5.1930
4.4	1.4816	8.9	2.1861	190	5.2470

Table D
A Short Table of Integration Formulas

1. $\int x^n \, dx = \dfrac{x^{n+1}}{n+1} + C \quad n \neq -1$

2. $\int e^{ax} \, dx = \dfrac{e^{ax}}{a} + C$

3. $\int \dfrac{1}{x} \, dx = \ln |x| + C$

4. $\int a^x \, dx = \dfrac{a^x}{\ln a} + C \quad a > 0$

5. $\int xe^{ax} \, dx = \dfrac{e^{ax}}{a^2}(ax - 1) + C$

6. $\int x^n e^{ax} \, dx = \dfrac{x^n e^{ax}}{a} - \dfrac{n}{a} \int x^{n-1} e^{ax} \, dx$

7. $\int \ln x \, dx = x \ln x - x + C$

8. $\int x^n \ln x \, dx = x^{n+1} \left[\dfrac{\ln x}{n+1} - \dfrac{1}{(n+1)^2} \right] \quad n \neq -1$

9. $\int (\ln x)^n \, dx = x(\ln x)^n - n \int (\ln x)^{n-1} \, dx \quad n \neq -1$

10. $\int \dfrac{1}{x^2 - a^2} \, dx = -\dfrac{1}{2a} \ln \left| \dfrac{x-a}{x+a} \right| + C$

11. $\int \dfrac{1}{a^2 - x^2} \, dx = \dfrac{1}{2a} \ln \left| \dfrac{a+x}{a-x} \right| + C$

12. $\int \dfrac{1}{\sqrt{x^2 - a^2}} \, dx = \ln |x + \sqrt{x^2 - a^2}| + C$

13. $\int \dfrac{1}{\sqrt{x^2 + a^2}} \, dx = \ln |x + \sqrt{x^2 + a^2}| + C$

14. $\int \dfrac{x}{ax + b} \, dx = \dfrac{x}{a} - \dfrac{b}{a^2} \ln |ax + b| + C$

15. $\int \dfrac{x}{\sqrt{ax + b}} \, dx = \dfrac{2(ax - 2b)}{3a^2} \sqrt{ax + b} + C$

16. $\int \dfrac{x}{(ax + b)^2} \, dx = \dfrac{b}{a^2(ax + b)} + \dfrac{1}{a^2} \ln |ax + b| + C$

17. $\int \dfrac{1}{x(ax + b)} \, dx = \dfrac{1}{b} \ln \left| \dfrac{x}{ax + b} \right| + C$

18. $\int \dfrac{1}{x(ax + b)^2} \, dx = \dfrac{1}{b(ax + b)} + \dfrac{1}{b^2} \ln \left| \dfrac{x}{ax + b} \right| + C$

19. $\int \sqrt{x^2 \pm a^2} \, dx = \dfrac{x}{2} \sqrt{x^2 \pm a^2} \pm \dfrac{a^2}{2} \ln |x + \sqrt{x^2 \pm a^2}| + C$

20. $\int \dfrac{1}{x^2 \sqrt{a^2 - x^2}} \, dx = \dfrac{-\sqrt{a^2 - x^2}}{a^2 x} + C$

Answers to Odd-Numbered Problems

1. $(1, 5)$, $(2, -1)$, $(-2, -1)$, and $(\frac{3}{2}, \frac{5}{2})$ satisfy the equation.

3. $P(-3, 2)$; $Q(0, 1)$; $R(1, 3)$; $S(4, 2)$; $T(-2, -2)$; $V(3, -1)$

5.
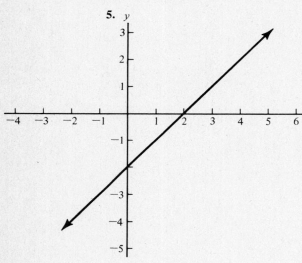

7.
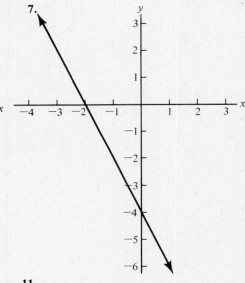

9.
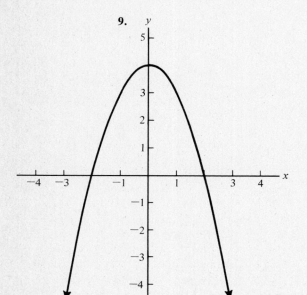

11.
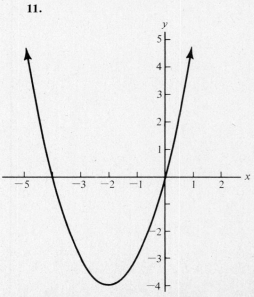

13.

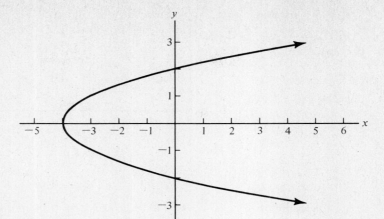

15.

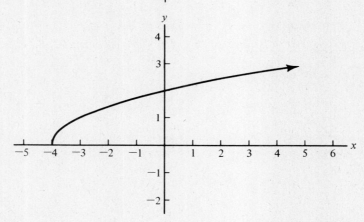

17. $S = 4h$

EXERCISE 1.2

1. a. $R = \{28, 20, 12, 4, -4\}$;
$f = \{(-5, 28), (-3, 20), (-1, 12), (1, 4), (3, -4)\}$
b. $R = \{2, 11, 26\}$; $f = \{(2, 2), (3, 11), (4, 26)\}$
c. $R = \{4.5, 7, 15\}$; $f = \{(.5, 4.5), (-2, 7), (-10, 15)\}$
d. $R = \{-\frac{1}{3}, -8, \frac{22}{3}\}$; $f = \{(0, -\frac{1}{3}), (1, -8), (3, \frac{22}{3})\}$

3. Domain $= \{x | x \text{ is any real number}\}$
Range $= \{y | y \leq 8\}$

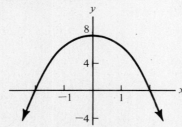

5. Domain = $\{x|x$ is any real number$\}$
Range = $\{y|y$ is any real number$\}$

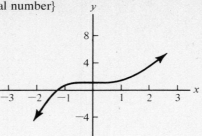

7. a. $f(7) = 17$ **b.** $f(\frac{3}{4}) = \frac{9}{2}$ **c.** $f(-3) = -3$ **d.** $f(a) = 2a + 3$
e. $f(a + 2) = 2a + 7$ **f.** $f(a + 2) - f(a) = 4$

9. a. $f(0) = 4$ **b.** $f(-1) = 6$ **c.** $f(6) = 34$ **d.** $f(1) = 4$
e. $f(1 + h) = h^2 + h + 4$ **f.** $f(1 + h) - f(1) = h^2 + h$

11. a. $f(0) = -1$ **b.** $f(1) = -2$ **c.** $f(-2) = -\frac{1}{2}$ **d.** $f(\frac{1}{2}) = -\frac{4}{3}$
e. $f(1 + h) = \dfrac{2}{h - 1}$ **f.** $f(1 + h) - f(1) = \dfrac{2h}{h - 1}$

13. a. 16.8 miles/gallon

15.

17. a. y represents parking lot fee

$$y = \begin{cases} 1.00 & 0 < t \le 1 \\ 1.50 & 1 < t \le 2 \\ 2.00 & 2 < t \le 3 \\ 2.50 & 3 < t \le 24 \end{cases}$$

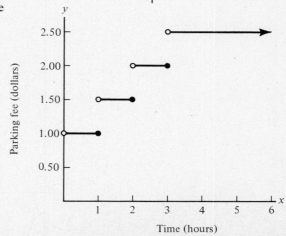

EXERCISE 1.3

1.

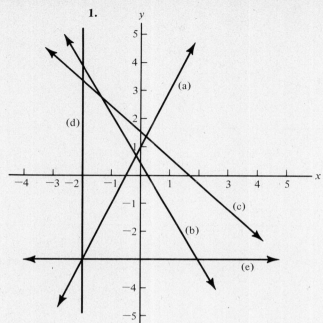

3. Slope $= -3$, x-intercept $= -\frac{2}{3}$, y-intercept $= -2$

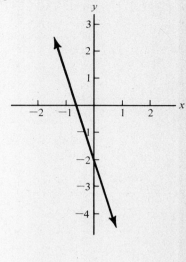

5. Slope $= \frac{1}{4}$, x-intercept $= -\frac{5}{2}$, y-intercept $= \frac{5}{8}$

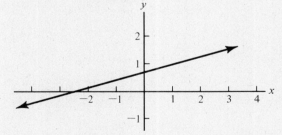

7. $2x - 3y = 24$

9. $y + x = 2$

11. $2y + 2x = 7$

13. $y = x + (d - c)$

15. $y = 4$

17. $3y - 2x = 3$

19. a. $D = 0.20M + 12$
b. $D = 0.25M + 5$
c. 140 miles

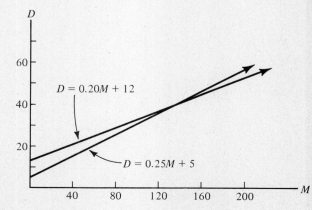

21. a. $A = 1000 + 180t \qquad 0 \le t \le \frac{1}{2}$
   b. $A = 1000 + 15t \qquad 0 \le t \le 6$
   c.

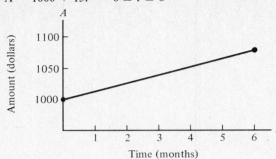

   d. $1060.00

EXERCISE 1.4

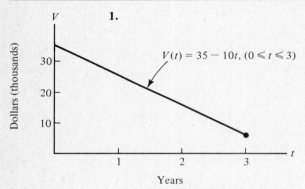

1. $V(t) = 35 - 10t, \ (0 \le t \le 3)$

3. $V = -800t + 4400$ (5 year life)
   $V = -1050t + 4400$ (4 year life)

5. a. $p = -2q + 35$ (q in thousands)
   $p = -0.002q + 35$ (q in units of one)
  b. $p = \$17.50$

7. $p = \dfrac{q}{80} + \dfrac{55}{4}$

9. $P = 55x - 220$
   break-even point $= (4, 0)$

11. a. s: sales in thousands of dollars
   S: salary in dollars
   Plan I: $S = 30s + 500$

   Plan II: $S = \begin{cases} 600 & s \le 15 \\ 50s - 150 & s > 15 \end{cases}$

  b. $\$\frac{10}{3}$ thousand $< s < \$\frac{110}{3}$ thousand

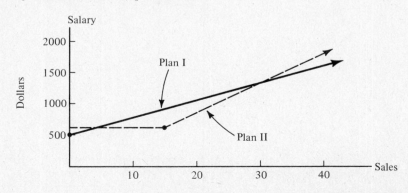

13. a. $36.00

    b. $P = 16x - 800,000$

    c. 50,000 units

15. $q = 50; p = 6$

17. $q = 3.1; p = 4.2$

EXERCISE 1.5

1.

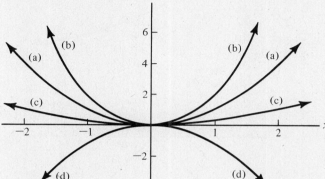

3.

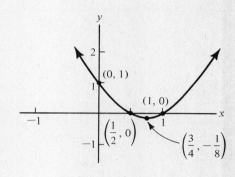

5.

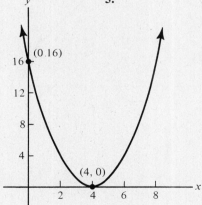

7.

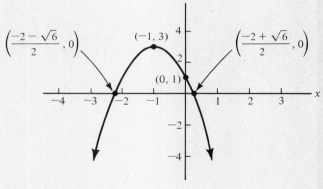

9. Room rate that maximizes profit is $39.50

11. x equals number of passengers in excess of 50

$$R(x) = \begin{cases} 10,000 + 200x & -50 \leq x \leq 0 \\ 10,000 + 150x - x^2 & x > 0 \end{cases}$$

EXERCISE 1.6

1. a, b, and e are polynomials.

3.

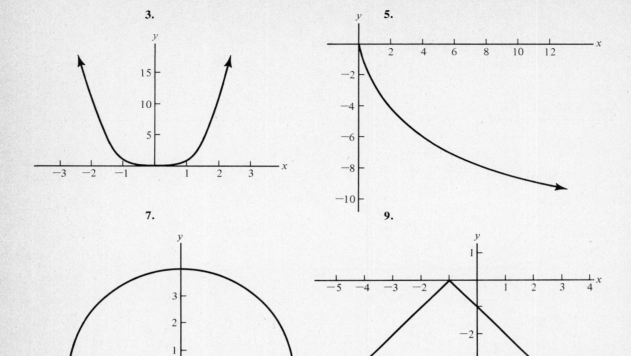

5.

7.

9.

EXERCISE 2.1

1. $\lim_{x \to 2} (5 - 4x) = -3$

x	y	x	y
1.0000	1.0000	3.0000	−7.0000
1.5000	−1.0000	2.5000	−5.0000
1.9000	−2.6000	2.1000	−3.4000
1.9900	−2.9600	2.0100	−3.0400
1.9990	−2.9960	2.0010	−3.0040
1.9999	−2.9996	2.0001	−3.0004

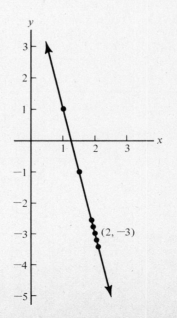

$(2, -3)$

3. $\lim\limits_{x\to 3}\dfrac{x^2-3x}{x^2-9}=0.5000$

x	y	x	y
2.0000	0.4000	4.0000	0.5714
2.5000	0.4545	3.5000	0.5385
2.9000	0.4915	3.1000	0.5082
2.9900	0.4992	3.0100	0.5008
2.9990	0.4999	3.0010	0.5001
2.9999	0.5000	3.0001	0.5000

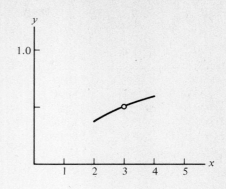

5. $\lim\limits_{x\to 0}\dfrac{x^2}{x}=0$

x	y	x	y
−1.0000	−1.0000	1.0000	1.0000
−0.5000	−5.0000	0.5000	0.5000
−0.1000	−0.1000	0.1000	0.1000
−0.0100	−0.0100	0.0100	0.0100
−0.0010	−0.0010	0.0010	0.0010
−0.0001	−0.0001	0.0001	0.0001

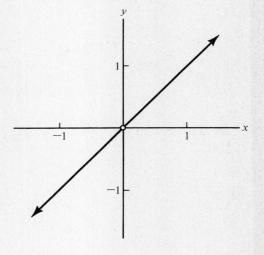

7. $\lim\limits_{x\to 0}\dfrac{x}{x^2}$ Does not exist

x	y	x	y
−1.0000	−1	1.0000	1
−0.5000	−2	0.5000	2
−0.1000	−10	0.1000	10
−0.0100	−100	0.0100	100
−0.0010	−1000	0.0010	1000
−0.0001	−10000	0.0001	10000

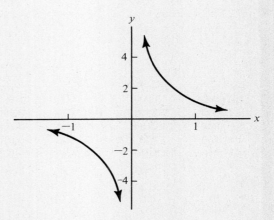

9. $\lim\limits_{x \to 1} \dfrac{x^3 - 1}{x - 1} = 3$

x	y	x	y
0.0000	1.0000	2.0000	7.0000
0.5000	1.7500	1.5000	4.7500
0.9000	2.7100	1.1000	3.3100
0.9900	2.9701	1.0100	3.0301
0.9990	2.9970	1.0010	3.0030
0.9999	2.9997	1.0001	3.0003

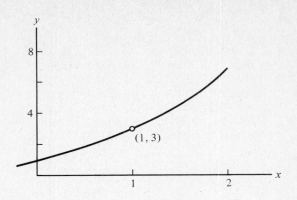

11. $\lim\limits_{x \to 0} \sqrt{x}$ Does not exist

x	y	x	y
−1.0000	not	1.0000	1.0000
−0.2500	defined	0.2500	0.5000
−0.0100		0.0100	0.1000
−0.0001		0.0001	0.0100

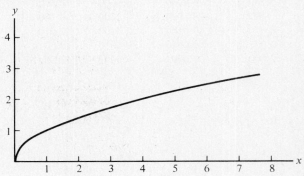

13. $\lim\limits_{x \to -1} = f(x)$ Does not exist

x	y	x	y
−2.0000	4.0000	0.0000	1.0000
−1.5000	2.2500	−0.5000	0.5000
−1.1000	1.2100	−0.9000	0.1000
−1.0100	1.0201	−0.9900	0.0100
−1.0010	1.0020	−0.9990	0.0010
−1.0001	1.0002	−0.9999	0.0001

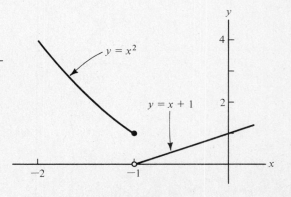

15. $\lim\limits_{x \to -1} \dfrac{x^2 - 1}{x + 1} = -2$

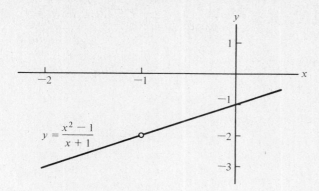

$y = \dfrac{x^2 - 1}{x + 1}$

17. $\lim\limits_{x \to 4} \dfrac{x - 4}{\sqrt{x} - 2} = 4$

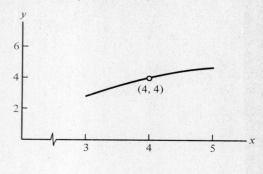

(4, 4)

EXERCISE 2.2

1. 26

3. $\frac{7}{3}$

5. 1

7. 12

9. 0

11. $\frac{2}{3}$

13. 2

15. Limit does not exist

17. $\frac{1}{3}$

EXERCISE 2.3

1. a. Discontinuous at $x = 2$ (B)
   c. Discontinuous at $x = -1$ (B)
   e. Discontinuous at $x = 3$ (D)
      Discontinuous at $x = -2$ (D)
   g. Continuous everywhere

   b. Continuous everywhere
   d. Discontinuous at $x = 2$ (B)
   f. Discontinuous at $x = -1$ (C)

3.

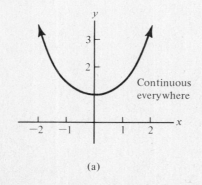

Continuous everywhere

(a)

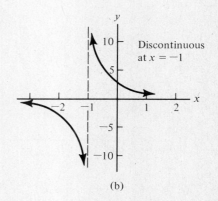

Discontinuous at $x = -1$

(b)

3. (continued)

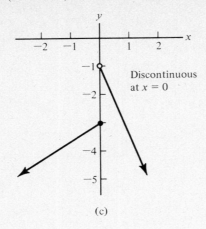

(c)

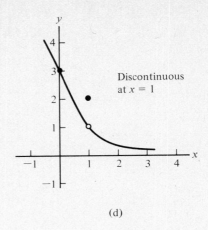

(d)

EXERCISE 3.1

1. a. $79.1 per year   **b.** $105.90 per year   **c.** $124.60 per year

3. a. 0.31   **b.** 0.46   **5. a.** Slope = 2   **b.** Slope = 2

7. a. Slope = -1 **b.** Slope = 0   **9. a.** Slope = 1   **b.** Slope = $\frac{1}{3}$

11. $2 - 2x - h$ **13.** $\dfrac{2x + h}{x^2(x^2 + 2xh + h^2)}$

15. $4x^3 + 6x^2h + 4xh^2 + h^3$

EXERCISE 3.2

1.

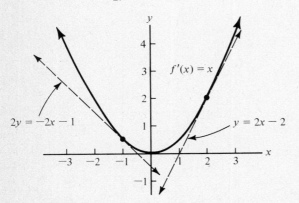

3.

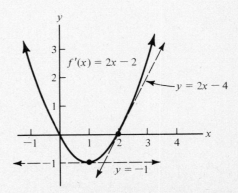

5.

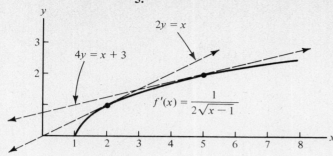

7.

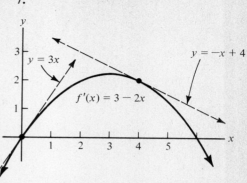

9. a. $(3, \frac{9}{2})$ **b.** $(-\frac{5}{2}, \frac{25}{8})$ **11. a.** $(3, 3)$ **b.** $(-\frac{1}{2}, \frac{5}{4})$

13. a. $(\frac{1}{3}, 3)$ and $(-\frac{1}{3}, -3)$ **b.** $(4, \frac{1}{4})$ and $(-4, -\frac{1}{4})$

EXERCISE 3.3

1. $4x^3 - 6x$

3. 0

5. $30x^{14} + 99x^8$

7. $-\dfrac{10}{x^3}$

9. $21x^2 + \dfrac{12}{x^4}$

11. $-\dfrac{78}{x^{14}}$

13. $\frac{4}{3}\sqrt[3]{x}$

15. $5 + \dfrac{8}{x^3}$

17. $\dfrac{5}{2}x^{3/2} + 12x^3 + \dfrac{1}{x^2}$

19. Slope $= -3$; tangent line, $y = -3x + 3$

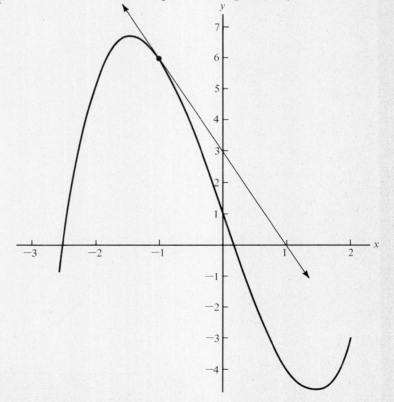

472

21. Slope $= -5$; tangent line, $y = -5x - 1$

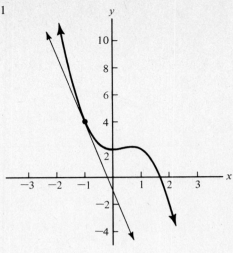

23. $(1, 2)$ **25.** $(2, \frac{1}{2})$ and $(-2, -\frac{1}{2})$

EXERCISE 3.4

1. $(3x + 2)(1) + (x - 5)(3) = 6x - 13$

3. $(x^3 - x + 3)(6) + (6x - 4)(3x^2 - 1) = 24x^3 - 12x^2 - 12x + 22$

5. $(x^4 + x^2 + 1)(6x^5) + (x^6 - 1)(4x^3 + 2x) = 10x^9 + 8x^7 + 6x^5 - 4x^3 - 2x$

7. $\dfrac{(x - 5)(2) - 2x(1)}{(x - 5)^2} = \dfrac{-10}{(x - 5)^2}$ **9.** $\dfrac{(3x + 4)(2x) - x^2(3)}{(3x + 4)^2} = \dfrac{3x^2 + 8x}{(3x + 4)^2}$

11. $\dfrac{(x - 4)(2) - (2x + 6)(1)}{(x - 4)^2} = \dfrac{-14}{(x - 4)^2}$ **13.** $\dfrac{(1 - x)(1) - (x + 1)(-1)}{(1 - x)^2} = \dfrac{2}{(1 - x)^2}$

15. $\dfrac{(x + 1)(1/2\sqrt{x}) - (\sqrt{x})(1)}{(x + 1)^2} = \dfrac{1 - x}{2\sqrt{x}(x + 1)^2}$

17. Slope $= -1$; tangent line, $y = -x + 4$

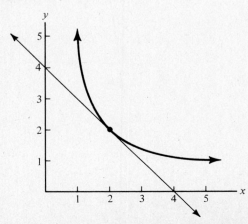

19. $(1, -1)$, $(-\frac{1}{9}, -\frac{13}{243})$ **21.** $(1, 8)$

23. $(4, 2)$, $(2, 0)$

EXERCISE 3.5

1. a. Quotient rule **b.** Sum-difference rule **c.** Sum-difference rule
   d. Product rule **e.** General power rule **f.** Product rule

3. $18(x^2 - 7x + 1)^5(2x - 7)$

5. $8\left(2 + \dfrac{1}{x}\right)^7\left(\dfrac{-1}{x^2}\right) = \dfrac{-8}{x^2}\left(2 + \dfrac{1}{x}\right)^7$

7. $\dfrac{1}{3}(6x - 10)^{-2/3}(6) = \dfrac{2}{\sqrt[3]{(6x - 10)^2}}$

9. $-\dfrac{5}{2}(7 - 2x^6)^{-3/2}(-12x^5) = \dfrac{30x^5}{\sqrt{(7 - 2x^6)^3}}$

11. $(7x + 2)(8)(x^4 + 5)^7(4x^3) + (x^4 + 5)^8(7) = (x^4 + 5)^7(231x^4 + 64x^3 + 35)$

13. $\dfrac{(x - 3)^2(6) - (6x + 4)(2)(x - 3)}{(x - 3)^4} = \dfrac{-6x - 26}{(x - 3)^3}$

15. $\dfrac{x^2(\frac{1}{2})(7x + 2)^{-1/2}(7) - \sqrt{7x + 2}(2x)}{x^4} = \dfrac{-21x - 8}{2x^3\sqrt{7x + 2}}$

17. $(x^4 + 2)\left(\dfrac{1}{2}\right)(3 - x^2)^{-1/2}(-2x) + (3 - x^2)^{1/2}(4x^3)$

$$= 4x^3\sqrt{3 - x^2} - \dfrac{x^5 + 2x}{\sqrt{3 - x^2}}$$

19. $(x^3 + 7x + 2)^3(2)(2x^4 - x^3 + 5)(8x - 3x^2)$
$$+ (2x^4 - x^3 + 5)^2(3)(x^3 + 7x + 2)^2(3x^2 + 7)$$

21. **23.** $\left(\frac{13}{3}, 3\right)$

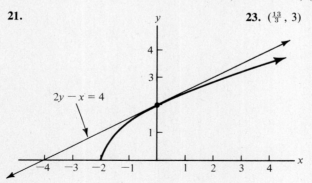

EXERCISE 3.6

1. $V(t) = 15t^2 - 12t + 9$ $V(2) = 45$ ft/sec
   $a(t) = 30t - 12$ $a(2) = 48$ ft/sec^2

3. $V(t) = 14t + \dfrac{1}{2\sqrt{t + 2}}$ $V(7) = \dfrac{589}{6}$ ft/sec

$a(t) = 14 - \dfrac{1}{4\sqrt{(t + 2)^3}}$ $a(7) = \dfrac{1511}{108}$ ft/sec^2

5. $V(t) = \dfrac{6}{(t + 3)^2}$ $V(0) = \dfrac{2}{3}$ ft/sec

$a(t) = \dfrac{-12}{(t + 3)^3}$ $a(0) = \dfrac{-12}{27}$ ft/sec

7. a. $R'(x) = 200 - 2x$ **b.**

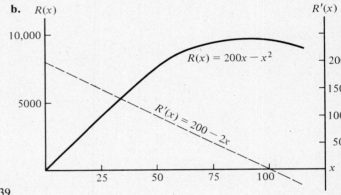

c. $R'(30) = \$140/\text{unit}$
$R(31) - R(30) = \$139$

9. $S'(x) = \dfrac{dS}{dx} = 0.01 - 0.002x$ $0 \le x \le 50$

EXERCISE 3.7

1. 0

3. $36x - 18$

5. $\dfrac{48}{x^4}$

7. $\dfrac{-5}{4} x^{-3/2} = \dfrac{-5}{4\sqrt{x^3}}$

9. $216(1 - 6x)$

11. $\dfrac{-1}{4\sqrt{(x - 1)^3}}$

13. $\dfrac{-4}{(1 + x)^3}$

15. $96x^6(x^4 + 1) + 36x^2(x^4 + 1)^2 - \dfrac{2}{x^3}$

17. $\dfrac{6x}{\sqrt{x^2 + 1}} - \dfrac{2x^3}{\sqrt{(x^2 + 1)^3}}$

19. $\dfrac{2}{x^3} - \dfrac{2}{9\sqrt[3]{x^5}}$

EXERCISE 3.8

1. $\dfrac{dy}{dx} = -\dfrac{5}{2}$

3. $\dfrac{dy}{dx} = \dfrac{3x^2 - 2x - 2y}{3 + 2x}$

5. $\dfrac{dy}{dx} = -\dfrac{x}{y}$

7. $\dfrac{dy}{dx} = \dfrac{4x - 1}{2y + 4}$

9. $\dfrac{dy}{dx} = \dfrac{-6xy^2}{5y^4 + 6x^2y}$

11. $\dfrac{dy}{dx} = \dfrac{14x - 3(x + y)^2}{3(x + y)^2 - 2y}$

13. $\dfrac{dy}{dx} = \dfrac{6y + 5}{1 - 6x}$

15. Slope $= \frac{3}{2}$; tangent line, $2y = 3x - 5$

17. Slope $= -\frac{3}{4}$; tangent line, $4y + 3x = 25$

EXERCISE 4.1

1.

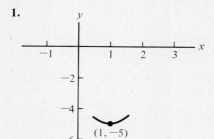

(1, −5)

3.

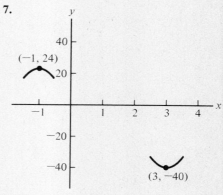

$\left(-\dfrac{5}{2}, -\dfrac{29}{4}\right)$

5.

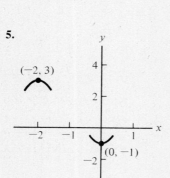

(−2, 3) (0, −1)

7.

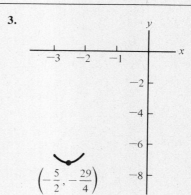

(−1, 24) (3, −40)

9.

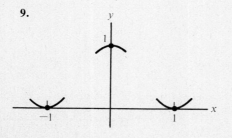

11.

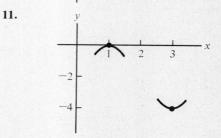

13.

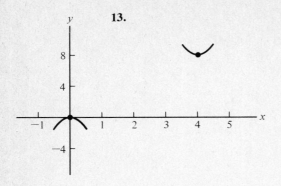

15.

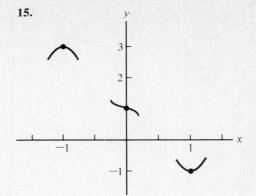

17. $(-1, 0)$, minimum
    $(1, 2)$, maximum

21. $(4, 4)$, maximum
    $(16, 0)$, minimum

23. $(1, 1)$, minimum
    $(5, \frac{49}{5})$, maximum

19. $(1, 1)$, maximum
    $(3, \frac{1}{3})$, minimum

EXERCISE 4.2

1. 3 ft by 3 ft, area = 9 ft²

5. $\frac{1}{2}$

9. 6 in. by 6 in. by 9 in.

13. $6.60

17. 65 trees per acre

21. 12.5%

3. 27.5 and 27.5

7. 30 ft by 60 ft

11. $\dfrac{25 \text{ in.}}{6}$ by $\dfrac{25 \text{ in.}}{6}$

15. $p = \$2.75$, $x = 25$

19. 5 batches of 80,000 books each

23.
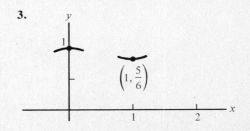

EXERCISE 4.3

1.

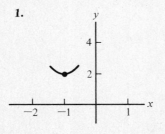

3.

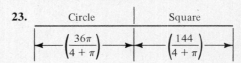

5.

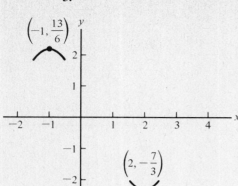

7.

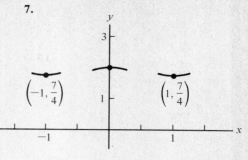

9.

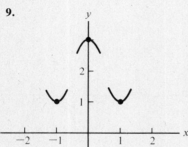

11.

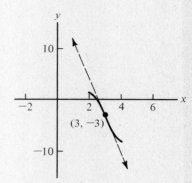

13.

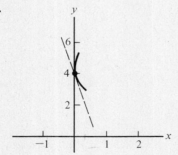

15.

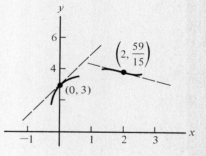

17.

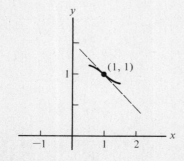

EXERCISE 4.4

1.

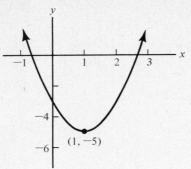

3.

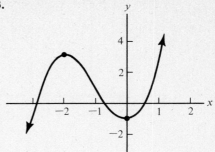

5.

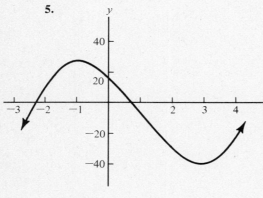

7.

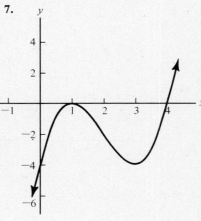

9.

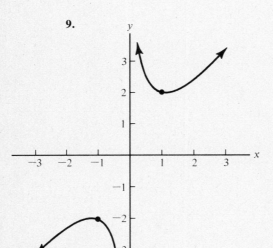

11.

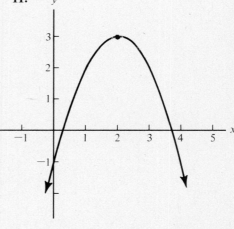

13.

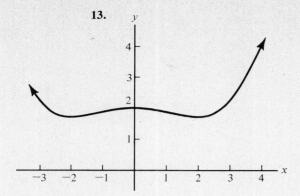

15.

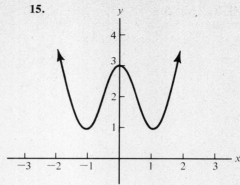

17.

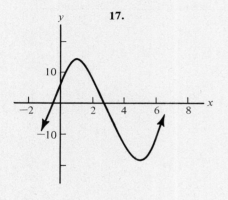

19.

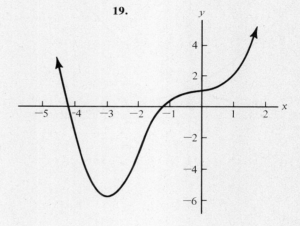

21.

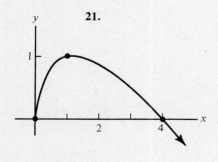

23.

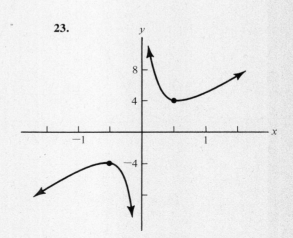

EXERCISE 5.1

1. $3472.88

3.

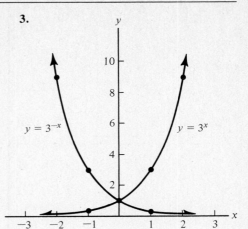

5.

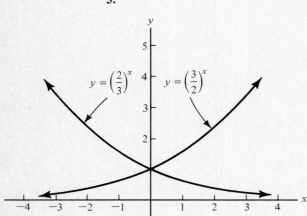

7. a. $y = 2^x$ **b.** $y = 9^x$ **c.** $y = (\frac{1}{3})^x$ **d.** $y = (\frac{1}{3})^x$ **e.** $y = (\frac{1}{8})^x$

9. a.   **b.**

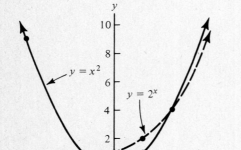

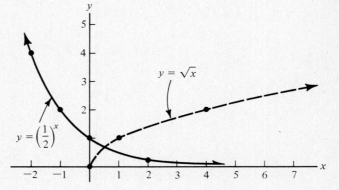

11. $2851.52

13. 1.61×10^6 bacteria after 5 hours
2.59×10^6 bacteria after 10 hours

15. a. $12,689.86
    b. $13,138.26

17. $788.49

EXERCISE 5.2

1.

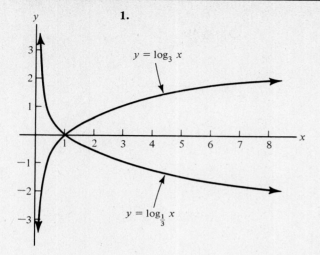

$y = \log_3 x$

$y = \log_{\frac{1}{3}} x$

3. $\log_{2/3} \frac{16}{81} = 4$

5. $\log_{5/4} \frac{64}{125} = -3$

7. $\log_x 20 = 3$

9. $\log_x \frac{5}{3} = -4$

11. $x = \frac{1}{8}$

13. $x = 5$

15. $x = \frac{3}{2}$

17. $x = \frac{1}{2}$

19. $x = 2$

21. $x = \frac{1}{2}$

23. $x = 7$

25. 2.1584

27. 2.4084

29. 1.9084

31. 3.1126

33. -2.8626

35. $x = -\frac{2}{3}$

37. $x = \frac{25}{8}$

EXERCISE 5.3

1.

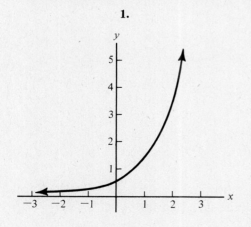

3.

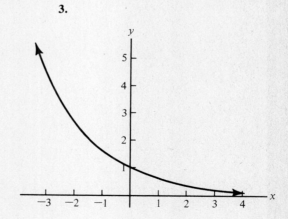

5. 4.9 billion

7. \$136 million

9.

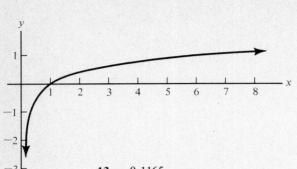

11.

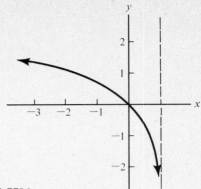

13. −0.1165

17. 1.5440

21. $x = 1.6610$

15. 8.7796

19. $x = 4.7737$

23. a. $1098   **b.** $813   **c.** $1721

EXERCISE 6.1

1. $2x \ln x + x$

3. $6x + \dfrac{1}{x}$

5. $\dfrac{1}{x} - \dfrac{1}{x(\ln x)^2}$

7. $\dfrac{4(\ln x)^3}{x}$

9. $\dfrac{-1 + \ln(1 - x)}{(1 - x)^2}$

11. $\dfrac{1}{x}$

13. $\dfrac{2}{(2x + 3) \ln 5}$

15. $\dfrac{1}{x + 1} - \dfrac{1}{x + 2} = \dfrac{1}{(x + 1)(x + 2)}$

17. $\dfrac{6}{x^3 + 3x}$

19. Slope = 4; $y = 4x - 8$

21. Slope = 0; $y = \dfrac{1}{e}$

23. Slope = -1; $y = x + 1$

25. $x = 1$; $y = 2(\ln 4) = 2.77$

27. $x = e$; $y = \dfrac{1}{e}$

29.

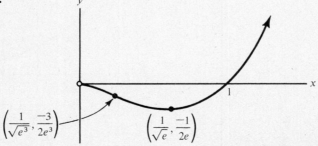

31.

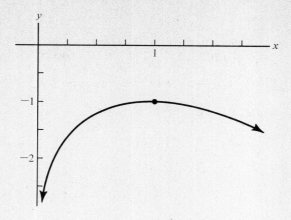

EXERCISE 6.2

1. $(2x + x^2)e^x$

3. $e^x - e^{-x}$

5. $(6x + 9)e^{(x^2+3x)}$

7. $\dfrac{1 + e^x}{2\sqrt{x + e^x}}$

9. $10^x(\ln 10)$

11. $\dfrac{e^x}{e^x + 1}$

13. $-(\ln 6)6^{2-x}$

15. Slope $= e$; $y = ex$

17. Slope $= -3e$; $y = -3ex - 2e$

19. Slope $= -1$; $y = -x + 1$

21. $x = -\ln 2 = -0.69$; $y = 2$

23. $x = 0$; $y = 1$

25.

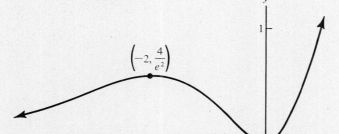

27.

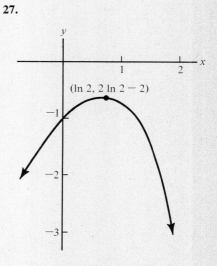

EXERCISE 6.3

3. a. $V(4) = \$205$ **b.** $\dfrac{V'(t)}{V(t)} = \dfrac{15t^{1/2}}{125 + 10t^{3/2}}$ **c.** $\dfrac{V'(4)}{V(4)} = \dfrac{6}{41}$

5. a. $S'(2) = -0.15e^{-0.02} = -\$0.15$ million per year

   b. $\dfrac{S'(t)}{S(t)} = \dfrac{-0.15e^{-0.01t}}{15e^{-0.01t}} = -.01$

7. a. $\dfrac{S'(1)}{S(1)} = 1$ or 100%

   b. $t = 2$ years

EXERCISE 7.1

1. a. $dy = 18x^2\,dx$
   b. $dy = 1.8$

3. a. $dy = (2x - 5)\,dx$
   b. $dy = (6 - 5)(1) = 1$

5. a. $dy = \dfrac{3}{(3 - 2x)^2}\,dx$

   b. $dy = -\frac{3}{2}$

7. a. $dy = \dfrac{1}{x + 1}\,dx$

   b. $dy = -\frac{1}{32}$

9. a. $dy = \dfrac{1 - \ln x}{x^2}\,dx$

   b. $dy = 0.30$

11. $26 \cong \sqrt{25} + \dfrac{1}{2\sqrt{25}}\,(1)$

   $\cong 5.10$

13. $\sqrt[3]{28} \cong \sqrt[3]{27} + \dfrac{1}{3(\sqrt[3]{27})^2} = \frac{82}{27}$

15. $\sqrt[4]{15.7} \cong \sqrt[4]{16} + \dfrac{-0.3}{4(\sqrt[4]{16})^3} = \frac{637}{370}$

17. $\dfrac{dA}{A} = \dfrac{2x\,dx}{x^2} = 2\left(\dfrac{dx}{x}\right) = \pm 0.02$

19. $\dfrac{dR}{R} = (1 - 1.5x)\dfrac{dx}{x} = (-2)(\pm 0.10) = \mp 0.20$

21. $dy = e^x dx; \ dy = e^2(-1) = -e^2$

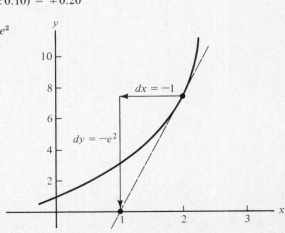

23. $dy = \dfrac{-1}{\sqrt[2]{1-x}}\, dx$; $dy = \dfrac{-1}{2}\,(-3) = \dfrac{3}{2}$

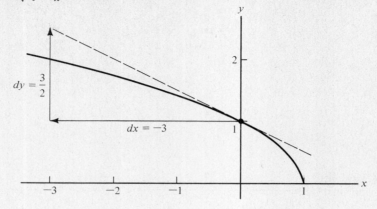

EXERCISE 7.2

1. a. $x_5 = 1.7100$ **b.** $x_3 = 0.5777$ **c.** $x_5 = 0.0699$

3. $t = 2.317$ **5.** $t = 1.164$

7. $t = 2.091$ **9.** $x = -1.145$; $y = -6.589$

11. $i = 13.0\%$ **13.** $i = 29\%$

15. $r = 0.41 = 41\%$

EXERCISE 8.1

1. $\dfrac{x^6}{6} + C$ **3.** $\dfrac{x^2}{2} + C$

5. $\dfrac{7x^3}{3} + 2x^2 - 9x + C$ **7.** $10x - \dfrac{5x^3}{3} + C$

9. $2x^4 - 2e^{2x} + 2x + C$ **11.** $x^5 + \dfrac{4x^{3/2}}{3} + C$

13. $6\sqrt{x} + 4x + C$ **15.** $\dfrac{-1}{x} - 5\ln|x| + C$

17. $3x^2 - 4\ln|x| + \dfrac{e^{2x}}{2} + C$ **19.** $\dfrac{x^3}{3} + \dfrac{3x^2}{2} - \dfrac{4}{x} + C$

21. $y = F(x) = \dfrac{2x^{3/2}}{3} + 3x - 43$ **23.** $y = F(x) = x^4 + \dfrac{1}{4x^2} + \dfrac{7}{4}$

25. $T(x) = 270\sqrt[3]{x}$; $T(27) = 810$ minutes

EXERCISE 8.2

1. Area $= \frac{27}{4}$ 3. Area $= \frac{117}{6}$

5. Area $= \frac{32}{3}$ 7. Area $= \frac{35}{6}$

9. Area $= \dfrac{e^3 - 1}{e}$ 11. Area $= \frac{8}{5}$

13. Area $= 12$ 15. Area $= \frac{101}{24}$

17. Area $= \frac{26}{3} - \ln 3 = 7.57$ 19. Area $= 10$

21. **a.** \$75,610   **b.** 4.58 years   23. 300 units

EXERCISE 8.3

1. Sum of the areas $= A_\text{I} + A_\text{II} + A_\text{III} = (\frac{1}{3})(\frac{1}{3}) + (\frac{1}{3})(\frac{2}{3}) + (\frac{1}{3})(\frac{3}{3}) = \frac{6}{9} = \frac{2}{3}$

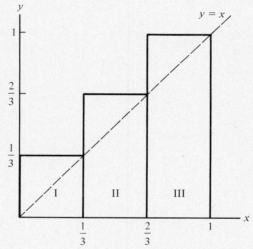

3. **a.** $A_1 + A_2 + A_3 + \cdots + A_n = \dfrac{4}{n^2} \dfrac{n(n + 1)}{2} = \dfrac{2(n + 1)}{n}$

   b. Area under curve $= \lim\limits_{n \to \infty} \dfrac{2(n + 1)}{n} = 2$

5. $\frac{5}{2}$ 7. $6 \ln 3$

9. $\frac{662}{3}$ 11. $\frac{1}{2}$

13. $\frac{45}{2}$ 15. 124

17. $\frac{35}{6}$ 19. $\frac{46}{3}$

21. Two solutions: $t_1 = -2$, $t_2 = 1$ 23. $t = \dfrac{\sqrt[3]{2}}{2}$

25. Area $= 2$ 27. Area $= \frac{1}{4}$

29. Area $= \frac{37}{12}$

31. Change in total revenue $= \$2400$ thousand $= \$2,400,000$

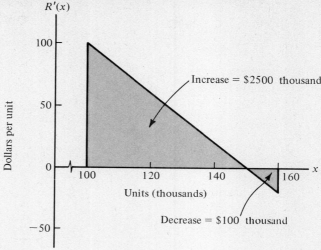

R'(x)

Increase = $2500 thousand

Decrease = $100 thousand

Dollars per unit

Units (thousands)

33. $\$\frac{20}{3}$ billion

EXERCISE 8.4

1. Area $= 12$

3. Area $= \dfrac{4e - 1}{4}$

5. Area $= e^2 - e - \ln 2$

7. Area $= \frac{1}{12}$

9. Area $= \frac{1}{3}$

11. Additional sales $= 300e^{0.25} - 367.5 = \$17.71$ million

13. a. Energy savings $= \dfrac{40}{.03} e^{0.3} - \dfrac{30}{.02} e^{0.2} - \left(\dfrac{40}{0.03} - \dfrac{30}{0.02}\right) = 134$ million kwh.

b. Dollar savings $= \dfrac{2}{0.09} e^{0.9} - \dfrac{1.5}{0.08} e^{0.8} - \left(\dfrac{2}{0.09} - \dfrac{1.5}{0.08}\right) = \$9.46$ million

EXERCISE 8.5

1. $20,000(e^{0.90-1}) = \$29,192.06$

3. $R = \left(\dfrac{1200}{52}\right)\left(\dfrac{1}{e^{1.6-1}}\right) = 5.84$ (Assuming employee's contribution is matched by employer.)

5. $R = \dfrac{2000}{3(1 - e^{-1.60})} = \$835.32$

7. a. $\$61.25$ million   **b.** $\$70.31$ million

EXERCISE 9.1

1. $\dfrac{(x-5)^4}{4} + C$

3. $e^{(x+2)} + C$

5. $\dfrac{(2x+1)^6}{6} + C$

7. $\ln |4x+3| + C$

9. $\dfrac{(x^2-1)^5}{10} + C$

11. $\dfrac{-e^{-x^2}}{2} + C$

13. $\dfrac{-5}{6(3x^2+7)} + C$

15. $\dfrac{(x^6+2)^2}{12} + C$

17. $\dfrac{-e^{(1-x^2)}}{2} + C$

19. $\ln(2+e^x) + C$

21. $\dfrac{e^{x^2}}{2} - \ln(x^2+1) + C$

23. $-e^{1/x} + C$

25. $\dfrac{2(x+3)^{5/2}}{5} - \dfrac{4(x+3)^{3/2}}{3} + C$

27. $\dfrac{e^4 - e}{2}$

29. $2 \ln 4 - \ln 2$

31. $\dfrac{\ln 4}{6}$

33. $\ln(2+e) - \ln 3 = \ln\left(\dfrac{2+e}{3}\right)$

EXERCISE 9.2

1. $\dfrac{xe^{2x}}{2} - \dfrac{e^{2x}}{4} + C$

3. $\dfrac{x^2 e^{4x}}{4} - \dfrac{xe^{4x}}{8} + \dfrac{e^{4x}}{32} + C$

5. $x \ln 3x - x + C$

7. $\frac{2}{3}x^{3/2} \ln x - \frac{4}{9}x^{3/2} + C$

9. $(x+1)\ln(x+1) - (x+1) + C$

11. $-(x+2)e^{-x} + C$

13. $\dfrac{-2}{e}$

15. $\dfrac{18 \ln 3 - 8 \ln 2 - 5}{2}$

17. $e - 2$

19. 1

EXERCISE 9.3

1. $\dfrac{x^4 \ln x}{4} - \dfrac{x^4}{16} + C$

3. $\dfrac{\ln |x + \sqrt{x^2+1}|}{3} + C$

5. $\dfrac{1}{16(4x+1)} + \dfrac{\ln |4x+1|}{16} + C$

7. $\dfrac{3x}{5} - \dfrac{6 \ln |5x+2|}{25} + C$

9. $\dfrac{-2\sqrt{4-9x^2}}{3x} + C$

11. $\dfrac{5^x}{\ln 5} + C$

13. $\frac{1}{4} \ln \left| \frac{x^2 - 1}{x^2 + 1} \right| + C$

15. $\frac{1}{6x^2 + 18} + \frac{1}{18} \ln \left| \frac{x^2}{x^2 + 3} \right| + C$ 17. $\frac{(x^2 - 10)}{3} \sqrt{x^2 + 5} + C$

19. $\frac{1}{4} \ln 3$ 21. $\frac{3}{2} - \frac{3}{4} \ln 3$

EXERCISE 9.4

1. $\frac{8}{3}$, fundamental theorem; $\frac{11}{4}$, trapezoidal rule; $\frac{8}{3}$, Simpson's rule

3. 1.72, fundamental theorem; 1.73, trapezoidal rule; 1.72, Simpson's rule

5. 1.87, fundamental theorem; 1.87, trapezoidal rule; 1.87, Simpson's rule

7. $\frac{\ln 2}{2} = 0.3466$, fundamental theorem; 0.3413, trapezoidal rule; 0.3467, Simpson's rule

9. 0.7825, trapezoidal rule; 0.7850, Simpson's rule

11. 2.382, trapezoidal rule; 2.378, Simpson's rule

13. 1.1356, trapezoidal rule; 1.1162, Simpson's rule

EXERCISE 9.5

1. $\frac{2}{3}$ 3. 2

5. $\frac{e^2}{2} = 3.69$ 7. 6

9. $\frac{1}{2}$ 11. $-\frac{1}{4}$

13. Integral is divergent   15. 0

17. $5.71 million

EXERCISE 10.1

1. a. 2 b. -11 c. -13 d. $h^2 + 4h + 2$ e. $h^2 + 4h$ f. $4 + h$ g. 4

3. a. -7 b. -29 c. $+1$ d. -1

5. a. 2 b. 2 c. $e + \frac{1}{e}$ d. $ae^b - be^a$

7, 9, 11.

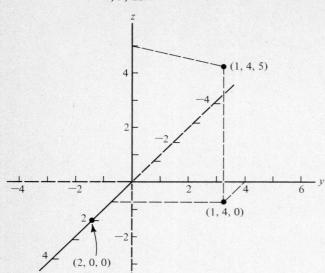

(1, 4, 5)

(1, 4, 0)

(2, 0, 0)

13.

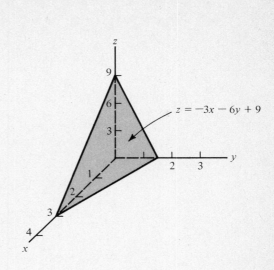

$z = -3x - 6y + 9$

15.

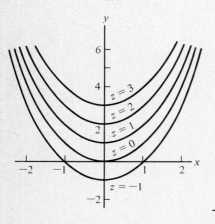

$z = 3$
$z = 2$
$z = 1$
$z = 0$
$z = -1$

17.

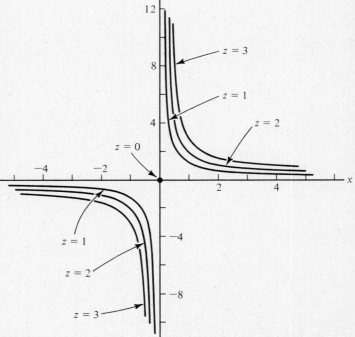

$z = 3$
$z = 1$
$z = 2$
$z = 0$
$z = 1$
$z = 2$
$z = 3$

19. $R_1 = 10p_1 - p_1^2 + 0.5p_1p_2$
$R_2 = 8p_2 + 0.4p_1p_2 - p_2^2$

21. $C = 3.50LW + 2LH + 2WH$
$C = \$1610$

EXERCISE 10.2

1. $f_x = 12x^3; f_y = -18y^2$

3. $f_x = \frac{1}{2}\sqrt{\frac{y}{x}}; f_y = \frac{1}{2}\sqrt{\frac{x}{y}}$

5. $f_x = \frac{-y}{x^2}; f_y = \frac{1}{x}$

7. $f_x = 6x(x^2 - y^2)^2; f_y = -6y(x^2 - y^2)^2$

9. $f_x = ye^{xy}; f_y = xe^{xy}$

11. $f_x = f_y = e^{x+y}$

13. Slope = 2

15. Slope = 1

17. $f_{xx} = 6; f_{yy} = -14; f_{xy} = f_{yx} = 2$

19. $f_{xx} = e^x - \frac{1}{x^2}; f_{yy} = \frac{-1}{y^2}; f_{xy} = f_{yx} = 0$

21. $f_{xx} = \frac{-y}{4\sqrt{x^3}}; f_{yy} = 0; f_{xy} = f_{yx} = \frac{1}{2\sqrt{x}}$

23. $\left.\frac{\partial S}{\partial T}\right|_{(70,16)} = 25,600 \frac{\text{dollars}}{°F}$ $\left.\frac{\partial S}{\partial T}\right|_{(90,16)} = 76,800 \frac{\text{dollars}}{°F}$

EXERCISE 10.3

1. (0, 0, 1), minimum

3. (0, 0, 3), saddle point

5. (4, −1, −12), minimum

7. (1, −4, 22), saddle point

9. (0, 1, 5), saddle point
(−2, 1, 9), maximum

11. (0, 0, −5), saddle point
(1, 1, −6), minimum

13. (2, 1, −18), minimum
(−1, 1, 9), saddle point

15. (0, 0, 4), saddle point
(6, 18, 112), maximum

17. $(\frac{4}{3}, \frac{9}{2}, 18)$, minimum

19. $p_1 = \$21.50; p_2 = \$30$

21. $x = y = w = 15$

23. 50 ft by 100 ft by 75 ft

EXERCISE 10.4

1. $x = y = 1; z_{max} = f(1, 1) = 1$

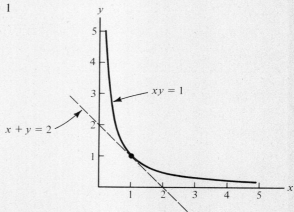

3. $x = \frac{2}{3}$; $y = \frac{1}{3}$; $z_{max} = \frac{4}{3}$ 5. $x = \frac{5}{2}$; $y = -1$; $z_{max} = 9$

7. $x = y = 1$; $z_{min} = e^3$ 9. 1600

11. -49 13. 30 ft by 60 ft

EXERCISE 10.5

1. $y = 0.63x + 0.77$ 3. $y = 0.62x + 0.96$

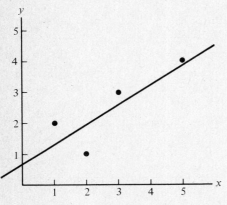   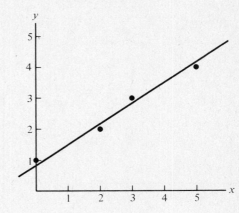

5. $y = -0.58x + 4.95$ 7. $y = -1.23x + 5.63$

9. $y = -1.5x + 7.50$ 11. $p = -0.34x + 3.38$

13. **a.** $y = 0.10x - 0.47$
    b. \$2.0 thousand

EXERCISE A

1. **a.** T **b.** F **c.** F **d.** T **e.** T **f.** T **g.** F **h.** F

3. **a.** $\{?\}, \{!\}, \{\$\}, \{?, !\}, \{?, \$\}, \{\$, !\}, \varnothing, A$ **b.** $\{?\}, \{!\}, \{\$\}, \{?, !\}, \{?, \$\}, \{\$, !\}, \varnothing$

5. $B \cap C = \varnothing$ 7. $\{0, 2, 3, 4\}$

9. $\varnothing$ 11. $\{0, 2, 3, 4, 5\}$

EXERCISE B

1. x^6 3. $-\frac{27}{64}$

5. $\dfrac{3}{x^4}$ 7. x^3

9. 2 11. $\dfrac{xy}{x + y}$

13. $\frac{7}{4}$ 15. x^4

17. $\dfrac{8x^2}{y^4}$

19. $8x$

21. $9x$

23. $\dfrac{5x^3}{3y^5}$

25. $(x + 4)(x - 4)$

27. $(x - 5)^2$

29. $(2x - 3)(x + 1)$

31. $(4x + 9y)(x + y)$

33. $2x(x - 4)(x^2 + 4x + 16)$

35. $\dfrac{3(x + 1)(x - 2)}{2}$

37. $\dfrac{x + 2}{x + 5}$

39. $\dfrac{x - 1}{x^2 + x}$

41. $\dfrac{2x^2 - 4x + 3}{(x + 2)(x - 2)}$

43. $x = 8$

45. $x = -3$

47. $x = \frac{4}{3}$

49. $x_1 = 0; \ x_2 = \frac{5}{2}$

51. $x_1 = \dfrac{3 + \sqrt{13}}{2}; \ x_2 = \dfrac{3 - \sqrt{13}}{2}$

53. $x_1 = \dfrac{-1 + \sqrt{21}}{10}; \ x_2 = \dfrac{-1 - \sqrt{21}}{10}$

55.

57.

Index

495

A Short Table of Integration Formulas

1. $\int x^n \, dx = \dfrac{x^{n+1}}{n+1} + C$

2. $\int e^{ax} \, dx = \dfrac{e^{ax}}{a} + C$

3. $\int \dfrac{1}{x} \, dx = \ln |x| + C$

4. $\int a^x \, dx = \dfrac{a^x}{\ln a} + C \qquad a > 0$

5. $\int xe^{ax} \, dx = \dfrac{e^{ax}}{a^2}(ax - 1) + C$

6. $\int x^n e^{ax} \, dx = \dfrac{x^n e^{ax}}{a} - \dfrac{n}{a} \int x^{n-1} e^{ax} \, dx$

7. $\int \ln x \, dx = x \ln x - x + C$

8. $\int x^n \ln x \, dx = x^{n+1} \left[\dfrac{\ln x}{n+1} - \dfrac{1}{(n+1)^2} \right] \qquad n \neq -1$

9. $\int (\ln x)^n \, dx = x(\ln x)^n - n \int (\ln x)^{n-1} \, dx \qquad n \neq -1$

10. $\int \dfrac{1}{x^2 - a^2} \, dx = -\dfrac{1}{2a} \ln \left| \dfrac{x + a}{x - a} \right| + C$

11. $\int \dfrac{1}{a^2 - x^2} \, dx = \dfrac{1}{2a} \ln \left| \dfrac{a + x}{a - x} \right| + C$

12. $\int \dfrac{1}{\sqrt{x^2 - a^2}} \, dx = \ln |x + \sqrt{x^2 - a^2}| + C$

13. $\int \dfrac{1}{\sqrt{x^2 + a^2}} \, dx = \ln |x + \sqrt{x^2 + a^2}| + C$

14. $\int \dfrac{x}{ax + b} \, dx = \dfrac{x}{a} - \dfrac{b}{a^2} \ln |ax + b| + C$

15. $\int \dfrac{x}{\sqrt{ax + b}} \, dx = \dfrac{2(ax - 2b)}{3a^2} \sqrt{ax + b} + C$

16. $\int \dfrac{x}{(ax + b)^2} \, dx = \dfrac{b}{a^2(ax + b)} + \dfrac{1}{a^2} \ln |ax + b| + C$

17. $\int \dfrac{1}{x(ax + b)} \, dx = \dfrac{1}{b} \ln \left| \dfrac{x}{ax + b} \right| + C$

18. $\int \dfrac{1}{x(ax + b)^2} \, dx = \dfrac{1}{b(ax + b)} + \dfrac{1}{b^2} \ln \left| \dfrac{x}{ax + b} \right| + C$

19. $\int \sqrt{x^2 \pm a^2} \, dx = \dfrac{x}{2} \sqrt{x^2 \pm a^2} \pm \dfrac{a^2}{2} \ln |x + \sqrt{x^2 \pm a^2}| + C$

20. $\int \dfrac{1}{x^2 \sqrt{a^2 - x^2}} \, dx = \dfrac{-\sqrt{a^2 - x^2}}{a^2 x} + C$